# Communications in Computer and Information Science 780

*Commenced Publication in 2007*
Founding and Former Series Editors:
Alfredo Cuzzocrea, Xiaoyong Du, Orhun Kara, Ting Liu, Dominik Ślęzak,
and Xiaokang Yang

More information about this series at http://www.springer.com/series/7899

Jian Chen · Thanaruk Theeramunkong
Thepchai Supnithi · Xijin Tang (Eds.)

# Knowledge and Systems Sciences

18th International Symposium, KSS 2017
Bangkok, Thailand, November 17–19, 2017
Proceedings

 Springer

*Editors*
Jian Chen
Tsinghua University
Beijing
China

Thanaruk Theeramunkong
Thammasat University
Bangkadi Muang Pathumthani
Thailand

Thepchai Supnithi
National Electronics
 and Computer Technology Center
Pathumthani
Thailand

Xijin Tang
Academy of Mathematics
 and Systems Science
Chinese Academy of Sciences
Beijing
China

ISSN 1865-0929          ISSN 1865-0937   (electronic)
Communications in Computer and Information Science
ISBN 978-981-10-6988-8          ISBN 978-981-10-6989-5   (eBook)
https://doi.org/10.1007/978-981-10-6989-5

Library of Congress Control Number: 2017957199

Printed on acid-free paper

This Springer imprint is published by Springer Nature
The registered company is Springer Nature Singapore Pte Ltd.
The registered company address is: 152 Beach Road, #21-01/04 Gateway East, Singapore 189721, Singapore

# Preface

The annual International Symposium on Knowledge and Systems Sciences aims to promote the exchange and interaction of knowledge across disciplines and borders to explore the new territories and new frontiers. With over 17 years of continuous endeavors, attempts to strictly define the knowledge science may be still ambitious, but a very tolerant, broad-based, and open-minded approach to the discipline can be taken. Knowledge science and systems science can complement and benefit each other methodologically.

The First International Symposium on Knowledge and Systems Sciences (KSS 2000) was initiated and organized by Japan Advanced Institute of Science and Technology (JAIST) in September of 2000. Since then, KSS 2001 (Dalian), KSS 2002 (Shanghai), KSS 2003 (Guangzhou), KSS 2004 (JAIST), KSS 2005 (Vienna), KSS 2006 (Beijing), KSS 2007 (JAIST), KSS 2008 (Guangzhou), KSS 2009 (Hong Kong), KSS 2010 (Xi'an), KSS 2011 (Hull), KSS 2012 (JAIST), KSS 2013 (Ningbo), KSS 2014 (Sapporo), KSS 2015 (Xi'an), and KSS 2016 (Kobe) have been held successfully, with contributions by many scientists and researchers from different countries. During the past 17 years, people interested in knowledge and systems sciences have become a community, and an international academic society has existed for 14 years.

This year KSS was held in Bangkok, Thailand, to provide opportunities for presenting interesting new research results, facilitating interdisciplinary discussions, and leading to knowledge transfer under the theme of "Artificial Intelligence and Information Systems for Knowledge, Technology and Service Management" during November 17–19, 2017. Five distinguished scholars delivered the keynote speeches reflecting these diverse features of KSS topics:

- Peter A. Gloor (MIT, USA), "From the Age of Emperors to the Age of Empathy"
- Yoshitsugu Hayashi (Chubu University, Japan), "Quality-of-Life (QOL) Based Urban Transport Planning Utilising ICT"
- Ryosuke Shibasaki (University of Tokyo, Japan), "Urban Computing Using Big Data from Mobile Network"
- Vilas Wuwongse (Mahidol University, Thailand), "An Evolution-Theoretical Approach to the Analysis of Social Systems"
- Minjie Zhang (University of Wollongong, Australia), "Multi-agent Solutions for Supply-Demand Management in Smart Grid Markets"

The organizers of KSS 2017 received 63 submissions, and finally 21 submissions were selected for the proceedings after a rigorous review process. The co-chairs of international Program Committee made the final decision for each submission based on the review reports from the referees, who came from Australia, China, France, Japan, New Zealand, Thailand, UK, and USA.

We received a lot of support and help from many people and organizations for KSS 2017. We would like to express our sincere thanks to the authors for their remarkable contributions, all the Technical Program Committee members for their time and expertise reviewing the papers under a very tight schedule, and the Springer for their professional help in the publication process. This is the second time that KSS proceedings are published as a CCIS volume by Springer after the success of the 2016 publication. We greatly appreciate our five distinguished scholars for accepting our invitation to deliver keynote speeches at the symposium. Last but not least, we are very indebted to the local organizers for their hard work.

We were happy to witness not only cross-cultural learning and integration at the conference, but also academic achievements and professionalism refined into the essence of knowledge and systems sciences.

November 2017

Jian Chen
Thanaruk Theeramunkong
Thepchai Supnithi
Xijin Tang

# Organization

## Organizer

International Society for Knowledge and Systems Sciences

## Host

Artificial Intelligence Association of Thailand (AIAT)

## General Chairs

| | |
|---|---|
| Jian Chen | Tsinghua University, China |
| Thanaruk Theeramunkong | SIIT, Thammasat University, Thailand |

## Program Committee Chairs

| | |
|---|---|
| Van-Nam Huynh | Japan Advanced Institute of Science and Technology, Japan |
| Thepchai Supnithi | NECTEC, Thailand |
| Xijin Tang | CAS Academy of Mathematics and Systems Science, China |
| Jiangning Wu | Dalian University of Technology, China |

## Technical Program Committee

| | |
|---|---|
| Quan Bai | Auckland University of Technology, New Zealand |
| Chutima Beokhaimook | Rangsit University, Thailand |
| Prachya Boonkwan | NECTEC, Thailand |
| Marut Buranarach | NECTEC, Thailand |
| Meng Cai | Xidian University, China and Harvard University, USA |
| Zhigang Cao | Beijing Jiaotong University, China |
| Hao Chen | Nankai University, China |
| Jindong Chen | China Academy of Aerospace Systems Science and Engineering, China |
| Yu-wang Chen | Manchester Business School, UK |
| Narumol Chumuang | Muban Chombueng Rajabhut University, Thailand |
| Zengru Di | Beijing Normal University, China |
| Yong Fang | CAS Academy of Mathematics and Systems Science, China |
| Serge Galam | Sciences Po and CNRS, France |

| | |
|---|---|
| Chonghui Guo | Dalian University of Technology, China |
| Kiyota Hashimoto | Prince of Songkla University, Thailand |
| Narit Hnoohom | Mahidol University, Thailand |
| Rui Hou | Guangdong University of Technology, China |
| Van-Nam Huynh | Japan Advanced Institute of Science and Technology, Japan |
| Peerasak Intarapaiboon | Thammasat University, Thailand |
| Chuleerat Jaruskulchai | Kasetsart University, Thailand |
| Bin Jia | Beijing Jiaotong University, China |
| Jiradett Kerdsri | Defense Technology Institute, Thailand |
| Rachada Kongkachandra | Thammasat University, Thailand |
| Cheng-Siong Lee | Monash University, Australia |
| Xianneng Li | Dalian University of Technology, China |
| Yongjian Li | Nankai Univertsity, China |
| Zhenpeng Li | Dali University, China |
| Bo Liu | CAS Academy of Mathematics and Systems Science, China |
| Dehai Liu | Dongbei University of Finance and Economics, China |
| Jiamou Liu | University of Auckland, New Zealand |
| Yijun Liu | CAS Institute of Science and Development, China |
| Tieju Ma | East China University of Science and Technology, China |
| Patiyuth Pramkeaw | King Mongkut's University of Technology Thonburi, Thailand |
| Mina Ryoke | University of Tsukuba, Japan |
| Pokpong Songmuang | Thammasat University, Thailand |
| Leilei Sun | Tsinghua University, China |
| Thepchai Supnithi | NECTEC, Thailand |
| Xijin Tang | CAS Academy of Mathematics and Systems Science, China |
| Jing Tian | Wuhan University of Technology, China |
| Haibo Wang | Texas A&M International University, USA |
| Mingzheng Wang | Dalian University of Technology, China |
| Yuwei Wang | CAS Institute of Automation, China |
| Cuiping Wei | Yangzhou University, China |
| Jiang Wu | Wuhan University, China |
| Jiangning Wu | Dalian University of Technology, China |
| Haoxiang Xia | Dalian University of Technology, China |
| Gang Xie | CAS Academy of Mathematics and Systems Science, China |
| Yun Xue | South China Normal University, China |
| Guangfei Yang | Dalian University of Technology, China |
| Worawut Yimyam | Phetchaburi Rajabhat University, Thailand |
| Thaweesak Yingthawornsuk | King Mongkut's University of Technology Thonburi, Thailand |

# Abstracts of Keynotes

# From the Age of Emperors to the Age of Empathy

Peter A. Gloor

Center for Collective Intelligence, Sloan School of Management,
Massachusetts Institute of Technology, USA
pgloor@mit.edu

**Abstract.** The age of imperial CEOs residing in the corner office is over, Mark Zuckerberg shares the same open office space with the rest of his Facebook employees. Today's Millennials do not want to be led by emperors high on testosterone and authority, but by leaders high on empathy and compassion. This talk is based on my new books "SwarmLeadership" and "Sociometrics". "SwarmLeadership" introduces a framework based on "social quantum physics", which explains how all living beings are connected through empathy in entanglement, and learning. To track empathy, entanglement, and learning we have developed "seven honest signals of collaboration" which can be used to measure empathy, entanglement, and learning on any level, from the global level on social media, inside the organization with e-mail, down to face-to-face entanglement using the body sensors of smartwatches. The talk will present the main concepts and the underlying algorithms and models, documenting them by numerous industry examples from our own work.

# Quality-of-Life (QOL) Based Urban Transport Planning Utilising ICT

Yoshitsugu Hayashi

Chubu University, Japan
y-hayashi@isc.chubu.ac.jp

**Abstract.** The main stream of this keynote speech is 1) to network the existing transport system TukTuk by ICT, 2) to evaluate the performance of network plan based on time serial profile of QOL(Quality of Life) versus $CO_2$ emission, 4) to propose options of policies combining infrastructure improvement by visualised 3D mappings, 5) to develop big data for planning.

# Urban Computing Using Big Data from Mobile Network

Ryosuke Shibasaki

Center for Spatial Information Science, University of Tokyo, Japan
shiba@csis.u-tokyo.ac.jp

**Abstract.** While a cellular phone or a smart phone becomes an indispensable item in our daily life. It is possible to utilize mobile information with privacy preservation in several innovative applications such as humanitarian aid, traff congestion solution, urban facility design, etc. This talk presents our recent research on urban computing with introduction of the applications and technologies for integrating and deep mining heterogeneous data. Examples include population behavior analysis for reasoning following large-scale disasters, tourist behavior analysis, and village building phenomena analysis.

# An Evolution-Theoretical Approach to the Analysis of Social Systems

Vilas Wuwongse

Mahidol University, Thailand
vilasw@gmail.com

**Abstract.** "Evolution" in Darwin's Theory of Evolution consists of two major mechanisms: mutation and natural selection. Mutation in essence means "change of organisms' properties." It is a mechanism to create new organisms and diversity. Natural selection is the process whereby organisms that better fit with other organisms and their surrounding environment tend to survive and reproduce. As organisms are also part of their environment, natural selection is essentially "interaction" among organisms as well as with their environment. If "change of organisms' properties" could be viewed as "fluctuation of organisms' states," then "evolution of a system" comprises "fluctuation of the states of system components" and "interaction among system components." This view of "evolution" could be applied to the analysis of various systems, not only the biological one. In this talk, it will be applied to the analysis of social systems, with an example of Thai social system.

# Multi-agent Solutions for Supply-Demand Management in Smart Grid Markets

Minjie Zhang

School of Computing and Information Technology,
University of Wollongong, Australia
minjie@uow.edu.au

**Abstract.** A smart grid market is a complex and dynamic market with various participators, including energy generators, general consumers, interruptible consumers, storage consumers, or even small renewable energy producers, such as solar systems and windmills. Moreover, different participators exhibit a variety of behaviours. For instance, the behaviours of solar and wind energy producers are closely related to the weather conditions, while some interruptible consumers can contribute extra energy to supply-demand balance. Besides, the large energy generators may produce variant quantities of energy from day to day. Due to the complexity and dynamics, it is of great challenge to manage supply-demand balance in the Smart Grid market.

Agent and multi-agent technologies offer potential solutions to the above challenge, by using the capabilities of intelligent modelling, management and group collaboration, in addition to the learning and self-organising abilities and autonomous decision making of individual agents. This talk will introduce our two new solutions in smart grid research, including (1) an agent-based broker model for power trading in smart grid markets; and (2) a load forecasting approach in smart grid market through customer behaviour learning.

# Contents

Dynamics of Brand Acceptance Influenced by the Spread of Promotive
Information in Social Media.................................... 1
   *Qian Pan, Haoxiang Xia, and Shuangling Luo*

Complex Network's Competitive Growth Model
of Degree-Characteristic Inheritance .......................... 12
   *Hualu Gu, Xianduan Yang, and Shouyang Wang*

An Alternative Fuzzy Linguistic Approach for Determining Criteria
Weights and Segmenting Consumers for New Product Development:
A Case Study................................................ 23
   *Sirin Suprasongsin, Van-Nam Huynh, and Pisal Yenradee*

MANDY: Towards a Smart Primary Care Chatbot Application........... 38
   *Lin Ni, Chenhao Lu, Niu Liu, and Jiamou Liu*

Sequence-Based Measure for Assessing Drug-Side Effect Causal Relation
from Electronic Medical Records .............................. 53
   *Tran-Thai Dang and Tu-Bao Ho*

A Multi-center Physiological Data Repository for SUDEP: Data Curation,
Data Conversion and Workflow ................................ 66
   *Wanchat Theeranaew, Bilal Zonjy, James McDonald, Farhad Kaffashi,
Samden Lhatoo, and Kenneth Loparo*

Concept Name Similarity Measure on SNOMED CT................. 76
   *Htet Htet Htun and Virach Sornlertlamvanich*

Comparative Study of Using Word Co-occurrence to Extract Disease
Symptoms from Web Documents .............................. 91
   *Chaveevan Pechsiri and Renu Sukharomana*

Forecasting the Duration of Network Public Opinions Caused
by the Failure of Public Policies: The Case of China................. 101
   *Ying Lian, Xuefan Dong, Ding Li, and Yijun Liu*

Modeling of Interdependent Critical Infrastructures Network
in Consideration of the Hierarchy ............................. 117
   *ChengHao Jin, LiLi Rong, and Kang Sun*

Emergency Attribute Significance Ranking Method Based
on Information Gain . . . . . . . . . . . . . . . . . . . . . . . . . . . . . . . . . .    129
    *Ning Wang, Haiyuan Liu, Huaiming Li, Yanzhang Wang, Qiuyan Zhong,*
    *and Xuehua Wang*

Predicting Hashtag Popularity of Social Emergency by a Robust Feature
Extraction Method. . . . . . . . . . . . . . . . . . . . . . . . . . . . . . . . . . . .    136
    *Qianqian Li and Ying Li*

Mining Online Customer Reviews for Products Aspect-Based Ranking . . . . .    150
    *Chonghui Guo, Zhonglian Du, and Xinyue Kou*

An Empirical Analysis of the Chronergy of the Impact of Web Search
Volume on the Premiere Box Office . . . . . . . . . . . . . . . . . . . . . . . . .    162
    *Ling Qu, Guangfei Yang, and Donghua Pan*

Societal Risk and Stock Market Volatility in China: A Causality Analysis . . .    175
    *Nuo Xu and Xijin Tang*

A New Hybrid Linear-Nonlinear Model Based on Decomposition
of Discrete Wavelet Transform for Time Series Forecasting . . . . . . . . . . . .    186
    *Warut Pannakkong and Van-Nam Huynh*

The Distribution Semantics of Extended Argumentation. . . . . . . . . . . . . . .    197
    *Nguyen Duy Hung*

An Ontology-Based Knowledge Framework for Software Testing . . . . . . . . .    212
    *Shanmuganathan Vasanthapriyan, Jing Tian, and Jianwen Xiang*

The Effect of Task Allocation Strategy on Knowledge Intensive Team
Performance Based on Computational Experiment. . . . . . . . . . . . . . . . . . .    227
    *Shaoni Wang, Yanzhong Dang, and Jiangning Wu*

A Kind of Investor-Friendly Dual-Trigger Contingent Convertible Bond . . . .    242
    *Wenhua Wang and Xuezhi Qin*

Analysis on Influencing Factors on Cultivation of Graduate Innovation
Ability Based on System Dynamics. . . . . . . . . . . . . . . . . . . . . . . . . . .    250
    *Bing Xiao and Vira Chankong*

**Author Index** . . . . . . . . . . . . . . . . . . . . . . . . . . . . . . . . . . . . . . .    267

# Dynamics of Brand Acceptance Influenced by the Spread of Promotive Information in Social Media

Qian Pan[1(✉)], Haoxiang Xia[1], and Shuangling Luo[2]

[1] Institute of Systems Engineering,
Dalian University of Technology, Dalian, China
lovelyrita@mail.dlut.edu.cn, hxxia@dlut.edu.cn
[2] Collaborative Innovation Center for Transport Studies,
Dalian Maritime University, Dalian, China
slluo@dlmu.edu.cn

**Abstract.** In this paper we propose an agent-based model that combine the Majority-Rule-based Voter model in opinion dynamics and the SI Model for information spreading to analyze the dynamics of brand acceptance in social media. We focus on two important parameters in diffusion dynamics: the decayed transmission rate ($\beta$) and the diffusion interval ($\theta$). When the system is stable, the order parameter of the system is the duration time ($\tau$). In the absence of opinion interaction, the simulation results indicate that, when a disadvantaged brand tries to occupy a large market share through social marketing approaches, it is always effective to let the opponent be the propaganda target. While with the Majority-Rule-based Voter model included, we observe that the opinion interaction could have a dual function, which show that a brand holding a small market share needs to adopt diverse marketing methods according to different population types.

**Keywords:** Social marketing · Dynamics of brand acceptance · Opinion dynamics · Diffusion dynamics

## 1 Introduction

Kotler and Zaltman [1] proposed the concept of "social marketing", which implies one can market a social perspective to groups to whom a commodity is sold. Nowadays this idea of social marketing becomes even more prevailing due to today's widespread of social media and online social networks. For example, in the summer of 2013, Coca-Cola Company launched Coca-Cola nickname bottle in the Chinese market. They printed cyber language on the bottle and spread across multiple social platforms, covering WeChat, renren.com, Douban.com and QZONE, winning the China Effie Creative Awards. Coca-Cola's success indicates social marketing is thriving in brand diffusion. Today, we can also see some social marketing model enterprise, like VANCL, Durex and Nike. This brings about a noticeable research issue how a company achieves a high-level brand acceptance through social marketing. To cope with this issue, it is deserved to explore the underlying dynamic process of brand acceptance

© Springer Nature Singapore Pte Ltd. 2017
J. Chen et al. (Eds.): KSS 2017, CCIS 780, pp. 1–11, 2017.
https://doi.org/10.1007/978-981-10-6989-5_1

in the context of social marketing. Such dynamics of brand acceptance can basically be modeled as process in which two groups of opinions (i.e. positive and negative) evolve within a population (i.e. the customers and potential customers). Much work has been done in the context of "opinion dynamics" [2–4].

Opinion dynamics is a sub-field of social dynamics, utilizing mathematical and physical models and the agent-based computational modeling tools, to explore the evolution of collective opinions in human population. Among different types of opinion-dynamics models, the voter model is a suitable form to depict the dynamics of brand acceptance in public. The academic inquires on the voter model can be traced back to 1970s; and plenty of researches have been conducted since then [5–7]. Frachebourg and Krapivsky [8] proposed a dynamical process of an ordered voter model in a regular lattice, and found that the ordering process relying on the dimensionality of the lattice. The voter-model-based opinion dynamics on all sorts of complex networks has become a focal topic of concern since the 2000s; and a series of endeavors have been conducted in this direction, e.g., [9]. As a variation of the standard voter model, the majority rule model concerns about the opinion dynamics under the situation of group discussions and debates. The Majority Rule model derives from the study of "herd behavior" in social psychology [2, 10, 11], which illustrates a general trend of attitudes and beliefs that individuals are willing to be consistent with the social organization in which they live. Galam and his colleagues gave extensive researches on the dynamics based on majority voting rules. Some intuitive results are given to explain some absurd social phenomena, such as the formation of dictatorship from hierarchical democratic majority-rule voting [12, 13]. In addition, Chen and Redner [14] investigated the majority rule in the fixed odd-sized population.

Besides the opinion-dynamic process, the acceptance of a brand is simultaneously related to the diffusion of relevant information and news about the brand. The study of such diffusion process involves in the diffusion modeling and dynamics. The diffusion dynamics are closely related to the epidemic dynamics of infectious diseases. The simplest epidemiological model is the susceptible–infected (SI) model in which individuals can only hold two discrete states, susceptible and infected. The other two well-known studies on the epidemiology models are SIS and SIR models [15]. The diffusion of information and opinions in a human population can to some extent be analogous to the infection of diseases; thus, we can borrow and revise the mathematical models of epidemiology to study the diffusion of opinions through contagions, e.g., [16, 17]. Rumors and information spreading phenomena are the prototypical examples of social contagion processes in which the infection mechanism can be considered of psychological origin. Also, Zanette [18] proposed an information diffusion model based on a small world network. The nodes in the model may be in one of three states: (1) Susceptible (S), where the individual has not yet known about the information, called "ignorant"; (2) Infected (I), where individuals who can spread rumors, information or knowledge; and (3) Refractory (R), where individuals who are adopters of the information but no longer spread it. The fast and efficient diffusion of information is often desired in social marketing. Thus, the connection between these modeling methods and commercial applications is also deserve attention. The epidemic models for information propagation may be used in marketing campaigns applying the viral marketing techniques [16, 19]. Information diffusion in multilayer networks have also gained much stressing in recent years [20–22]. For instance, Wang et al. [23] have investigated the asymmetrical interrelation between two process account for the spreading of an epidemic through a physical-contact network, and the spreading of information awareness on a communication network.

As mentioned above, both opinion dynamics and information diffusion dynamics have been comprehensively studied. However, the key characteristics of dynamics of brand acceptance cannot been fully addressed if simply applying either an opinion-dynamic model or an information-diffusion model to the situation of brand reputation formation. Factually, the dynamics of brand acceptance is essentially an interwoven dynamics of endogenous opinion-dynamics disturbed by an information-diffusion process. However, the dual dynamics of opinion-evolution and information-diffusion has not been well-studied in literature. Therefore, in this paper we propose an agent-based model that combine the Majority-Rule-based Voter model in opinion dynamics and the SI Model for information diffusion to analyze the dynamics of brand acceptance in social media. To coherent with this aim, the rest of the paper is structured as follows. The model is elaborated in the next section. Then, we present and discuss the results of simulation experiments in different situations and parameter settings. The implications for social marketing strategies will then be discussed. The paper is finally concluded in the last section.

## 2 Model Description

In this paper, we choose a Majority-Rule-based Voter model as the base model, which has been extensively studied [24, 25]. According to the Social Influence Theory [26, 27], there are three main factors influencing the formation of individual opinion: the individual's own views, neighboring individuals' collective views and the external influence from the government policy and the mass media. The first two factors are taken into account in the Majority-Rule-based Voter model adopted in this work, in which the collective influences of neighbors obey the Majority rule. Thus, the opinion of an individual is determined by a function that combines his or her own view and that of the majority of neighbors. A conviction power ($\varphi$) can then be defined to measure the extent to which one sticks to his\her own opinion under the influence of others.

In the proposed model, we use $O = U_{i=1}^{N} o_i$ as the set of opinions. Alike to the original voter model, this Majority-Rule-based Voter model uses discrete values (i.e. 0 or 1) to represent individuals' opinions. Opinions randomly distributed in the grid. The initial density of positive opinion is $\rho_A(0)$, so the density of negative opinion is $1 - \rho_A(0)$. The update rule is defined as follows:

$$O_{i,t+1} = \text{sgm}\left( \varphi * O_{i,1} + (1 - \varphi) * \text{sgn}\left( \sum_j O_{j,t} - 5 \right) - 0.5 \right) \qquad (1)$$

where,

$$sgn(x) = \begin{cases} 1, & x \geq 0 \\ 0, & x < 0 \end{cases} \qquad (2)$$

In Eq. (1), $O_{i,t}$ denotes the opinion of individual $i$ at time step $t$. When majority of neighbors (equal to or greater than 5) hold positive opinion, the neighbors' opinion is taken as 1, otherwise taken as 0. When conviction power $\varphi$ is small, it indicates that the individual in the group is easy to be affected by others; $\varphi = 0.5$, the individual's opinion is influenced by its own intrinsic view and neighbors' view with the same extent, called rational in this paper. While when $\varphi$ is much bigger, people always stick to its own opinion and very stubborn. They will not change under whatever circumstances.

The previous majority-rule model is combined with an information diffusion model. The brand competes for approvals from different types attitudes is similar to two diseases compete for the same population of hosts because one disease kills hosts before the other can infect them. In some epidemiology study is the behavior of competing virus [28, 29]. So in order to simulate the competition between brands based on SI Model, we use A and B express a certain brand's supporters (hold opinion 1) and opponents (hold opinion 0). A given node has only two states in the message propagation process: inactive and active. Initially, the brand is in disadvantaged situation in the market, holding only a small percentage of market share. There are a large number of opponents B in the group and a small number of supporters A. They are all in inactive states, then a seed node are randomly chosen to become diffusion source from the supporters. Therefore, there are four states in the diffusion dynamics: ActiveA, ActiveBA, InactiveA and InactiveB. In the diffusion process, only disadvantaged brand can take social marketing approaches to attract customers. The opponents will not strike back. Thus, the dynamics is asymmetric. Finally, acquiring a new idea or being convinced that a new information is grounded may need time and exposure to more than one source of information, which memory has an important role [30]. Therefore, we define the persuaded active nodes: activeBA, individuals changing from opponents to supporters, have decayed spreading rate $\beta$. While the convincing active nodes (activeA) have a larger diffusion rate $\alpha$. We also assume that the neighbors around the seed node do not influence its opinion. Figure 1 illustrates the diffusion process.

- activeA and activeBA—iffusion source, those who are convinced for the adoption of one brand, usully fascinated about his choice for that they also try to convince others to take the same choice, which is analogous to "infective state" in epidemics;
- inactiveA and inactiveB—susceptible, those who have not yet convinced to spread their attitudes.

For example, when an activeA agent interact with an inactiveA agent, the inactiveA node will change into acticeA state at bigger probability $\alpha$. Or an inactiveB agent interact with an active node, the inactiveB agent may change to activeBA state at smaller probability $\beta$.

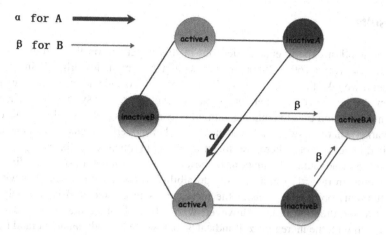

**Fig. 1.** Illustration of the diffusion epidemics Model: inactiveA is the agent holding the positive attitude for the brand but do not have the transmission capacity, while inactiveB agent holds the negative attitude. Similarly, active node is willing to transmit its opinion to the neighbor. An inactiveA node can acquire the news and transmission capacity at probability $\alpha$. Nodes in state inactiveB can be contagious with the competitive attitude at probability $\beta$. The network structure is two-dimensional lattice.

The interaction diagram shown in Fig. 1 with the following processes:

- activeA + inactiveA $\rightarrow$ 2activeA with initial convincing rate $\alpha$;
- activeA + inactiveB $\rightarrow$ active + activeBA with initial convincing rate $\alpha$;
- activeBA + inactiveB $\rightarrow$ activeBA + active with decayed rate $\beta$;
- activeBA + inactiveB $\rightarrow$ 2activeBA with decayed rate $\beta$.

Based on these rules, we show that the combination of opinion dynamics and the information spreading to analyze the dynamics of brand acceptance in social media. Under different initial conditions, how does the diffusion interval ($\theta$) and the decayed diffusion rate ($\beta$) influence the dynamics of brand acceptance. The order parameter is the duration time $\tau$, which denotes duration time steps when the system reaches to a stable state. In this paper, when all inactive nodes become active nodes, or the proportion of active individuals fluctuates slightly near a value, which indicates that the system has reached a stable state [31]. The case that the diffusion has not carried out smoothly is out of discussion. Then we firstly investigate the situation when it only exists the diffusion process. Afterwards, we consider opinion interactions in situations wherein the diffusion process also exists. The model proposed in this paper assumes that: (1) all individuals with the same opinion have a constant transmission rate; (2) the propagation of the message is intermittent and recurrent; (3) the opinion evolution and the news diffusion happened simultaneously.

## 3  Results

All the simulation experiments carried out in this paper are on the two-dimensional lattice of N = 160 * 160 = 25600 agents with a Moore neighborhood. In the first experiment, we take the "diffusion-only" situation into consideration. In this paper, we focus on the special case when the initial market share is asymmetric. When the disadvantaged brand tries to compete with other competitors that have already occupied large market share, how to advertising? Whether advertisements can achieve the anticipated effect or not? Therefore, to set up the experiments, we fix the $\rho_A = 0.3$ to represent the disadvantaged situation. It shows obviously in the log-log coordinates in Fig. 2. Diffusion rate $\alpha(\beta)$ indicates the possibility to convince the neighbor with the same (different) opinion to transfer the message. With a fixed $\alpha$, $\beta$ is inversely proportional to the duration time $\tau$ However, with a fixed $\beta$, the duration time does not change much with the increased $\alpha$. It indicates that, when a disadvantaged brand tries to occupy a large market share through social marketing approaches, it is always effective to let the opponent be the propaganda target.

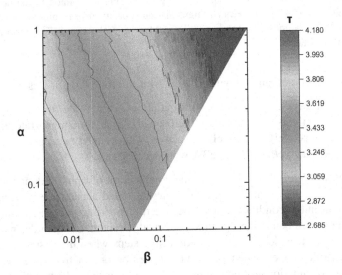

**Fig. 2.** Variation of duration time for different values of $\alpha$ and $\beta$, also $\alpha > \beta$ represent the competitive relationship between different opinions. The color bar illustrates the value of duration time. Warmer color represents larger value. (Color figure online)

According to the results, we then fixed $\alpha = 1$ in the following experiments for simplicity. In Fig. 3, $\theta = 0.1, 0.4$ and $0.8$, represents different diffusion intensities. The result shows the duration time decreases exponentially as the transmission rate increases, and such trend remains the same under various diffusion intervals.

However, in real world, brand acceptance is influenced by two factors: brand reputation and social marketing activities intended to strengthen the competitiveness of the brand. Therefore, we consider opinion interactions in situations wherein the diffusion process also exists, reaching some results quite different from those in the above

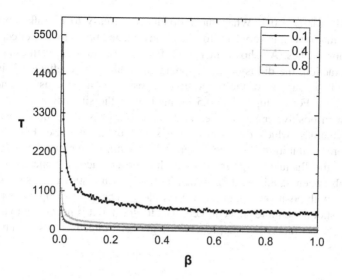

**Fig. 3.** Variation of duration time with transmission rate β for different values of θ = 0.1, 0.4 and 0.8.

discussed situations. Figure 4 shows for $\varphi = 0.4$, a group of individuals are more likely to adopt the neighbor's opinion; $\varphi = 0.5$, neighbors and their own views have the equal impact; with $\varphi = 0.6$, the group is very stubborn, which their views are not easy to be changed. Therefore, the results could be quite different with different group types.

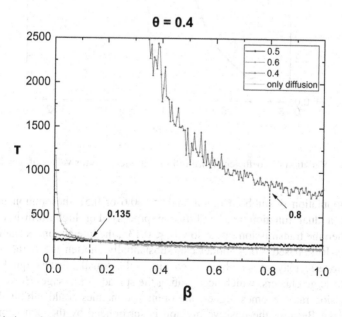

**Fig. 4.** Variation of duration time with transmission rate β for different values of $\varphi = 0.4$, 0.5 and 0.6.

Especially for $\varphi = 0.4$, When the individuals are easy to be influenced by the neighboring majority, it is very hard for the disadvantaged brand to take effective social marketing approaches. As shown in Fig. 5, for $\theta \geq 0.285$, due to the inhibition of opinion interaction, the diffusion can't spread smoothly. The coefficient P is used to uniform results in support of the proportion of positive opinion consensus appears in 100 simulations. For example, P = 0.5 means half of the simulation experiments are consensus with positive opinion. Therefore, there is a phase transition process in the diffusion dynamics, which differs from the single dynamics case. For $\theta \geq 0.285$, although there is inhibition of opinion interaction, the diffusion process could also carry out. Despite the fluctuant experimental results, the coefficient P increases with the increased $\theta$ between 0.285 and 0.32. When $\theta \geq 0.32$, simulation results steadily reach to consensus with positive opinion. When people are change their opinions easily, the propagation should be more concentrated. It indicates that it is necessary to adopt high frequency social marketing approach to get an effective result under such circumstances.

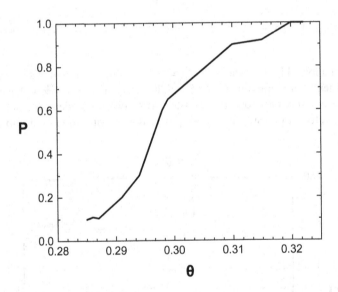

**Fig. 5.** Variation of coefficient P with diffusion interval $\theta$ for values of $\varphi = 0.4$.

If the population is stubborn or rational ($\tau = 0.6$ or $0.5$), the opinion interaction could have a dual function in the diffusion process. For high diffusion intensity $\theta = 0.4$, when the transmission rate is low $\beta < 0.135$, the interaction mechanism from the Majority-Rule-based Voter model can substantially shorten the duration time of diffusion process, playing an obvious positive role. Small clusters quickly gather together into large clusters, which accelerating the spread of message. However, when the transmission rate becomes higher, the opinion dynamics could inhibit the propagation process. Because the positive opinion is surrounded by the same opinion, the clusters inhibit the diffusion process. This implies that, a disadvantaged brand should take high-frequency and low-intensity marketing strategy to get better outcomes from activities. However, for $\varphi = 0.4$, $\tau$ fluctuates, the effectiveness of social marketing is with great uncertainty influenced by opinion interaction. When $\beta > 0.8$, the results tend to be stable. Therefore, according to the result shown in Fig. 4, if individuals in a population always change their minds easily, it's hard for the diffusion process to carry out smoothly. Thus a disadvantaged brand should take high-frequency and high-intensity marketing approaches to get better outcomes from marketing activities, if it faces a population who easily change their opinions when influenced by others.

## 4   Conclusion

In this paper, we attempt to disclose the implications for social marketing strategies through a series of computational simulations combining the Majority-Rule-based Voter model with SI Model. We firstly investigate the situation wherein it exists only the diffusion process. Results show that, when a disadvantaged brand tries to occupy a large market share through social marketing approaches, it is always effective to let the opponent be the propaganda target. The duration time decreases exponentially as the transmission rate increases, and such trend remains the same under various situations of different diffusion intervals. The results are quite different when we put opinion interaction together with the diffusion process. When the population is stubborn or rational, the opinion interaction could have a dual function in the diffusion process. For high diffusion intensity, when the decayed diffusion rate is low, the opinion interaction process within the population can substantially shorten the duration time of diffusion process, thus playing an obvious positive role, while the high decayed diffusion rate could inhibit the diffusion process. This implies that, a disadvantaged brand should take high-frequency and lower intensity marketing strategies to get better outcomes from activities. However, if individuals tend to accept others' opinions easily, which the opinion interactions could have a strong destabilization impact on the diffusion process, a disadvantaged brand should take high-frequency and high-intensity social marketing approaches correspondingly.

To achieve effective marketing outcomes, generally, a brand holding a small market share needs to adopt diverse marketing approaches according to different population types. To a certain degree, it can be argued that a stubborn or rational population is in favor of a disadvantaged brand, because once individuals change their opinions when they are influenced by the information, they will not easily change back. In such situations, therefore, the high-frequency, low-intensity marketing activities should be effective enough in achieving larger brand acceptance. If individuals tend to easily accept others' opinions, however, a disadvantaged brand should take high-frequency, high-intensity social marketing activities.

**Acknowledgements.** This work is partly supported by the National Natural Science Foundation of China under grant Nos. 71401024 and 71371040, respectively.

# References

1. Kotler, P., Zaltman, G.: Social marketing: an approach to planned social change. J. Mark. **35**, 3–12 (1971)
2. Castellano, C., Fortunato, S., Loreto, V.: Statistical physics of social dynamics. Reviews of Modern Physics, 81, 556–591 (4AD)
3. Holyst, J.A., Kacperski, K., Schweitzer, F.: Social impact models of opinion dynamics. Ann. Rev. Computat. Phys. **9**, 253–273 (2001)
4. Stauffer, D.: Sociophysics simulations II: opinion dynamics. AIP Conf. Proc. **779**, 56–68 (2005)
5. Dornic, I., Chaté, H., Chave, J., Hinrichsen, H.: Critical coarsening without surface tension: the universality class of the voter model. Phys. Rev. Lett. **87**, 045701 (2001)
6. Holley, R.A., Richard, T.M.: Ergodic theorems for weakly interacting infinite systems and the voter model. Ann. Probab. **2**, 347–370 (1975)
7. Lambiotte, R., Saramäki, J., Blondel, V.D.: Dynamics of latent voters. Phys. Rev. E – Stat. Nonlinear Soft Matter Phys. **79**(4) (2009)
8. Frachebourg, L., Krapivsky, P.: Exact results for kinetics of catalytic reactions. Phys. Rev. E **53**, R3009–R3012 (1996)
9. Castellano, C., Vilone, D., Vespignani, A.: Incomplete ordering of the voter model on small-world networks. Europhys. Lett. **63**, 153–158 (2003)
10. Cox, J.T.: Coalescing random walks and voter model consensus times on the torus in Zd. Ann. Probab. **17**, 1333–1366 (1989)
11. Wenglinsky, M., Milgram, S.: Obedience to authority: an experimental view. Contemp. Sociol. **4**, 613 (1975)
12. Galam, S.: Real space renormalization group and totalitarian paradox of majority rule voting. Phys. A: Stat. Mech. Appl. **285**, 66–76 (2000)
13. Galam, S., Wonczak, S.: Dictatorship from majority rule voting. Eur. Phys. J. B **18**, 183–186 (2000)
14. Chen, P., Redner, S.: Majority rule dynamics in finite dimensions. Phys. Rev. E – Stat. Nonlinear Soft Matter Phys. **71**(3) (2005)
15. Hethcote, H.W.: The mathematics of infectious diseases. SIAM rev. **42**, 599–653 (2000)
16. Moreno, Y., Nekovee, M., Pacheco, A.F.: Dynamics of rumor spreading in complex networks. Phys. Rev. E – Stat. Nonlinear Soft Matter Phys. **69**(62) (2004)
17. Gronlund, A., Holme, P.: A network-based threshold model for the spreading of fads in society and markets. Adv. Complex Syst. **8**, 261–273 (2005)
18. Zanette, D.H.: Dynamics of rumor propagation on small-world networks. Phys. Rev. E – Stat. Nonlinear Soft Matter Phys. **65**(4) (2002)
19. Leskovec, J., Adamic, L.A., Huberman, B.A.: The dynamics of viral marketing. ACM Trans. Web **1**, 5 (2007)
20. Kivelä, M., Arenas, A., Barthelemy, M., Gleeson, J.P., Moreno, Y., Porter, M.A.: Multilayer networks. J. Complex Netw. **2**(3), 203–271 (2014)
21. Boccaletti, S., Bianconi, G., et al.: The structure and dynamics of multilayer networks. Struct. Dyn. Multilayer Netw. **544**(1), 1–122 (2014)
22. Shao, J., Buldyrev, S.V., Havlin, S., Stanley, H.E.: Cascade of failures in coupled network systems with multiple support-dependence relations. Phys. Rev. E – Stat. Nonlinear Soft Matter Phys., vol. 83, (2011)
23. Wang, W., et al.: Asymmetrically interacting spreading dynamics on complex layered networks. Sci. rep. **4**, 5097 (2014)

24. Galam, S.: The september 11 attack: a percolation of individual passive support. Eur. Phys. J. B **26**, 269–272 (2002)
25. Mobilia, M., Redner, S.: Majority versus minority dynamics: phase transition in an interacting two-state spin system. Phys. rev. E Stat. Nonlinear Soft Matter Phys. **68**, 046106 (2003)
26. Lewenstein, M., Nowak, A., Latane, B.: Statistical mechanics of social impact. Phys. Rev. A **45**, 763–776 (1992)
27. Schweitzer, F., Holyst, J.A.: Modelling collective opinion formation by means of active brownian particles. Eur. Phys. J. B **732**, 1–10 (2000)
28. Castillo-Chavez, C., Huang, W., Li, J.: Competitive exclusion in gonorrhea models and other sexually transmitted diseases. SIAM J. Appl. Math. **56**, 494–508 (1996)
29. Andreasen, V., Lin, J., Levin, S.A.: The dynamics of cocirculating influenza strains conferring partial cross-immunity. J. Math. Biol. **35**, 825–842 (1997)
30. Dodds, P.S., Watts, D.J.: A generalized model of social and biological contagion. J. Theor. Biol. **232**, 587–604 (2005)
31. Pastor-Satorras, R., Vespignani, A.: Epidemic dynamics and endemic states in complex networks. Phys. Rev. E **63**, 066117 (2001)

# Complex Network's Competitive Growth Model of Degree-Characteristic Inheritance

Hualu Gu[1,2,3], Xianduan Yang[1,4], and Shouyang Wang[1,2,4(✉)]

[1] University of Chinese Academy of Sciences, Beijing 100190, China
{guhualu15,yangxianduan15}@mails.ucas.ac.cn
[2] School of Economics and Management,
University of Chinese Academy of Sciences, Beijing 100190, China
[3] Technology and Engineering Center for Space Utilization,
Chinese Academy of Sciences, Beijing 100094, China
[4] Academy of Mathematics and Systems Science,
Chinese Academy of Sciences, Beijing 100190, China
sywang@amss.ac.cn

**Abstract.** Complex network is a kind of network between regular network and stochastic network. Inspired by biological evolution, we introduced resource competition and genetic inheritance into the growth process of network, and proposed a new growth model with the priority connection of scale-free network. Emulated analysis shows that the network model of competitive growth is no longer power-law, but it obeys exponential distribution. The competitive growth model of degree-characteristic inheritance is negative skew. And it shows a linear relationship between the logarithm of average degree and the network size, which is also proved by the mathematical deduction. In addition, the average clustering coefficient of network decreases with the increase of the genetic coefficient, while the average path length increases with the increase of the inherited coefficient. The whole model is topologically tunable. Different combinations of parameters can produce network models with different properties.

**Keywords:** Exponential distribution · Inheritance · Competition · Average degree · Clustering coefficient · Shortest path

## 1 Introduction

Gaussian distribution and Pareto distribution is the two major golden rules of natural evolution and the development of human society. Gaussian distribution is essentially independent. Under the condition of the law of large numbers, a large number of homogeneous independent events lead to Gaussian distribution. Pareto distribution is also called power-law distribution and its essence is positive feedback mechanism. When the event is no longer independent, the individuals' evolution development will have an impact on themselves and other similar things. This leads to Pareto distribution [1]. A large number of homogeneous or relevant things can be seen as a system. There are complex relationships among individuals in the system. It is significantly meaningful for us to study these relationships and further identify the systematical laws.

© Springer Nature Singapore Pte Ltd. 2017
J. Chen et al. (Eds.): KSS 2017, CCIS 780, pp. 12–22, 2017.
https://doi.org/10.1007/978-981-10-6989-5_2

Complex systems can be described by the network. Combining the network with complex systems has become a new discipline. The network's node represents individuals in the system, and the edge between different nodes represents the relationship among individuals. Boolean network can only reflect the relationship's existence or not, while the weighted network can further describe the intensity and direction of the influence [2]. So far, many classic network models have been developed, such as regular network, small-world network, stochastic network, scale-free network, hierarchical network, and deterministic network, etc. [3]. These models have achieved their goals while describing the relevant systems. But there are still many problems. The regular network cannot reflect the huge difference of the nodes' degree distribution [4]. The degree distribution calculated by small-world network [5] is not the same with the actual network. And the network cannot grow independently. The scale-free network has no clustering characteristic because the connective probability between nodes is constant [6].

BA scale-free network model has the characteristics of growth and prior connection. The prior connection makes the network obey the power-law distribution [7, 8]. It can explain the Matthew effect well. But it still has the following problems. Firstly, the network size is linearly and infinitely growing [9]. Secondly, the network's structure has no effect on the growth. Thirdly, the new nodes are not relevant to the original nodes. Every node is isolated. Fourthly, the old nodes occupy most resources. The nodes with lowest degree always occupy the largest space, which does not match many phenomena in reality. The new nodes are discriminated, and the node degree is the power-law function of time. Fifthly, the network's clustering coefficient and the average degree is hard to adjust [10]. It is not suitable for the fickle complex environmental analysis.

In view of the above problems, this paper introduced the mechanism of inheritance and competition into the traditional scale-free network model. A new model named inherited competitive growth network is proposed. Then we analyze the different properties shown by the new models in detail.

## 2  Model Construction

We assume that all nodes are greedy. Their basic goal is to obtain the maximum connective degree. But connection costs at each time. Every new node is generated by two nodes that have the nearest degree. We call the two nodes parent nodes. All new nodes tend to connect to the points with larger degree regardless of the bigness or smallness of the initial connective degree [11]. However, the old nodes tend to connect the new nodes whose parent nodes have large degree. Under the same condition, the weak parents' sub-node is more likely to be rejected. The followings are the calculation steps to construct the inherited competitive growth network.

Step 1 is to initialize the network. Firstly, we construct an initial network with $n_0$ number of nodes. There are three kinds of structural form that include complete isolated network, stochastic network, and interconnected network.

Step 2 is heredity and inheritance. We assume that the network's growth rate is $r$. Every time we randomly select $n_{new}$ numbers of node as parent nodes. Then all parent nodes are ordered according to their respective node degree. $n$ represents the network size and $N$ represents the number of iterative times. There is the following relationship.

$$n_{new} = 2 * r * n \tag{1}$$

$$n = n_0 * (1+r)^N \tag{2}$$

According to the order, each two adjacent nodes generate a sub-node uninterruptedly. $k_{pi}$ represents the degree of parent node $i$. There are two cases when the sub-node is connected for the first time. One is the constant value and the other is to generate the initial degree in accordance with the inherited coefficient $h$.

$$k_{new} = \begin{cases} h * k_p, h \in (0,1), m_0 \leq h * k_p \\ m_0, m_0 > h * k_p \end{cases} \tag{3}$$

$$k_p = \frac{k_{p1} + k_{p2}}{2} \tag{4}$$

Step 3 is prior connection. At each step, the sub-node will be unconditionally connected to its two parent nodes to ensure that the network is connected, and then they will independently and uninterruptedly connected to nodes in the original network. The number of connections is determined by the initial connective degree obtained in step 2, and the probability that the sub-node is connected with the node $i$ is:

$$\Pi_i = \frac{k_i + 1}{\sum_j (k_j + 1)} \tag{5}$$

Step 4 is rejecting to connect. The old nodes have the right to reject the connection with the new node. The new nodes always tend to choose nodes with larger degree. However, the marginal value brought for the old nodes by the new node decreases as the old nodes' degree increase. So the old nodes tend to reject. For those new nodes with large parent nodes, the old nodes value its parent nodes' extensive connection relationships. So the old nodes will be more likely to connect with them. The old nodes with smaller degree have lower rejection probability due to their poor condition. The following is a rejection function. $v$ is the adjustment coefficient:

$$if \left( k_{old} - k_p * rand * v \right) > k_{avg}, reject \tag{6}$$

Step 5, for nodes that are rejected to connect, there are two possibilities: finding the next node that can be connected continuously or giving up. The nodes that have been rejected for many times will die because they do not obtain enough resources. This results in the low growth rate of network. For one initial network with a large difference in degree distribution, the new node will be removed once it is born.

## 3  Simulation and Analysis

### 3.1  Degree Distribution of Competitive Growth Network

Firstly, we only consider the mechanism of competition. The inherited coefficient is zero, and those rejected nodes can continue to connect in order that the degree of new nodes is constant. The network's degree distribution is shown in Fig. 1. In experiment of different number of iterations, the earliest nodes joining in the network have largest degree. They are independent and away from other nodes in the Fig. 1(a). The larger degree of initial network's node is and the smaller the threshold value is, the smaller chances of connecting new nodes have. Assuming that is the system's minimum protection net, the degree of initial node is regarded as a ladder. Then, if the threshold value is too small, it can't play the role of protection. With the advance of the time step, the ladder will be increasingly high and narrow. Figure 2 is the scatter diagram between the logarithm of the probability of new node's degree distribution and the degree. The scatter diagram obviously shows a concave, indicating that the new node's degree does not comply with the power-law distribution.

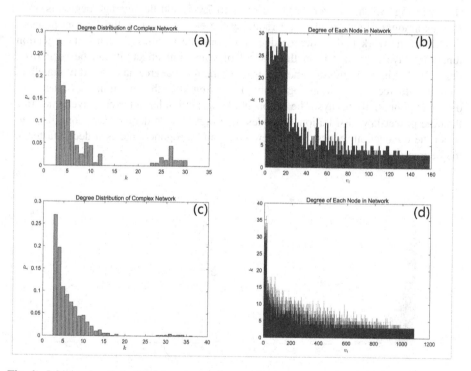

**Fig. 1.** Initial network is full interconnected. $n_0 = 20, h = 0, m_0 = 3$. (a) Degree distribution histogram at $N = 20$; (b) Degree of each node at $N = 20$; (c) Degree distribution histogram at $N = 40$; (d) Degree of each node at $N = 40$.

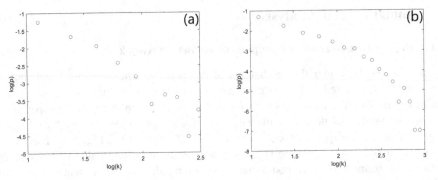

**Fig. 2.** (a) Logarithm of degree distribution' probability and degree at $N = 20$; (b) Logarithm of degree distribution' probability and degree at $N = 40$.

Then, it needs to be considered whether there is exponential relationship between the probability of new node's degree distribution and the degree or not in the situation of competitive mechanism. Continuing to expand the number of iterative times, it is shown in Fig. 3. The network size increases to 7345, and the average degree is 6.03. A large amount of data is good for reducing the error caused by the random process. The initial network node's degree does not continue to grow, because the rejection function plays a good role in the respect of constraint effect. It can be seen from Fig. 3(b) that linear regression method is used, the node degree has a good fitting effect with the degree after taking logarithm, indicating that the distribution of network degree exponentially decays. The exponential distribution has a faster convergence rate than the power-law distribution, so nodes with higher node degrees are more likely to reject the connection request of the new node, and arrogance makes it decay faster in the degree distribution.

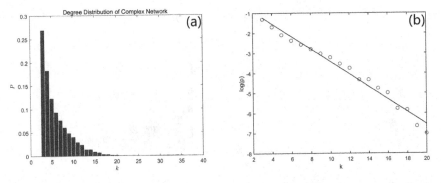

**Fig. 3.** (a) Degree distribution histogram at $N = 60$; (b) Logarithm of degree distribution' probability and degree at $N = 60$, and $log(p) = -0.301 - 0.312 * k$

Furthermore, the connection rejection means missing a connection opportunity, and the initial degree of the new node is variable, which is shown in Fig. 4. After 50 iterations, the two ends of the network' node degree distribution are different. In Fig. 4(b), the right-hand part is concave. In Fig. 4(d), the curve has a better fitting effect indicating that it is still exponential distribution, but fitting effect is not as good as Fig. 3. In Fig. 4(b), the left-hand part does not obey the power-law distribution or the exponential distribution. It is continuously tested by the Gaussian distribution (also called normal distribution). The significance of the K-S test is less than 0.05, so it does not support the assumption of normal distribution.

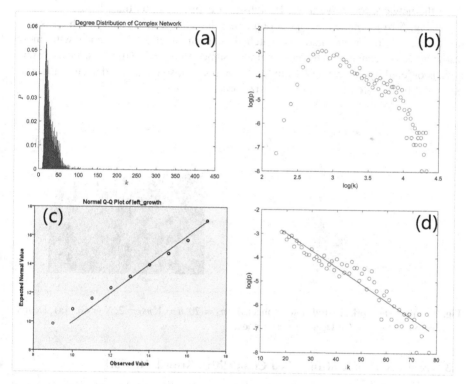

**Fig. 4.** Initial network is full interconnected. $n_0 = 20, h = 0, m_0 = 3, N = 50$. (a) Degree distribution histogram; (b) Logarithm of degree distribution' probability and degree; (c) Q-Q plot of nodes' degree in the left-hand of picture a; (d) Logarithm of degree distribution' probability and degree in the right-hand of picture a.

## 3.2 Degree Distribution of Inherited Competitive Growth Network

After introducing the inherited mechanism into the model, the new node completely inherits the degree characteristic of parent node. Figure 5(a) shows that the distribution of network node degree is negative skew and all nodes' degree is very large. In the experiment, this phenomenon is universal and it is not related with the network's initial state. We shall explain the reason for this phenomenon next. The growth of the node with the largest node degree will be constrained under the influence of rejection function, so it will miss the opportunity of being connected and fall into the range with lower degree under the same conditions. Moreover, the new node's characteristic of prior connection will force the old node degree to continue to grow. There are both the power of resistance in front and thrust behind. A large number of nodes gradually cluster under the threshold value of rejection function, and therefore the threshold value of rejection function is the key factor that determines the degree position with the largest degree distribution probability. On the other hand, the joining of new nodes increases the network's overall average degree. The threshold value of rejection function becomes larger with the increase of the number of iteration, which will further drive the network to evolve towards high-level cooperation on the whole.

In Fig. 5(b) the node's degree arrangement is dentate, the reason of which is that the degree of the parent node is matched in each iteration. The node with more resources is associated with the node with the same resource. Randomly selected parent node is ordered by the degree $k$ from the large to small, so the initial degree of the new node will be dentate according to the size order.

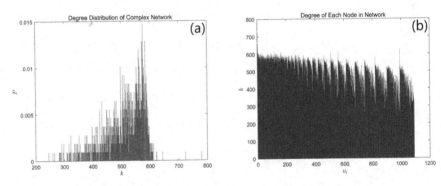

**Fig. 5.** Initial network is full interconnected. $n_0 = 20, h = 1, m_0 = 3, N = 40$. (a) Degree distribution histogram; (b) Degree of each node.

### 3.3    Average Degree of Inherited Competitive Growth Network

Figure 6 shows the relationship between the network's average degree and the number of iteration under the influence of different inherited coefficients. When the inherited coefficient is small, the network average degree will quickly converge to a stable point, and the maximum value of convergence will be determined by the threshold value. When the inherited coefficient increases, the average degree gradually increases. When the inherited coefficient is 0.8, it can be seen from the picture that the average degree is linearly related to the number of iteration. When the sub-node completely inherits the degree characteristic from its parent node, which is shown in Fig. 6(b), the fitting curve of the average degree and the number of iteration is concave. At the same time, the network scale is also concave. We have known that network size is the exponential function of the number of iteration. So is there similar relationship between the average degree and the iteration number?

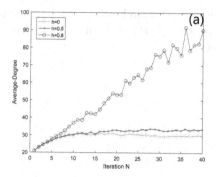
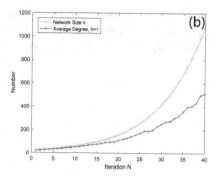

**Fig. 6.** (a) The relation graph of average degree of the network and the iteration number; (b) The relation graph of the network size and average degree and the iteration number.

In the following, the mathematical formula will be used to deduce the relationship between the average degree and the network size. Assuming that the total degree of the network is $K$ and each time step it iterates for one time, the difference equation can be obtained:

$$K_{t+1} - K_t = 2 * r * K_t \tag{7}$$

After many times of iteration, we can get the total degree:

$$K_t = K_0 * (1 + 2 * r)^t \tag{8}$$

On the other hand, the total degree is the product of the total number of nodes and the average degree of the network.

$$K = n * k_{avg} \tag{9}$$

We can deduce the following equation after combining formula (2), (8) and (9):

$$ln(k_{avg}) = \left(\frac{ln(1 + 2 * r)}{ln(1 + r)} - 1\right) * ln(n) + ln(k_0) - \frac{ln(1 + 2 * r)}{ln(1 + r)} * ln(n_0) \tag{10}$$

The coefficient of $ln(n)$ is a function of $r$.

$$f(r) = \frac{ln(1 + 2 * r) - ln(1 + r)}{ln(1 + r)}, 0 < r < 0.5 \tag{11}$$

In the domain interval, there is $0 < f < 1$. After two derivations, it can be proved that $f$ is the monotonically decreasing function of $r$ in the domain interval.

After the experimental test and taking logarithm of the average degree and the total network node numbers, the fitting curve conforms to the linear relationship, which further explains the correctness of the double logarithmic model. Thus, the average degree is also an exponential function of the number of iteration. In Fig. 7, the slope of the fitting curve is determined by the growth rate of network scale. The slope will become smaller with the increase of the growth rate of network scale, which is consistent with the above mathematical deduction.

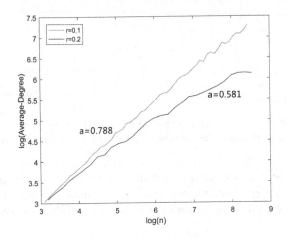

**Fig. 7.** The relation graph of average degree of the network and the logarithm of network size. The slope of the blue curve at $r = 0.1$ is 0.788 while the slope of the red curve at $r = 0.2$ is 0.581. (Color figure online)

### 3.4   Average Clustering Coefficient and Average Path Length of the Inherited Competitive Growth Network

The clustering coefficient of the computer network is shown in Fig. 8. Regardless of how the clustering coefficient changes, the clustering coefficient decrease with the increase of the number of iteration. When the inherited coefficient is 1, the average clustering coefficient is linearly related to the number of iteration. The path length among nodes can be obtained by using the Dijkstra algorithm to calculate the shortest

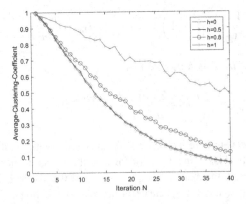

**Fig. 8.** The relation graph of average clustering coefficient and iteration number in the inherited competitive growth network.

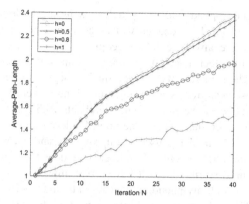

**Fig. 9.** The relation graph of average path length and iteration number in the inherited competitive growth network.

path. Figure 9 shows the relationship between the network's average path length and the number of iteration. It can be seen that the network's average path is short and conforms to the characteristic of the small-world network. No matter how the inherited coefficient changes, the average path length increases with the increase of the number of iteration. Interestingly, when the inherited coefficient is 1, the average path length is linearly related to the number of iteration.

# 4 Conclusion

In this paper, a new complex network model is proposed. The traditional scale-free network model adopts the "prior connection" principle. The new node has the connection choice right, but the old node can only passively accept it. Based on inheriting the scale-free model, the competitive growth network holds the view that all nodes are greedy, and the old node has the rejection right of connection. On this basis, the inherited mechanism is introduced, and we proposed the complex network's competitive growth model of degree-characteristic inheritance and solved the problem of where the new node comes from, while the previous research focuses on where the new node goes. The model is topologically tunable, and different parameters produce network models with different properties. When the inherited coefficient is zero and the adjustment coefficient of rejection function is infinite, this model will be the traditional scale-free network model; when the inherited coefficient is zero and the rejection function is not zero, this model will become a competitive growth model.

From the experimental results, it can be found that the degree distribution of the competitive growth network is no longer power-law; however, it complies with the exponential distribution. The completely inherited competitive growth network is negative skew. And the average degree is linearly related to the network's logarithm, which is also proved by the mathematical derivation. In addition, the average clustering coefficient of network decreases with the increase of inherited coefficient, and the average path length increases with the increase of inherited coefficient.

This model can be applied into many fields, including biological evolution, the World Wide Web, and the balanced growth of economy, etc. Like the problem of queuing to eat, there are always a lot of people queuing in the front of delicious restaurants; the herding effect makes the number of consumers exceed the merchant's reception capacity. Then, merchants' service quality begins to decline, and they are more willing to choose customers with strong purchasing power. At the same time, the rest of customers will choose the surrounding merchants with common taste. The other example is the gap between the rich and the poor. Society should provide people with ladder and safety net. The ladder is used to improve life and the safety net is used to prevent the poor falling into the abyss. However, how does the higher threshold value affect the entrepreneurial spirits? Does it have a bad impact on the systematical evolution? How does the variability phenomenon caused by the inherited process affect the network structure? These interesting problems need to be further studied.

# References

1. Andriani, P., McKelvey, B.: From Gaussian to Paretian thinking: causes and implications of power laws in organizations. Organ. Sci. **20**(6), 1053–1071 (2009). doi:10.1287/orsc.1090. 0481
2. Liu, W.Y., Liu, B.: Study on congestion control for complex network based on weighted routing strategy. Syst. Eng.-Theory Pract. **35**(4), 1063–1068 (2015)
3. Lv, J.H.: Mathematical models and synchronization criterions of complex dynamical networks. Syst. Eng.-Theory Pract. **24**(4), 17–22 (2004)
4. Wang, X.F., Li, X., Chen, G.R.: Theory of Complex Networks and its Application. Tsinghua University Press, Beijing (2009)
5. Watts, D.J., Strogatz, S.H.: Collective dynamics of small-world network. Nature **393**(6684), 440–442 (1998). doi:10.1038/30918
6. Shekatkar, S.M., Ambika, G.: Complex networks with scale-free nature and hierarchical modularity. Eur. Phys. J. B **88**(9), 1–7 (2015). doi:10.1140/epjb/e2015-60501-y
7. Santos, F.C., Pacheco, J.M.: Scale-free networks provide a unifying framework for the emergence of cooperation. Phys. Rev. Lett. **95**, 098104 (2005). doi:10.1103/PhysRevLett. 95.098104
8. Barabasi, A.L., Albert, R.: Emergence of scaling in random networks. Science **286**, 509–512 (1999). doi:10.1126/science.286.5439.509
9. Li, J., Wang, B.H.: Growing complex network model with acceleratingly increasing number of nodes. Acta Phys. Sinica **55**(8), 4051–4057 (2006)
10. Wang, Z., Yao, H.: Repeated snowdrift game on tunable scale-free networks. Syst. Eng.-Theory Pract. **36**(1), 121–126 (2016)
11. Holme, P., Kim, B.J.: Growing scale-free networks with tunable clustering. Phys. Rev. E Stat. Nonlinear Soft Matter Phys. **65**(2), 026107 (2002). doi:10.1103/PhysRevE.65. 026107

# An Alternative Fuzzy Linguistic Approach for Determining Criteria Weights and Segmenting Consumers for New Product Development: A Case Study

Sirin Suprasongsin[1,2]($\boxtimes$), Van-Nam Huynh[1], and Pisal Yenradee[2]

[1] Japan Advanced Institute of Science and Technology, Nomi, Ishikawa, Japan
{sirin,huynh}@jaist.ac.jp
[2] Sirindhorn International Institute of Technology, Thammasat University,
Bangkok, Pathumthani, Thailand
pisal@siit.tu.ac.th

**Abstract.** In new product development, identifying a product concept (criterion) and a market segment is a critically important issue. To do so, it is necessary to deal with uncertainties in an evaluation process. In this paper, we introduce (a) an alternative function principle based approach for coping with inter-relation uncertainties in human criteria assessment and heterogeneity of decision makers, in determining criteria weights for a multiple criteria group decision making problem (b) an alternative intersected area based approach in segmenting consumers for new product development. A case study is used to illustrate the discussed approaches. The proposed approaches are helpful to alleviate the tediousness of the mathematical operations and also make the fuzzy linguistic approach for a multiple criteria group decision making problem more convenient and practical to be used.

**Keywords:** Function principle · Pascal triangular graded mean approach · Linguistic term set · Unknown criteria weights · Market segmentation

## 1 Introduction

New product development (NPD) project generally composes of many processes ranging from product-concept identification through product launch [11]. As stated by Calatone et al. [4], initially prioritizing the product concepts can help manager to eliminate the risky product concepts at the beginning stage before significant investment are made and opportunity cost incurred. It is also found that initial prioritizing had the highest correlation with new product prior to commercialization resulting in resource consumption [21]. Thus, prioritizing product concepts is a very important task in NPD project [20]. However, in developing a new product, knowing which product concepts are the most and the least important is not enough to succeed in a highly competitive market.

© Springer Nature Singapore Pte Ltd. 2017
J. Chen et al. (Eds.): KSS 2017, CCIS 780, pp. 23–37, 2017.
https://doi.org/10.1007/978-981-10-6989-5_3

Identifying a market segment for a particular product concept is also a key to the success. Market segmentation helps companies to find the homogeneous market segment and expand their market [32]. The result of market segments is expected to create a similar purchasing behavior [13]. With its importance, companies need to evaluate the result of segmentation and select the most suitable one(s). The selection process might be difficult since it may affect other decisions associated with marketing strategy [33].

To investigate what to produce and whom to purchase, a questionnaire survey is widely accepted to be used for gathering evaluation information. However, analyzing the data from questionnaire survey is complicated since there are various uncertainties involved in the evaluation process. Basically, two of the most common uncertainties in evaluation process for a group decision making problem are (1) the uncertainty associated with the criteria assessment of DMs (2) the uncertainty associated with the heterogeneity of DMs. Each DM has different characteristics, interests, and knowledge. Thus, the relative importance of each DM is not equal and ambiguity. In dealing with the uncertainty, fuzzy set theory [35], rough set theory, [24], intuitionistic fuzzy set [1], interval-valued fuzzy set [17], interval-valued intuitionistic fuzzy set [2], and other advanced mathematical theories are exploited in many research areas. In addition, linguistic terms are usually used in information assessment since human beings are usually more comfortable when they provide a linguistic information, rather than a crisp number [9].

In this paper, we focus on the approach to deal with the inter-relation fuzzy data, taking uncertainties in criteria assessment and in heterogeneous DMs into consideration. This approach would be helpful for DMs to prioritize product concept which is the most and the least important to customers. In addition, we also focus on the approach to segment which customers will purchase a product according to a particular concept. In other words, with the use of fuzzy theory and linguistic assessment, what product concepts to be used and their target customers are identified in this study.

To do so, firstly, the relative importances of product concepts (criteria), with respect to the relative importances of DMs, need to be determined. Typically, there have been witnessed that fuzzy criteria assessments and fuzzy DM weights are usually individually converted to a scalar unit before combining them together. For example, Qi et al. [25] separately utilized the generalized cross-entropy measure to determine a set of criteria weights and a set of DMs weights. Then, the overall set of criteria weights is determined. Ye [34] proposed the entropy weights-based correlation coefficient for an intuitionistic fuzzy set (IFSs) and an interval-valued intuitionistic fuzzy sets (IVIFSs) in order to determine criteria weights, while the expert weights are given beforehand. Farhadinia [14], made use of a concept of entropy measure of hesitant fuzzy linguistic term sets in dealing with completely unknown criteria weight. However, such an information conversion may lead to the loss of information caused by many approximation processes. Thus, to consistently maintain the nature of human beings, delaying the transformation of linguistic term to a scalar unit is needed.

To deal with inter-relation fuzzy data, Zadeh [36] introduced an extension principle to combine the fuzzy information. Later, Chen [5] proposed a function principle to alleviate the complexity of mathematical operation of the extension principle and also proved that the principle does not change the type of membership function under arithmetical operation of fuzzy numbers. The difference between these two principles is illustrated by Hsieh [18], as shown in Fig. 1. Besides, Chen [6] also showed how to treat the operations of fuzzy numbers with step form by using function principle. Due to its simplicity and effectiveness, function principle is used in this study as an operation of multiplication and addition of fuzzy numbers.

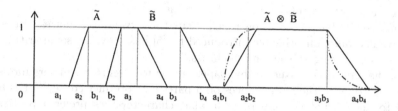

**Fig. 1.** The fuzzy multiplication operation of function principle (-) and extension principle (.....), introduced by Hsieh [18]

According to the limitation of human perception, the output of fuzzy process is transformed to be a single scalar quantity, not a collection of membership values [15,30]. This transformation is generally regarded as a defuzzifying process. Defined by Ross [28], the defuzzification results from reducing a fuzzy set to a crisp single-valued quantity, or to a crisp set. Various methods in defuzzifying are studied over time such as center of gravity (COG), mean of maxima (MeOM), mean max membership, weighted fuzzy mean (WFM), graded mean integration representation, pascal triangular graded mean, and so on. For the sake of simplicity, in this paper, pascal triangular graded mean approach is applied to defuzzify the data.

This paper is organized as follows. Section 2 begins with a problem statement. Then, a fundamental knowledge of function principle and pascal triangular graded mean approach are reviewed. In Sect. 3, a new evaluation approach for determining criteria weight is introduced. A case study is discussed in this section. Then, a new linguistic based approach for segmenting consumers is illustrated in Sect. 4. Finally, the conclusion is presented in Sect. 5.

## 2    Preliminaries

### 2.1    Fuzzy Linguistic Approach for MCDM Problems

This subsection describes a multiple criteria group decision making problem in determining a set of criteria weights through a case study taken from Wangsukjai export limited company. In this paper, only qualitative data are taken into account.

To begin with, let us define a set of evaluation grades in evaluating a subjective criteria assessment and a relative importance of DMs. Here, the notion of criteria infers to product concepts. The evaluation grade is defined as

$$S = \{S_{-l}, \ldots, S_0, \ldots, S_l\} \tag{1}$$

where $S_l$ is called the evaluation grade at a level that criterion $k$ is evaluated. That is, $S$ provides a complete set of distinct standards for assessing qualitative data [19]. As suggested by Miller [23], most of human beings can perceive only seven evaluation grades. For the sake of simplicity, this paper considers only five evaluation grades or $l = 2$ and the same set $S$ are applied into every evaluation assessment. To determine a set of criteria weights, it is necessary to aggregate fuzzy criteria assessments and fuzzy DMs weights.

Thus, the problem here is how to aggregate those two uncertainties in a rational way. The function principle introduced by [5] has provided some arithmetic operations to aggregate such uncertainties.

Another concerned issue in this paper is how to classify consumers into segments according to their preferences on a particular product concept. It is beneficial to know what consumers with what characteristics prefer a particular product concept significantly more than average consumers. This information can lead to the development of effective advertisement and promotion to attract consumers in that segment to purchase the product. In this paper, a market segment and a consumer segment have the same meaning, which is a group of consumers who apparently prefer the criteria (product concepts) to average consumers. The term of more preferable can be evaluated from their provided linguistic evaluation grade. If a consumer's grade is higher than a fuzzy-number (FN) criterion weight and their FNs do not intersect, a consumer is segmented into that criterion (product concept). For instance,

- Mr. A provides a linguistic evaluation grade for criterion 1 as $s_2$, which has a set of FNs = (0.75, 1, 1)
- FN weight of criterion 1 is (0.119, 0.281, 0.548)

Then, the evaluation grade of Mr. A to criterion 1 is higher than FN weight of criterion 1. Furthermore, it also does not intersect, since the highest FN value of criterion 1 (0.548) is less than the lowest FN value of his evaluation grade (0.75). This can be interpreted that Mr. A is very fond of criterion 1. If the product has been launched with this concept (criterion 1), he is likely to purchase it.

However, if there is a case that the weight and the grade have a partial intersection area, how can we segment this consumer? Thus, a new fuzzy linguistic based approach in segmenting consumers by considering the intersection area of FN criteria weights and linguistic evaluation grades, is proposed in this paper.

## 2.2 Fuzzy Arithmetical Operations Under Function Principle

The function principle aims to simplify the calculation of extension principle proposed by Zadeh [36]. The arithmetical operations under function principle are described below.

Let assume that $\tilde{A} = (a_1, a_2, a_3)$ and $\tilde{B} = (b_1, b_2, b_3)$ are two sets of triangular fuzzy numbers. Then the four arithmetical operations can perform on triangular fuzzy numbers.

1. The addition of $\tilde{A}$ and $\tilde{B}$
   $\tilde{A} + \tilde{B} = (a_1, a_2, a_3) + (b_1, b_2, b_3) = (a_1 + b_1, a_2 + b_2, a_3 + b_3)$
2. The subtraction of $\tilde{A}$ and $\tilde{B}$
   $\tilde{A} - \tilde{B} = (a_1, a_2, a_3) - (b_1, b_2, b_3) = (a_1 - b_1, a_2 - b_2, a_3 - b_3)$
3. The multiplication of $\tilde{A}$ and $\tilde{B}$ is $\tilde{A} \times \tilde{B} = (c_1, c_2, c_3)$
   where $T = a_1b_2, a_1b_3, a_3b_1, a_3b_3$; $c_1 = \min T, c_2 = a_2b_2, c_3 = \max T$
   However, if $a_1, a_2, a_3, b_1, b_2, b_3$ are non-zero positive real numbers, then
   $\tilde{A} \times \tilde{B} = (a_1, a_2, a_3) \times (b_1, b_2, b_3) = (a_1b_1, a_2b_2, a_3b_3)$
4. The division of $\tilde{A}$ and $\tilde{B}$ is $\frac{\tilde{A}}{\tilde{B}} = (c_1, c_2, c_3)$ where $T = \frac{a_1}{b_2}, \frac{a_1}{b_3}, \frac{a_3}{b_1}, \frac{a_3}{b_3}$
   $c_1 = \min T, c_2 = \frac{a_2}{b_2}, c_3 = \max T$
   However, if $a_1, a_2, a_3, b_1, b_2, b_3$ are non-zero positive real numbers, then
   $\frac{\tilde{A}}{\tilde{B}} = (a_1, a_2, a_3) \div (b_1, b_2, b_3) = (\frac{a_1}{b_1}, \frac{a_2}{b_2}, \frac{a_3}{b_3})$

By applying fuzzy operation under function principle, two fuzzy data can be aggregated. Then, let us defuzzify fuzzy numbers to a scalar unit for the ease of human cognition.

## 2.3 Graded Mean Integration Representation Approach

Graded mean integration representation approach is introduced by Chen and Hsieh [7] in order to transform a triangular fuzzy number into a crisp number [22]. For more details, see [10]. In 2006, Chen et al. [8] proposed properties of the representation of fuzzy numbers and the multiplication of fuzzy numbers under Extension Principle by using Graded Mean Integration Representation method. The generalized form can be obtained as follows.

Let assume that $L^{-1}$ and $R^{-1}$ are inverse functions of function L and R, respectively and the graded mean h-level of generalized fuzzy number $A = (a_1, a_2, a_3 : w)$ is $\frac{h[L^{-1}(h) + R^{-1}]}{2}$. Then the defuzzified value P(A) based on the integral value of graded mean h-level can be defined using Eq. 2

$$P(A) = \frac{\int_0^h [\frac{L^{-1}(h) + R^{-1}(h)}{2}] dh}{\int_0^w h\, dh} \tag{2}$$

where $h$ is in between 0 and $w, 0 < w \leq 1$. Interestingly, the representation of fuzzy number can be generalized as shown in Eqs. 3 and 4. For example, if $A = (a_1, a_2, a_3)$ is a triangular fuzzy number, then

$$P(A) = \frac{1}{2} \frac{\int_0^1 \int h[a_1 + h(a_2 - a_1) - h(a_3 - a_2)] dh}{\int_0^1 h\, dh} \tag{3}$$

$$P(A) = \frac{a_1 + 4a_2 + a_3}{6} \tag{4}$$

## 2.4   Pascal Triangular Graded Mean Approach

Pascal triangular graded mean approach is an extension of graded mean integration representation approach. In 2013, Babu [3] indicated that there is no significant difference between graded mean integration representation approach and pascal triangular graded mean approach. Their mean and variance are not significantly different. Therefore, it can be said that a pascal triangular graded mean approach is an alternative approach of the graded mean integration representation approach. Based on this approach, the coefficient of fuzzy numbers of pascal's triangles and the simple probability approach are used to formulate the formula, as simplified in Eq. 5.

$$P(A) = \frac{a_1 + 2a_2 + a_3}{4} \tag{5}$$

# 3   New Evaluation Approach in Determining a Set of Fuzzy Number (FN) Criteria Weights

## 3.1   A Case Study of Soy Milk Beverage

In this section, we will show how the function principle and the pascal triangular graded mean approach work in practice through a case study. A case study is the development of a new food product at Wangsukjai export limited company, an international company in a health beverage market segment. To keep capturing a high growth trend in health food product, an instant soy milk powder was proposed. To increase the product value, a marketing survey has been conducted, making use of the fuzzy linguistic based evaluation approach. The purposes of the survey are

1. To determine a relative importance of a product concept (criterion)
2. To segment consumers for a particular product concept.

In this paper, a structure of the problem can be summarized in Table 1. According to the Table, the notion of $j, k, i$ represents the indice of a set of triangular fuzzy values $(f_j)$, criteria $(c_k)$, and DMs $(r_i)$. Note that from now on, decision makers (DMs) will be called respondents (RS).

**Table 1.** A structure of the problem

| RS$(r_i)$ | Weight of RS | | | Criterion 1 $(c_1)$ | | | Criterion 2 $(c_2)$ | | | $\cdots$ | | | | | Criterion k$(c_k)$ |
|---|---|---|---|---|---|---|---|---|---|---|---|---|---|---|---|
| | $a_{i1}$ | $a_{i2}$ | $a_{i3}$ | $b_{i11}$ | $b_{i21}$ | $b_{i31}$ | $b_{i12}$ | $b_{i22}$ | $b_{i32}$ | $\cdots$ | $\cdots$ | $\cdots$ | $\cdots$ | $\cdots$ | $b_{ijk}$ |
| $r_1$ | $[a_{11}]$ | $[a_{12}]$ | $[a_{13}]$ | $[a_{11}b_{111}]$ | $[a_{12}b_{121}]$ | $[a_{13}b_{131}]$ | $\cdots$ | $\cdots$ | $\cdots$ | | $\cdots$ | $\cdots$ | $\cdots$ | $\cdots$ | $b_{1jk}$ |
| $r_2$ | $[a_{21}]$ | $[a_{22}]$ | $[a_{23}]$ | $[a_{21}b_{211}]$ | $[a_{21}b_{321}]$ | $[a_{21}b_{331}]$ | : | : | : | : | : | : | : | : | $b_{2jk}$ |
| : | : | : | : | : | : | : | : | : | : | : | : | : | : | : | : |
| : | : | : | : | : | : | : | : | : | : | : | : | : | : | : | : |
| $r_i$ | $a_{1j}$ | $\cdots$ | $a_{ij}$ | $[a_{i1}b_{i11}]$ | $[a_{i2}b_{i21}]$ | $[a_{i3}b_{i31}]$ | $\cdots$ | $\cdots$ | $\cdots$ | $\cdots$ | $\cdots$ | $\cdots$ | $\cdots$ | $\cdots$ | $[a_{ij}b_{ijk}]$ |

Within a framework of the case study, 30 respondents are participated in the evaluation assessment. Those respondents are 15 males and 15 females, and are categorized based on their interests in soy milk consumption. The consumer interests are classified by [31], which are taste conscious, health conscious, dieter, and natural-lover conscious. Four product concepts (criteria) are selected for increasing product value denoted by $c_k (k = 1, 2, 3, 4)$, as described further in Table 2.

**Table 2.** Description of criteria

| Criteria ($c_k$) | Example of components of criteria |
|---|---|
| $c_1$: variety of flavor | Coco-malt, thai-tea, coffee, banana, and strawberry |
| $c_2$: for a specific group | A product for woman, man, eldery, dieter, and muscle builder |
| $c_3$: health additive | Vitamin, omega, collagen, gluta, Q-10 |
| $c_4$: added condiment | Jelly, basil seed, tofu sheet, soy custard |

Five linguistic terms are used to represent the subjective judgment of RS in assessing the importance of each criterion and RS weights. Note that the RS weights are considered from how often respondent consumes a soy milk beverage. A set of linguistic terms $s_l$ $(l = 2)$ and their triangular fuzzy numbers are summarized in Table 3.

**Table 3.** A set of triangular fuzzy numbers for relative consumption frequency of DMs and relative importance of criteria assessment

| Label ($s_l$) | Linguistic term | | Triangular fuzzy number |
|---|---|---|---|
| | Weight of respondent | Criteria assessment | |
| $s_{-2}$ | Rarely | Unimportant | (0, 0, 0.25) |
| $s_{-1}$ | Slightly often | Weakly important | (0, 0.25, 0.5) |
| $s_0$ | Often | Moderately important | (0.25, 0.5, 0.75) |
| $s_1$ | Very often | Very important | (0.5, 0.75, 1) |
| $s_2$ | Extremely often | Extremely important | (0.75, 1, 1) |

## 3.2  Proposed Approach in Criteria-Weights Determination

To begin with, let us interpret the data shown in Table 4. For example, R1 represents a female who is a taste and natural-lover conscious. She **often** ($s_0$) consumes soy milk beverage. She thinks that having a variety of flavor and a health added supplement are **extremely important** ($s_2$) to develop a soy milk product. Adding some condiments is also **moderately important** $s_0$), while developing a product for a specific group is **weakly important**($s_{-1}$).

The steps to determine a set of FN criteria weights are as follows.

First, let us transform the linguistic information shown in Table 4 into fuzzy numbers shown in Table 5 by using semantic information provided in Table 3. For example, a set of criteria assessment of R1 can be represented by

$$\{s_2, s_{-1}, s_2, s_0\} = \{(0.75, 1, 1), (0, 0.25, 0.5), (0.75, 1, 1), (0.25, 0.5, 0.75)\}$$

Second, apply the fuzzy arithmetic operations under function principle proposed by Chen [5] to determine criteria weights with respect to consumption frequency of RS. Multiplication operation is the first process to integrate criteria assessment and consumption frequency of RS. For example, see Table 6. Assessment of R1 to criterion 1 with respect to consumption frequency can be computed as

$$\begin{aligned}
\text{Assessment of R1 to criterion } 1 &= \{(a_{i1}b_{i11}), (a_{i2}b_{i21}), (a_{i3}b_{i31})\} \\
&= \{(0.25 \times 0.75), (0.5 \times 1), (0.75 \times 1)\} \\
&= \{0.188, 0.500, 0.750\}
\end{aligned}$$

where $a_{ij}$ is a consumption frequency of respondent $i$ with fuzzy value $j$, while $b_{ijk}$ represents assessment of criterion $k$ with fuzzy value $j$ by respondent $i$.

Third, the addition operation is exploited to combine the importance of individual criterion based on all RS' opinions, as shown in the row of "Total" in Table 6. For example, a set of triangular fuzzy numbers of criterion 1, with respect to all 30 RS, is equal to

$$\begin{aligned}
C1 = \ &\{(0.188 + 0.000 + \ldots + 0.000), (0.500 + 0.188 + \ldots 0.188), \\
&(0.750 + 0.500 + \cdots + 0.500)\} = \{2.688, 8.438, 16.438\}
\end{aligned}$$

According to an aggregation of criteria fuzzy numbers shown above, it is necessary to normalize them into the scales based on Table 3. Based on a provided maximum linguistic value $(s_2) = \{0.75, 1, 1\}$, the fuzzy numbers criteria weights are normalized, as presented in the row of "Criteria weights" in Table 6. To do so, the process is explained as follows.

$$\begin{aligned}
\text{Normalized weight of Criterion } 1 &= \left\{\frac{2.688}{n \times 0.75}, \frac{8.438}{n \times 1}, \frac{8.438}{n \times 1}\right\} \\
&= \{0.119, 0.281, 0.548\}
\end{aligned}$$

where $n$ infers to the total number of RS, which in this case, is equal to 30. After obtaining normalized FN criteria weights, pascal triangular graded mean approach is used to defuzzify fuzzy numbers to a scalar value according to Eq. 5. Then, a set of criteria percentages can be obtained, $c_k = \{21\%, 27\%, 30\%, 22\%\}$.

$$\text{Weight of Variety of flavor } (C_1) = \frac{(0.119 + 2 \times 0.281 + 0.548)}{4} = 21\%$$

At this point, we can notice that this approach not only yields a set of scalar criteria weights, but also provides an additional information of FN criteria weights (shown in Table 6, row "criteria weight"). In other words, without loss of information, the proposed approach can consistently maintain the flexibility for DMs in making a decision on fuzzy linguistic information.

**Table 4.** Information of respondents

| Respondent | Gender | Consumer conscious based on [31]* | Consumption frequency | $C_1$ | $C_2$ | $C_3$ | $C_4$ |
|---|---|---|---|---|---|---|---|
| R1 | Female | T, N | $s_0$ | $s_2$ | $s_{-1}$ | $s_2$ | $s_0$ |
| R2 | Male | H | $s_1$ | $s_{-1}$ | $s_1$ | $s_2$ | $s_{-1}$ |
| R3 | Male | H, D | $s_0$ | $s_0$ | $s_2$ | $s_1$ | $s_{-1}$ |
| R4 | Female | T, N | $s_2$ | $s_0$ | $s_2$ | $s_1$ | $s_2$ |
| R5 | Male | N | $s_{-1}$ | $s_{-1}$ | $s_1$ | $s_1$ | $s_2$ |
| R6 | Female | H | $s_2$ | $s_1$ | $s_2$ | $s_0$ | $s_1$ |
| R7 | Female | D | $s_1$ | $s_1$ | $s_2$ | $s_1$ | $s_0$ |
| R8 | Male | H,N | $s_{-2}$ | $s_0$ | $s_{-1}$ | $s_0$ | $s_{-1}$ |
| R9 | Male | T | $s_0$ | $s_2$ | $s_1$ | $s_1$ | $s_2$ |
| R10 | Female | H, D | $s_{-1}$ | $s_0$ | $s_1$ | $s_2$ | $s_1$ |
| R11 | Female | N | $s_2$ | $s_{-1}$ | $s_{-1}$ | $s_2$ | $s_0$ |
| R12 | Female | T, D, N | $s_1$ | $s_{-2}$ | $s_0$ | $s_1$ | $s_1$ |
| R13 | Female | H | $s_0$ | $s_{-1}$ | $s_1$ | $s_1$ | $s_{-1}$ |
| R14 | Male | N | $s_1$ | $s_1$ | $s_{-1}$ | $s_2$ | $s_{-2}$ |
| R15 | Male | N | $s_{-1}$ | $s_0$ | $s_0$ | $s_1$ | $s_{-1}$ |
| R16 | Female | T, H | $s_0$ | $s_2$ | $s_{-1}$ | $s_1$ | $s_0$ |
| R17 | Female | D | $s_2$ | $s_0$ | $s_2$ | $s_0$ | $s_2$ |
| R18 | Female | D, N | $s_{-2}$ | $s_{-2}$ | $s_0$ | $s_2$ | $s_1$ |
| R19 | Male | N | $s_0$ | $s_{-1}$ | $s_{-1}$ | $s_0$ | $s_{-1}$ |
| R20 | Male | D | $s_{-1}$ | $s_0$ | $s_1$ | $s_2$ | $s_{-2}$ |
| R21 | Male | H, D | $s_0$ | $s_2$ | $s_2$ | $s_1$ | $s_0$ |
| R22 | Male | H | $s_1$ | $s_{-1}$ | $s_0$ | $s_2$ | $s_{-1}$ |
| R23 | Male | H | $s_{-1}$ | $s_0$ | $s_1$ | $s_2$ | $s_1$ |
| R24 | Male | T, N | $s_0$ | $s_1$ | $s_0$ | $s_0$ | $s_0$ |
| R25 | Female | D, N | $s_{-1}$ | $s_0$ | $s_1$ | $s_2$ | $s_{-1}$ |
| R26 | Male | T | $s_1$ | $s_2$ | $s_1$ | $s_2$ | $s_2$ |
| R27 | Male | H, N | $s_2$ | $s_{-2}$ | $s_1$ | $s_0$ | $s_{-1}$ |
| R28 | Female | T | $s_{-1}$ | $s_{-1}$ | $s_2$ | $s_0$ | $s_{-2}$ |
| R29 | Female | D, N | $s_0$ | $s_1$ | $s_2$ | $s_1$ | $s_1$ |
| R30 | Female | H, N | $s_1$ | $s_{-1}$ | $s_0$ | $s_2$ | $s_0$ |

Note*: T = Taste conscious, H = Health conscious, D = Dieter, N = Natural-lover conscious

**Table 5.** Triangular fuzzy numbers of consumption frequency and criteria assessment

| Respondent (R_i) | Frequency | | | C1 | | | C2 | | | C3 | | | C4 | | |
|---|---|---|---|---|---|---|---|---|---|---|---|---|---|---|---|
| | $a_{i1}$ | $a_{i2}$ | $a_{i3}$ | $b_{i11}$ | $b_{i21}$ | $b_{i31}$ | $b_{i12}$ | $b_{i22}$ | $b_{i32}$ | $b_{i13}$ | $b_{i23}$ | $b_{i33}$ | $b_{i14}$ | $b_{i24}$ | $b_{i34}$ |
| R1 | 0.25 | 0.5 | 0.75 | 0.75 | 1 | 1 | 0 | 0.25 | 0.5 | 0.75 | 1 | 1 | 0.25 | 0.5 | 0.75 |
| R2 | 0.5 | 0.75 | 1 | 0 | 0.25 | 0.5 | 0.5 | 0.75 | 1 | 0.75 | 1 | 1 | 0 | 0.25 | 0.5 |
| : | : | : | : | : | : | : | : | : | : | : | : | : | : | : | : |
| R30 | 0.5 | 0.75 | 1 | 0 | 0.25 | 0.5 | 0.25 | 0.5 | 0.75 | 0.75 | 1 | 1 | 0.25 | 0.5 | 0.75 |

**Table 6.** Criteria weights in a form of triangular fuzzy numbers

| Respondent (R_i) | C1 | | | C2 | | | C3 | | | C4 | | |
|---|---|---|---|---|---|---|---|---|---|---|---|---|
| | $a_{i1}b_{i11}$ | $a_{i2}b_{i21}$ | $a_{i3}b_{i31}$ | $a_{i1}b_{i12}$ | $a_{i2}b_{i22}$ | $a_{i3}b_{i32}$ | $a_{i1}b_{i13}$ | $a_{i2}b_{i23}$ | $a_{i3}b_{i33}$ | $a_{i1}b_{i14}$ | $a_{i2}b_{i24}$ | $a_{i3}b_{i34}$ |
| R1 | 0.188 | 0.500 | 0.750 | 0.000 | 0.125 | 0.375 | 0.188 | 0.500 | 0.750 | 0.063 | 0.250 | 0.563 |
| R2 | 0.000 | 0.188 | 0.500 | 0.250 | 0.563 | 1.000 | 0.375 | 0.750 | 1.000 | 0.000 | 0.188 | 0.500 |
| : | : | : | : | : | : | : | : | : | : | : | : | : |
| R30 | 0.000 | 0.188 | 0.500 | 0.125 | 0.375 | 0.750 | 0.375 | 0.750 | 1.000 | 0.125 | 0.375 | 0.750 |
| Total | 2.688 | 8.438 | 16.438 | 4.188 | 11.375 | 19.375 | 4.938 | 12.875 | 21.438 | 3.313 | 9.313 | 17.063 |
| Criteria weights | 0.119 | 0.281 | 0.548 | 0.186 | 0.379 | 0.646 | 0.219 | 0.429 | 0.715 | 0.147 | 0.310 | 0.569 |

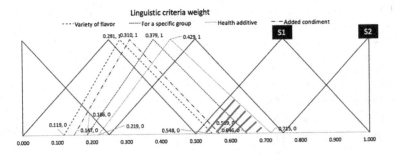

**Fig. 2.** FN criteria weights lied on linguistic evaluation grades

## 3.3 A Comparative Study

In the previous part, function principle and pascal triangular graded mean approach are applied to a case study. As a comparative study, let us analyze the provided linguistic information in Table 4, by defuzzifying linguistic information of both consumption frequency and criteria assessment to a scalar value by Eq. 5 before multiplying them together. The result is presented in Table 7. Then, multiplication and addition operations are used to obtain the set of criteria weights, as presented in Table 8.

It is worth to note here that although these two approaches provide the same set of scalar criteria weights, the proposed approach provides more information, i.e. a set of FN criteria weights, which can be further used in other analyses such as segmenting consumers.

**Table 7.** A scalar unit of RS weights (frequency) and criteria assessment $(c_k)$ obtained from pascal defuzzification

| Respondent | Frequency | $C_1$ | $C_2$ | $C_3$ | $C_4$ |
|---|---|---|---|---|---|
| R1 | 0.500 | 0.938 | 0.250 | 0.938 | 0.500 |
| R2 | 0.750 | 0.250 | 0.750 | 0.938 | 0.250 |
| : | : | : | : | : | : |
| R30 | 0.750 | 0.250 | 0.500 | 0.938 | 0.500 |

**Table 8.** Criteria weights by a scalar approach

| Respondent | $C_1$ | $C_2$ | $C_3$ | $C_4$ |
|---|---|---|---|---|
| R1 | 0.469 | 0.125 | 0.469 | 0.250 |
| R2 | 0.188 | 0.563 | 0.703 | 0.188 |
| : | : | : | : | : |
| R30 | 0.1888 | 0.375 | 0.703 | 0.375 |
| Total | 8.281 | 10.840 | 12.375 | 9.023 |
| Criteria weights | 21% | 27% | 30% | 22% |

# 4  New Fuzzy Linguistic Approach in Segmenting Consumers

## 4.1  Proposed Approach in Segmentation

The concept of this new approach is mainly based on the intersected area under the graph of FN criterion weight $(c_k)$ and linguistic evaluation grade $(s_l)$. The amount of intersected area, denoted by $\beta$, is used to judge whether a respondent will be included in the consumer segment. Let assume that if $\beta$ is less than 5% of linguistic evaluation grade $(s_l)$, a respondent who provides an evaluation grade $l$ is segmented to that criterion $k$. Note that 5% is called the cut-off point.

To exemplify, let us see the result of fuzzy numbers shown in Fig. 2. According to the graph, two analyses are performed to classify whether a respondent who evaluates $s_1$ and/or $s_2$, is segmented to that referred criterion $k$. The analyses are illustrated below. A case of criterion 3 is used as an example.

1. Linguistic evaluation grade $s_2$: In this case, a respondent who provides $s_2$ for criterion 3, he/she is segmented to criterion 3, because the FN criterion 3 weight (0.219, 0.429, 0.715) does not intersect with the linguistic evaluation grade $s_2$ (0.75, 1, 1).
2. Linguistic evaluation grade $s_1$: Noticing from the highlighted area in Fig. 2, $\beta$ of FN criterion 3 weight and linguistic evaluation grade $(s_1)$ is 14.69%, which is greater than 5%. Consequently, a respondent who provides $s_1$ for criterion 3 is not a consumer that belongs to the health additive $(c_3)$ segment.

Therefore, it can be concluded that respondents who provide $s_2$ is segmented to criterion 3. The results of the overall segmentation are presented in Table 9.

From Table 9, only criteria 1 and 4 include respondents who provide both $s_1$ and $s_2$. For criteria 2 and 3, only $s_2$ is taken into account.

Although, there are numerically advanced fuzzy approaches in segmenting consumers, i.e., 2-tuple fuzzy linguistic representation [29], fuzzy CODAS (COmbinative Distance-based ASsessment) method [16], Best Worst method [26], this proposed approach is a simply alternative approach based on FN criteria weights.

**Table 9.** The result of concerned linguistic evaluation grade, $s_l$ based on intersection area, $\beta$

| Product concept (criterion $k$) | Linguistic evaluation grade ($s_l$) | |
|---|---|---|
| | $s_1$ | $s_2$ |
| Variety of flavor ($c_1$) | ✓ | ✓ |
| For a specific group ($c_2$) | – | ✓ |
| Health additive ($c_3$) | – | ✓ |
| Added condiment ($c_4$) | ✓ | ✓ |

## 4.2  Result of Consumer Segmentation

The results are presented in Tables 10, 11, 12, 13, 14, 15, 16 and 17. In Tables 10, 11 and 12, it means that the respondent with "✓" feels that the criterion is significantly more important than the average degree of importance of that criteria derived from all respondents. Then, the detailed information of respondents are presented separately based on each criterion in Tables 13, 14, 15 and 16. Table 17 summarizes consumer segmentation. It can be concluded that a majority of respondents who significantly prefer the health additive than the average are from health conscious and natural-lover groups. A major respondents who prefer product for a specific group comes from dieter group. The respondents who prefer product with added condiment mainly come from natural-lover group. As expected, respondents who prefer product with varieties of flavor are mainly from taste conscious group. In addition, according to the gender percentage, males are likely to be more neutral than females. In other words, females are more likely to be attracted by the criteria (product concepts, e.g. health additive, for a specific group, added condiment, and variety of flavor) than males.

**Table 10.** RS1–10 who fond of criterion $k$

| Respondent | $C_1$ | $C_2$ | $C_3$ | $C_4$ |
|---|---|---|---|---|
| R1 | ✓ | | ✓ | |
| R2 | | | ✓ | |
| R3 | | ✓ | | |
| R4 | | ✓ | | ✓ |
| R5 | | | | ✓ |
| R6 | ✓ | ✓ | | ✓ |
| R7 | ✓ | ✓ | | |
| R8 | | | | |
| R9 | ✓ | | | ✓ |
| R10 | | | ✓ | ✓ |

**Table 11.** RS11–20 who fond of criterion $k$

| Respondent | $C_1$ | $C_2$ | $C_3$ | $C_4$ |
|---|---|---|---|---|
| R11 | | | ✓ | |
| R12 | | | | ✓ |
| R13 | | | | |
| R14 | ✓ | ✓ | | |
| R15 | | | | |
| R16 | ✓ | | | |
| R17 | | ✓ | | ✓ |
| R18 | | | ✓ | ✓ |
| R19 | | | | ✓ |
| R20 | | | | |

**Table 12.** RS21–30 who fond of criterion $k$

| Respondent | $C_1$ | $C_2$ | $C_3$ | $C_4$ |
|---|---|---|---|---|
| R21 | ✓ | ✓ | | |
| R22 | | | ✓ | |
| R23 | | | ✓ | ✓ |
| R24 | ✓ | | | |
| R25 | | | | |
| R26 | ✓ | | | ✓ |
| R27 | | | | |
| R28 | | ✓ | | |
| R29 | ✓ | ✓ | | ✓ |
| R30 | | | ✓ | |

**Table 13.** Variety of flavor ($C_1$)   **Table 14.** For a specific group ($C_2$)   **Table 15.** Health additive ($C_3$)

| Variety of flavor | | |
|---|---|---|
| Respondent | Gender | Conscious* |
| R1 | F | T, N |
| R6 | F | H, N |
| R7 | F | D |
| R9 | M | T |
| R14 | M | N |
| R16 | F | T, H |
| R21 | M | H, D |
| R24 | F | T |
| R26 | M | T |
| R29 | F | D |

| For a specific group | | |
|---|---|---|
| Respondent | Gender | Conscious* |
| R3 | M | H |
| R4 | F | T, N |
| R6 | F | H, N |
| R7 | F | D |
| R17 | F | D |
| R21 | M | H, D |
| R28 | F | T |
| R29 | F | D |

| Health additive | | |
|---|---|---|
| Respondent | Gender | Conscious* |
| R1 | F | T, N |
| R2 | M | H |
| R10 | F | H, D |
| R11 | F | T, N |
| R14 | M | N |
| R18 | F | D, N |
| R22 | M | H |
| R23 | M | H |
| R30 | F | H, N |

**Table 16.** Added condiment ($C_4$)

| Added condiment | | |
|---|---|---|
| Respondent | Gender | Conscious* |
| R4 | F | T, N |
| R5 | M | N |
| R6 | F | H, N |
| R9 | M | T |
| R10 | F | H, D |
| R12 | F | D, N |
| R17 | F | D |
| R18 | F | D, N |
| R19 | M | N |
| R23 | M | H |
| R26 | M | T |
| R29 | F | D, N |

**Table 17.** Summary of segmentation

| Criteria (product concepts) | Gender (%) | | Consumer conscious (%) based on [31]* | | |
|---|---|---|---|---|---|
| | Male | Female | Taste | Health | Dieter | Natural-lover |
| Health additive | 44.4 | 55.6 | 22.2 | 55.6 | 22.2 | 55.6 |
| For a specific group | 25.0 | 75.0 | 25.0 | 37.5 | 50.0 | 37.5 |
| Added condiment | 41.7 | 58.3 | 33.3 | 25.0 | 41.7 | 58.3 |
| Variety of flavor | 40.0 | 60.0 | 50.0 | 30.0 | 30.0 | 40.0 |

Typically, most companies are trying to gain insight in the attitude and behavior of their consumers that form the market share, which is usually represented by percentage [12]. Thus, it is worth addressing here that understanding the consumer interest is very valuable in further positioning the market and making an advertisement.

# 5   Concluding Remarks

This paper proposes an alternative fuzzy linguistic approach to determine criteria weights and segment consumers in new product development. To demonstrate the proposed approaches, a case study is also provided. Function principle and pascal triangular graded mean approach are applied to determine a set of FN criteria weights and a set of scalar criteria weights. In addition, a result of FN criteria weights is further used in segmenting consumers to a particular criterion (product concept) by considering the intersected area, $\beta$. From a practitioner point of view, DMs who want to keep a sense of linguistic information and consistently maintain fuzzy information in a decision process for determining product concepts for a particular consumer, may prefer simple and effective approaches as proposed in this paper to apply in their decision problem.

The approaches proposed can also be extended to apply to the uncertain assessment problem where each respondent may provide more than one linguistic term to each criterion. For future work, the decision model proposed by [27] may be applied to cope with this problem, but further research is required.

# References

1. Atanassov, K.T.: Intuitionistic fuzzy sets. Fuzzy Sets Syst. **20**(1), 87–96 (1986)
2. Atanassov, K.T.: More on intuitionistic fuzzy sets. Fuzzy Sets Syst. **33**(1), 37–45 (1989)
3. Babu, S.K.: Statistical optimization for generalised fuzzy number. Int. J. Mod. Eng. Res. (IJMER) **1**(3), 647–651 (2013)
4. Calantone, R.J., Benedetto, C.A., Schmidt, J.B.: Using the analytic hierarchy process in new product screening. J. Prod. Innov. Manage **16**(1), 65–76 (1999)
5. Chen, S.H.: Operations on fuzzy numbers with function principal. J. Manag. Sci. **6**(1), 13–26 (1985)
6. Chen, S.H.: Operations of fuzzy numbers with step form membership function using function principle. Inf. Sci. **108**(1–4), 149–155 (1998)
7. Chen, S.H., Hsieh, C.H.: Graded mean representation of generalized fuzzy numbers. In: Proceeding of Conference on Fuzzy Theory and Its Applications (1998)
8. Chen, S.H., Wang, S.T., Chang, S.M.: Some properties of graded mean integration representation of lr type fuzzy numbers. Tamsui Oxf. J. Math. Sci. **22**(2), 185 (2006)
9. Chen, Y.H., Wang, T.C., Wu, C.Y.: Multi-criteria decision making with fuzzy linguistic preference relations. Appl. Math. Model. **35**(3), 1322–1330 (2011)
10. Chou, C.C.: The canonical representation of multiplication operation on triangular fuzzy numbers. Comput. Math. Appl. **45**(10–11), 1601–1610 (2003)
11. Cooper, R.G., Kleinschmidt, E.J.: An investigation into the new product process: steps, deficiencies, and impact. J. Prod. Innov. Manage **3**(2), 71–85 (1986)
12. De Ruyter, K., Scholl, N.: Positioning qualitative market research: reflections from theory and practice. Qual. Market Res.: Int. J. **1**(1), 7–14 (1998)
13. Dibb, S.: Market segmentation success-making it happen. Strateg. Dir. **26**(9) (2010)
14. Farhadinia, B.: Multiple criteria decision-making methods with completely unknown weights in hesitant fuzzy linguistic term setting. Knowl.-Based Syst. **93**, 135–144 (2016)
15. Fortemps, P., Roubens, M.: Ranking and defuzzification methods based on area compensation. Fuzzy Sets Syst. **82**(3), 319–330 (1996)
16. Ghorabaee, M.K., Amiri, M., Zavadskas, E.K., Hooshmand, R., Antuchevičienė, J.: Fuzzy extension of the codas method for multi-criteria market segment evaluation. J. Bus. Econ. Manag. **18**(1), 1–19 (2017)
17. Gorzalczany, B.: Approximate inference with interval-valued fuzzy sets-an outline. In: Proceedings of the Polish Symposium on Interval and Fuzzy Mathematics, pp. 89–95 (1983)
18. Hsieh, C.H.: Optimization of fuzzy production inventory models. Inf. Sci. **146**(1), 29–40 (2002)
19. Huynh, V.N., Nakamori, Y., Ho, T.B., Murai, T.: Multiple-attribute decision making under uncertainty: the evidential reasoning approach revisited. IEEE Trans. Syst. Man Cybern.-Part A: Syst. Hum. **36**(4), 804–822 (2006)

20. Huynh, V.N., Nakamori, Y.: A linguistic screening evaluation model in new product development. IEEE Trans. Eng. Manag. **58**(1), 165–175 (2011)
21. Lin, C.T., Chen, C.T.: A fuzzy-logic-based approach for new product go/nogo decision at the front end. IEEE Trans. Syst. Man Cybern.-Part A: Syst. Hum. **34**(1), 132–142 (2004)
22. Lo, C.C., Chen, D.Y., Tsai, C.F., Chao, K.M.: Service selection based on fuzzy topsis method. In: IEEE 24th International Conference on Advanced Information Networking and Applications Workshops (WAINA), pp. 367–372. IEEE (2010)
23. Miller, G.A.: The magical number seven, plus or minus two: some limits on our capacity for processing information. Psychol. Rev. **63**(2), 81 (1956)
24. Pawlak, Z.: Rough sets. Int. J. Parallel Prog. **11**(5), 341–356 (1982)
25. Qi, X., Liang, C., Zhang, J.: Generalized cross-entropy based group decision making with unknown expert and attribute weights under interval-valued intuitionistic fuzzy environment. Comput. Ind. Eng. **79**, 52–64 (2015)
26. Rezaei, J., Wang, J., Tavasszy, L.: Linking supplier development to supplier segmentation using best worst method. Expert Syst. Appl. **42**(23), 9152–9164 (2015)
27. Rodriguez, R.M., Martinez, L., Herrera, F.: Hesitant fuzzy linguistic term sets for decision making. IEEE Trans. Fuzzy Syst. **20**(1), 109–119 (2012)
28. Ross, T.J.: Fuzzy Logic with Engineering Applications. Wiley, Hoboken (2009)
29. de Oliveira Moura Santos, L.F., Osiro, L., Lima, R.H.P.: A model based on 2-tuple fuzzy linguistic representation and analytic hierarchy process for supplier segmentation using qualitative and quantitative criteria. Expert Syst. Appl. **79**, 53–64 (2017)
30. Suprasongsin, S., Huynh, V.N., Yenradee, P.: Optimization of supplier selection and order allocation under fuzzy demand in fuzzy lead time. In: Chen, J., Nakamori, Y., Yue, W., Tang, X. (eds.) KSS 2016. CCIS, vol. 660, pp. 182–195. Springer, Singapore (2016). doi:10.1007/978-981-10-2857-1_16
31. Wansink, B., Park, S.B., Sonka, S.T., Morganosky, M.: How soy labeling influences preference and taste. Int. Food Agribus. Manag. Rev. **3**, 81 (2000)
32. Wedel, M., Kamakura, W.A.: Market Segmentation: Conceptual and Methodological Foundations, vol. 8. Springer Science & Business Media, New York (2012). doi:10.1007/978-1-4615-4651-1
33. Wind, Y., Thomas, R.J.: Segmenting industrial markets. Wharton School, University of Pennsylvania, Marketing Department (1984)
34. Ye, J.: Multiple attribute group decision-making methods with completely unknown weights in intuitionistic fuzzy setting and interval-valued intuitionistic fuzzy setting. Group Decis. Negot. **22**(2), 173–188 (2013)
35. Zadeh, L.A.: Fuzzy sets. Inf. Control **8**(3), 338–353 (1965)
36. Zadeh, L.A.: The concept of a linguistic variable and its application to approximate reasoning. Inf. Sci. **8**(3), 199–249 (1975)

# MANDY: Towards a Smart Primary Care Chatbot Application

Lin Ni[✉], Chenhao Lu, Niu Liu, and Jiamou Liu

Department of Computer Science, The University of Auckland,
Auckland, New Zealand
lni600@aucklanduni.ac.nz, jiamou.liu@auckland.ac.nz

**Abstract.** The paper reports on a proof-of-concept of Mandy, a primary care chatbot system created to assist healthcare staffs by automating the patient intake process. The chatbot interacts with a patient by carrying out an interview, understanding their chief complaints in natural language, and submitting reports to the doctors for further analysis. The system provides a mobile-app front end for the patients, a diagnostic unit, and a doctor's interface for accessing patient records. The diagnostic unit consists of three main modules: An analysis engine for understanding patients symptom descriptions, a symptom-to-cause mapper for reasoning about potential causes, and a question generator for deriving further interview questions. The system combines data-driven natural language processing capability with knowledge-driven diagnostic capability. We evaluate our proof-of-concept on benchmark case studies and compare the system with existing medical chatbots.

**Keywords:** Medicare chatbot · Patient interview · Natural language processing · AI and healthcare

## 1  Introduction

Patients arriving at a primary care service sometimes need to wait for a long time before being advised by a doctor [26]. This is often due to high workload and limited resources at the primary care service [7]. To facilitate the process, nurses and other health care staffs usually take the role of patient intake. An incoming patient would be first greeted by a receptionist who carries out an intake inquiry. The receptionist would typically be someone who has a certain level of medical proficiency, and the inquiry involves collecting patient information and understanding the symptoms of the patient. A brief report is generated as outcome of this inquiry to narrow down the causes of the symptoms, so that the doctor may then use minimum effort to perform differential diagnosis [30].

This seemingly robust system still has many shortcomings: Firstly, the medical staffs who carry out patient intake interviews are expected to acquire a good level of medical expertise; this limits the pool of potential candidates and increases the personnel cost. Secondly, at times the staffs need to meet

© Springer Nature Singapore Pte Ltd. 2017
J. Chen et al. (Eds.): KSS 2017, CCIS 780, pp. 38–52, 2017.
https://doi.org/10.1007/978-981-10-6989-5_4

the demand of a large number of patients and quickly attend to each individual; this increases the risk of losing crucial information in the interview reports. Thirdly, if the intake interview relies on standardized forms or questionnaires, the questions to patients would not be sufficiently personalized to reflect the specific symptoms of individuals, reducing the effectiveness of the interview.

The goal of this paper is to harness the power of artificial intelligence to automate the patient intake process, so that patients receive timely, cost-effective, and personalized healthcare services. To this end, we introduce Mandy, a mobile chatbot who interacts with patients using natural language, understands patient symptoms, performs preliminary differential diagnosis and generates reports.

Despite vast technological advancement, present-day clinics still very much rely on healthcare staff to handle patient intake and carry out initial interviews in a manual way [17]. On the other hand, it is widely viewed that data mining and AI may offer unprecedented opportunities and broad prospects in health [16]. Efforts have been made to deploy humanoid robots (e.g., "Pepper in Belgian hospitals[1]) in hospitals. However, a robot is expensive (e.g. Pepper comes with a price tag of £28000) and would not be able to efficiently cope with a large amount of people. Many industry giants are increasingly investing in AI-enhanced medical diagnosis tools. Notable products include Google DeepMind Health[2], IBM Watson Health[3] and Baidu's Melody[4]. The ambitious goal of these platforms is to allow AI to access and process vast amount of lab test results and genomic data for precision-driven medical diagnosis and predictions.

The novelty of Mandy lies in the fact that it is not directed at precise diagnosis and prediction, but rather, Mandy simply provides a humanized interface to welcome patients and understand their needs, and provide valuable information to physicians for further inquiry. In this way, the system aims to free up the time of healthcare staffs for more meaningful interactions with patients, and help to enable physicians to operate more efficiently.

Mandy is an integrated system that provides a range of functionalities: (1) Mandy provides a patient-end mobile application that pro-actively collects patient narratives of illness and register background information; this may take place at an arbitrary time before the doctor's appointment and at an arbitrary location. (2) Mandy is equipped with natural language processing (NLP) modules that understand patients' lay language, process the patient symptoms, and generate interview questions. (3) Based on interactions during the interview, Mandy will generate a report for the doctor regarding the patient's symptoms and likely causes. (4) Mandy also provides a doctor-end desk-top application for the doctors to check their patients' records and interview reports.

The potential benefits of Mandy are many-fold. Firstly, the system aims to reduce the workload of medical staffs by automating the patient intake process,

---

[1] http://www.bbc.com/news/technology-36528253.
[2] https://deepmind.com/applied/deepmind-health/, 2017.
[3] https://www.ibm.com/watson/health/, 2017.
[4] http://research.baidu.com/baidus-melody-ai-powered-conversational-bot-doctors-patients/, 2016.

and providing initial reporting to doctors. Secondly, Mandy provides personalized intake service to the patients by understanding their symptom descriptions and generating corresponding questions during the intake interview. Thirdly, by interacting with a chatbot, the patient avoids the need to express his health concerns out loud to people other than the doctor. This also reduces the likelihood of patients not seeking medical help due to shyness or cultural boundaries [28]. Furthermore, many studies have shown that patients tend to be more honest when facing a robot rather than a human health staff [1]. So Mandy is likely to collect truthful information about the patients.

**Paper Organization.** Section 2 presents related work and identifies insufficiencies with existing AI technology in terms of patient interviews. Section 3 describes system design and core algorithmic modules of Mandy. Section 4 evaluates a proof-of-concept of Mandy by test cases and discusses the results. Finally, Sect. 5 lists some future works which can further improve Mandy.

## 2    Problem Identification and Related Work

The "overcrowding" issue or long waiting time at emergency units of hospitals and other primary care services has been a world wide challenge [3,10,25]. To cope with the increasing population and an ever increasing demands of patients, a number of countries have implemented targets for reducing waiting time at the healthcare providers, e.g., New Zealand has implemented a "6-hours target" for the waiting time of patients at emergency department since 2009 [15].

Existing patient interview support applications often take the form of expert systems. A common challenge faced by all these applications is the ambiguity and diversity of patient answers. As a result, traditional expert systems usually fail to deliver effective decision support and lacks the flexibility that suits individual needs [14]. An example of AI-driven intake interview assistance system is provided by Warren in [30]. The system sets up complicated rules based on clinical experts' experience and medical knowledge. However, it does not demonstrate capabilities on personalizing the questions to patients and is not able to learn about the individual nature of patients. To apply the system, a clinic needs to provide necessary staffs with sufficient medical background to operate the system. The complicated interface of the system also requires considerable training time, which all adds extra costs to the health provider.

Numerous clinical decision support systems (CDSS) have employed AI technologies in various ways: MYCIN [29] is widely recognized as one of the very first rule-based expert systems that were used for diagnosing infectious diseases. It specialized in bacterial infections and it has been adapted as NEOMYCIN, a teaching and learning platform [8]. Other systems such as INTERNIST-I [22] used a much larger medical knowledge base – obtained from hospital case records – to assist medical personnel in diagnosing internal conditions the patient may have. The system has learning capability on patients' medical history to deliver more accurate results. CDSS technologies have been rapidly developed in the last 10 years. A recent study identified 192 commercially available CDSS applications in existence [18]. One of the more well-known achievements in this area is from IBM's Watson Health [12]. The system seamlessly combines natural language processing, dynamic learning and hypothesis generation and evaluation to provide useful systems in many key areas such as oncology, genomics, and medical imaging. We remark that most of the CDSS systems are designed to be used by the specialists but not the patients themselves.

Natural language processing has become a prevalent technology and formed an integral part of many IT applications; examples of which include e.g., Siri[5] and Cortana[6]. Chatbot Systems and Spoken Dialogue Systems (SDS) respond with comprehensible sentences and elaborately constructed paragraphs to communicate with the user, which has been adopted in medical domain. The well-known ELIZA [31] was designed to act roughly as psychotherapists. More recently, Florence Bot is a chatbot that reminds patients to take pills regularly[7]. Your.MD[8] and HealthTap[9] are miniature doctors. Studies have verified that SDS could help intervening human habits, to help patients quit smoking [23], or affect their dietary behaviour [9] and physical activity [11]. Others application also used SDS for chronic illness monitor systems, e.g. for hypertensive diabetic [6]. Medical counseling and education is another area which often requires the delivery of SDS [4,5,13]. Among them, Mandy resembles the user experiment of Your.MD the most. Your.MD constructs a Bayesian network with massive medical knowledge to compute the most likely cause of an indisposition. On the other hand, Your.MD has different purpose from Mandy as it is not meant to assist doctors.

## 3    System Design and Implementation

Figure 1 illustrates the architecture of Mandy. The patient interacts with Mandy through a mobile chatbot. All algorithms are executed and all data are processed in a web services (cloud). This means that all sentences to and from the patients are generated and analyzed in the cloud, respectively. After the intake interview, Mandy scores the patient's record and generate a report regarding the patient's conditions. The doctor can then login into the e-health information management system to access the personalized reports generated for the patient.

Mandy's logic flow simulates a well-established *clinical reasoning process* for differential diagnosis, which consists of a series of well-defined steps [19,24,27]. These steps are guidelines for medical inquiries by a practitioner:

1. *Data acquisition*: Collect patient's history and symptoms, which forms the basis for the initial diagnostic reasoning.
2. *Problem representation*: Summarize the chief complaints of the patient.
3. *Developing differential diagnosis*: Come up with the hypotheses list base on the data acquired.

---

5 http://www.imore.com/siri.

6 https://www.microsoft.com/en/mobile/experiences/cortana/.

7 Florence Bot, https://florence.chat/.

8 Your.MD, https://www.your.md/.

9 HealthTap, https://www.healthtap.com/.

4. *Prioritizing differential diagnosis*: Decide which should be the leading one among the hypotheses list.
5. *Testing hypothesis*: If additional data is required to confirm the hypotheses, order lab tests to take place.
6. *Review and re-prioritize differential diagnosis*: Rule out some diseases and then try to determine the cause of the symptoms. If a diagnosis cannot be drawn, go back to step 3.
7. *Test new hypotheses*: Repeat the process until a diagnosis is produced.

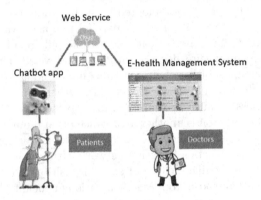

**Fig. 1.** An illustration of the application scenario and system architecture of Mandy.

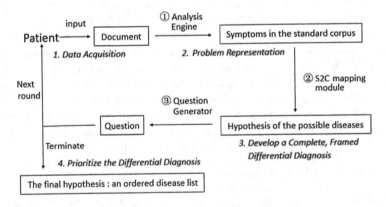

**Fig. 2.** Main procedure

Figure 2 illustrates the main control flow of Mandy. It simulates Steps 1–4 of the clinical reasoning process above. Mandy starts by asking the patient's chief complaint. After the patient inputs a text in natural language, the *analysis engine* extracts the symptoms in a standard corpus from the patient description text. In this way, the system gets an accurate problem representation. Then, the *symptoms-to-cause (S2C) mapping module* comes up with a list of hypothetic diseases based on the symptoms provided by the patient's complaint. The system ranks the possibility of the hypothetic diseases. If there is enough information for proposing the final hypothesis list, the procedure will terminate; Otherwise, the *question generator* will produce another question for the patient and repeats the procedure back to the analysis engine.

We next describe the key data structures and algorithmic modules. The internal algorithms of Mandy rely on the following sets:

A *symptom* is a subjective, observable condition that is abnormal and reflects the existence of certain diseases. For ease of terminology, we abuse the notion including also *signs*, which are states objectively measured by others. A patient *feature* is a fact reflecting the patients, age, gender, geographical and demographical information and life styles (e.g. smoking, alcoholic). Mandy uses a set $S$ of words representing standard symptoms and patient features that are extracted from an external knowledge base.

A *disease* is a medical condition that is associated with a set of symptoms. Mandy also uses a set $L$ of standard diseases. The connection between $S$ and $L$ is captured by a matching function $f : L \to 2^S$ where each disease $\ell \in L$ is associated with a subset $f(\ell) \subseteq S$.

*Example 1.* For the diseases "allergies" and "asthma", we have:

$f$(allergies) = {sneezing, runny nose, stuffy nose, cough, postnasal drip, itchy nose, itchy eyes, itchy throat, watery eyes, dry skin, scaly skin, wheezing, shortness of breath, chest tightness}

$f$(asthma) = {cough, difficulty breathing, chest tightness, shortness of breath, wheezing, whistling sound when exhaling, frequent colds, difficulty speaking, breathless}

**Module I: Analysis Engine.** The *analysis engine* understands user's natural language input and extracts a set of symptoms and features from the set $S$. To implement an effective mechanism that can handle arbitrary language input from the user, we apply Google's *word embedding* algorithm word2vec to map words into a vector space to capture their semantic similarities [2,20,21]. There are two scenarios that word embedding plays a key role: Firstly, when patients describe symptoms in a lay language, the analysis engine picks up keywords and constructs bags of words that represent all patients' symptoms. The algorithm analyzes the most likely diseases by comparing the similarity of the patient's symptoms and all common disease's symptoms. Secondly, when the input words do not appear in the standard corpus $S$, the analysis engine computes words similarity using a word2vec model, which is pre-trained on a large dataset of medical documents. The words similarity will allow the system to find symptoms in the set $S$ that best align with the input description (Fig. 3).

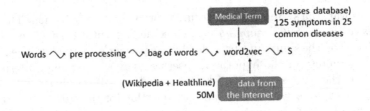

**Fig. 3.** The algorithmic process of the analysis engine in Mandy.

*Example 2.* We give two symptoms with their top-10 similar words:

```
rash: blisters, itchy, scabies, bumps, hives, ringworm, scaly, bite,
    flaky, planus
nausea and vomiting: dizziness, abdominal pain, nausea, drowsiness,
    lightheadedness, cramps, sleepiness, vertigo, weakness, bloating
```

**Module II: S2C Mapping.** This module takes the set $S' \subseteq S$ of patient's symptoms (output of the analysis engine) as input and computes a hypothesized disease $\ell \in L$ that corresponds to $S'$. We propose an algorithm, named *Positive-Negative Matching Feature Count* $(P - N)MFC$, to compare the similarity between $S$ and $f(\ell)$ for all $\ell \in L$. The algorithm runs the following steps: Suppose that we have a set $S_+$ of positive symptoms of the patient and a set $S_-$ of negative symptoms. Suppose also that the set of diseases $L$ is $\{d_1, d_2, \ldots\}$ and let $S_{d_i} = f(d_i)$ be the set of symptoms corresponding to $d_i$. For every $d_i \in L$:

1. Calculate $S_+ \cap S_{d_i}$, and let $n_i^+ = |S_+ \cap S_{d_i}|$.
2. Calculate $S_- \cap S_{d_i}$, and let $n_i^- = |S_- \cap S_{d_i}|$.
3. Calculate $\sigma_i = (n_i^+ - n_i^-)$, this is the similarity value of the patient's symptoms with each disease's symptom.

The $(P - N)MFC$ algorithm selects $d_i \in L$ that has the highest $\sigma_i$ value has the next hypothesis.

**Module III: Question Generator.** This module takes a list of hypothesized diseases $C \subseteq L$ as input, and generates a new question with a most likely symptom for the patient to confirm. Unless Mandy has obtained enough information to derive a diagnosis, the system will continue to pose new questions to the patient. Note that the input list of hypotheses is the result obtained from S2C Mapping Module; element in the list are ordered by the likelihood according to the current patient info. The output is a symptom that Mandy selects from the knowledge base which represent the most likely symptom the patient has. Mandy will form a question that asks the patient to confirm or reject this symptom. The detailed steps of the algorithm is as follows:

1. Update $S_+$ and $S_-$ according to the patients input.
2. If $S_+$ has a new element, perform the $(P - N)MFC$ algorithm to get the most likely disease $\ell \in L$. If $f(\ell) \setminus S_+ \neq \varnothing$, randomly choose one such symptom in $f(\ell)$ but not in $S_+$ and ask about it in the next question.

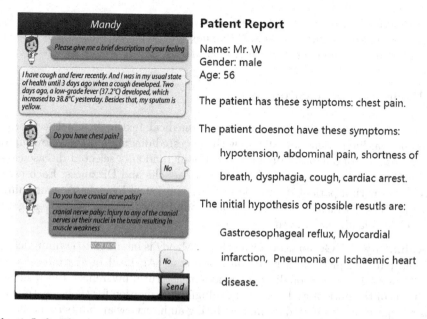

**Patient Report**

Name: Mr. W
Gender: male
Age: 56

The patient has these symptoms: chest pain.

The patient doesnot have these symptoms:

hypotension, abdominal pain, shortness of

breath, dysphagia, cough, cardiac arrest.

The initial hypothesis of possible resutls are:

Gastroesophageal reflux, Myocardial

infarction, Pneumonia or Ischaemic heart

disease.

**Fig. 4.** Left: The app user interface; Whenever users encounter obscure medical terms, the relevant explanation from Merriam-Webster Dictionary can be viewed by clicking the dialog box. Right: The generated initial interview outcome report.

3. If $f(\ell)$ does not contain any symptom not in $S_+$, the system will analyze patient's input, then choose the most similar symptom in our standard corpus, and use it in the next question.
4. Once the system has got enough information from the patient, it will generate a diagnosis result, list top-most possible diseases which are related to the patient's symptoms.

We deploy a proof-of-concept of Mandy on an Amazon Web Services Cloud[10]. It provides services for both the mobile app version[11] (see Fig. 4) and the PC version[12]. Knowledge about symptoms and diseases is constructed based on external sources[13]. In this proof-of-concept, we select 25 common diseases. The dataset for word2vec to train a word embedding consists of crawled entries from the Internet. Firstly, on Wikipedia[14], the crawler dredges data from the main page of "disease" and visit each medical terminology using hyperlinks. To collect

---

[10] https://aws.amazon.com/.
[11] https://github.com/lni600/Mandy.git.
[12] http://13.54.91.140:8080/HealthWebApp/ To log in, the user needs to input 'admin' as both Username and Password.
[13] E.g. online databases such as http://www.diseasesdatabase.com.
[14] https://en.wikipedia.org/.

more colloquial sentences, we also crawled data from Healthline[15]. The collected dataset contains approximately 20,000 web pages on Wikipedia and about 10,000 web pages on Healthline with a size of $\approx$50 MB.

## 4   Performance Evaluation

We extracted case studies from a standard medical textbook which contains numerous real-life patient complaints with suggested diagnosis [27]. We evaluate the performance of our proof-of-concepts on four randomly selected disease categories: Chest Pain, Respiratory Infections, Headache and Dizziness. Each case study starts with a patient description and then a list of hypotheses containing valid hypotheses which can be viewed as ground truth results. We investigate the result of Mandy on 11 such case studies.

**1. Evaluating the Generated Questions.** Mandy is intended to communicate with the patients just like a real healthcare staff. An ideal intake interviewer should pose a list of personalized questions that truthfully reflect the medical conditions of the patient and lead to meaningful information for their treatment. Thus the questions generated by Mandy during an interview amounts to a crucial criterion for its effectiveness.

From the patient description, we recognize main symptoms. We then input only the first symptom to the system and check if the system can generate high-quality questions. We regard the questions which covered the other symptoms as "high-quality" since they are sufficient and important for the doctors to come up with the hypothesis list.

*Example 3.* One case study includes the following patient description: *"Mrs. G is a 68-year-old woman with a history of hypertension who arrives at the emergency department by ambulance complaining of chest pain that has lasted 6 hours. Two hours after eating, moderate (5/10) chest discomfort developed. She describes it as a burning sensation beginning in her mid chest and radiating to her back. She initially attributed the pain to heartburn and used antacids. Despite multiple doses over 3 hours, there was no relief. Over the last hour, the pain became very severe (10/10) with radiation to her back and arms. The pain is associated with diaphoresis and shortness of breath. The patient takes enalapril for hypertension. She lives alone, is fairy sedentary, and smokes 1 pack of Cigarettes each day. She has an 80 pack year smoking history."* The symptoms extracted from the text are **severe chest pain** associated with **diaphoresis** and **shortness of breath**. To evaluate the generated questions, we only provide "severe chest pain", and see if "diaphoresis" and "shortness of breath" will be asked by Mandy.

After we input "severe chest pain" as the answer to the first question, Mandy generated the following interaction. The answers to the questions were obtained from understanding the text description above:

---

[15] http://www.healthline.com/.

Mandy: Do you have dysphagia?     Answer: no
Mandy: Do you have hypotension?     Answer: no
Mandy: Do you have cardiac arrest?     Answer: no
Mandy: Do you have hyperhidrosis?     Answer: yes
Mandy: Do you have fever?     Answer: no
Mandy: Do you have abdominal pain?     Answer: no
Mandy: Do you have shortness of breath?     Answer: yes
Mandy: Do you have nausea and vomiting?     Answer: no
Mandy: Do you have productive cough?     Answer: no
Mandy: Do you have any other symptoms?     Answer: no

Among the 9 questions symptoms, two match exactly as our expected questions. Thus we evaluate the accuracy as: **Question Accuracy** = matched questions ÷ expected questions = 2/2 = 100%.

Using the same approach, we calculate question accuracy for six test cases (the other five test cases are all single symptom cases, so they cannot be used to evaluate the question accuracy). See Table 2. Among the six cases, two are from each of Chest Pain and Respiratory Issues, and a single case is from each of Dizziness and Headache. Besides the case for Dizziness which only asks 2 high-quality questions out of the expected 3 ones, the question accuracies for the other cases are all 100%.

**2. The Performance of the Diagnosis Module.** Another natural evaluation criterion is the diagnosis capability of Mandy. For this, we input the entire paragraph of patient description into the system as the answer to the first question. We then answer subsequent questions manually based on understanding of the case description. When the system has no more questions for the patient, we check if the output hypothesis list from the system matches with the ground truth hypotheses from the book (Table 1).

**Table 1.** Diagnostic hypotheses for Mr.W.

| Diagnostic hypotheses | Clinical clues | Important tests |
| --- | --- | --- |
| **Leading hypothesis** | | |
| Stable angina | Substernal chest pressure with exertion | Exercise tolerance test Angiogram |
| **Active alternative—** | **Most common** | |
| GERD | Symptoms of heartburn chronic nature | EGD Esophageal pH monitoring |
| **Active alternative** | | |
| Musculoskeletal disorders | History of injury or specific musculoskeletal chest pain syndrome | Physical exam Response to treatment |

EGD, esophagogastroduodenoscopy; GERD, gastroesophageal reflux disease.

*Example 4.* The patient Mr.W complained that he felt chest pain with squeezing, sub-sternal pressure while climbing stairs. The only symptom recognized is chest pain. The diagnostic hypotheses including stable angina, GERD, and Musculoskeletal disorders from the guide book Table 8-1 [27] are shown here.

The hypotheses report from our system shows that one out of the four hypotheses is matched with the guide book (GERD). Another hypothesis "Myocardial infarction" (MI) from our system shares the same disease category with "stable angina" from the guide book. We regard MI as correct because it is close enough and "stable angina" does not exist in our current disease corpus.

**Table 2.** Question and prediction accuracy of Mandy over the case studies.

| Disease category | Question accuracy | Prediction accuracy |
| --- | --- | --- |
| Respiratory issues | 100% | 100% |
| Chest Pain | 100% | 64% |
| Headache | 100% | 25% |
| Dizziness | 66.7% | 14% |

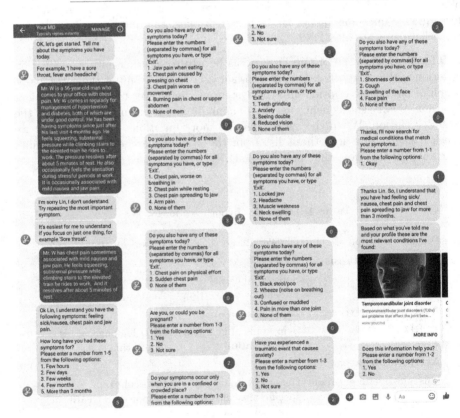

**Fig. 5.** Mr.W's case on Your.MD

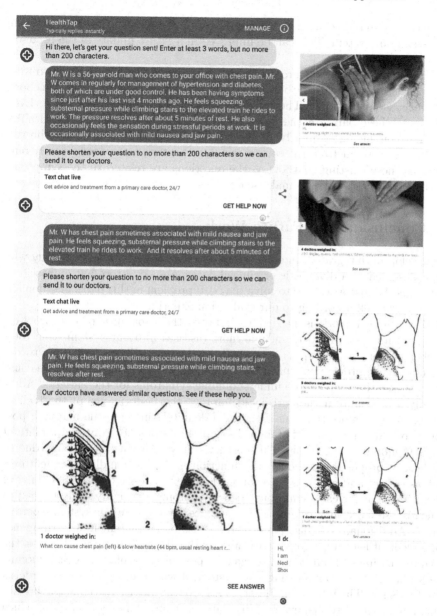

**Fig. 6.** Mr.W's case on HealthTap

Therefore we conclude that the final accuracy of our system for this case is: **_Prediction Accuracy_** = matched hypotheses from our system/ diagnostic hypotheses in guide book = 2/3 = 67%.

Following the same approach, we calculate all the prediction accuracy for the 11 test cases. See Table 2. The low prediction accuracies for Dizziness and

Headache are mainly caused by the lack of training data and knowledge in brain diseases in our system. This can be improved in a future update of the proof-of-concept.

To further evaluate our proof-of-concept, we input Mr.W's case on two well-known existing medical chatbots Your.MD and HealthTap from the Facebook Messenger Bots Platform. The conversations are shown in Figs. 5 and 6. Even including "chest pain" in the description, the results provided by HealthTap were not convincing. Similarly, after Your.MD checked 30 symptoms, the two conclusions were far from the correct one. On this test case, Mandy clearly outperforms these existing chatbots as the questions are related to the symptoms and the hypotheses list also make sense.

## 5    Conclusion and Future Work

We develop an integrated, intelligent and interactive system called Mandy who is not designed as a diagnostic or clinical decision-making tool but an assistant to doctors. We use word2vec to solve the NLP problem in this particular domain which works well according to our evaluation experiments.

Much further work is needed to improve the capability of our proof-of-concept. Firstly, we need to include more diseases into our system. The total number of human diseases is over 1000, so there is still a lot of work to do. Secondly, a symptom synonym thesauri should also be produced. Then we could generate questions with more understandable symptoms for the patients. Additionally, the symptom thesauri could improve the performance of the trained model. Because more than one target will definitely increase the mapping possibility of patient's doc and standard symptom corpus. Thirdly, the update of our S2C module is necessary. Currently, we only deal with symptoms, due to the lack of proper data, though the function is able to handle other features, such as gender and bad habits. Another data set we desired, is the S2C kind of file with weight for each symptom. Some symptoms are highly likely to lead to some disease more than others. This could greatly improve our system's performance. Additionally, we also plan to add a case-based Incremental learning and reinforcement learning algorithm to enhance the diagnosis accuracy. Besides, the separate modules in our structure make it possible to replace our S2C module like a plug-in with another diagnosis system, if which also provides an ordered hypothesis list. The last but not least, to achieve the ambitious goal, chatting like a human, we need to acquire real life Patient-Doctor conversation data, which will give us more confidence to provide smarter interaction.

**Acknowledgement.** The first author is partially funded by a scholarship offered by Precision Driven Health in New Zealand, a public-private research partnership aimed at improving health outcomes through data science. Initial progress of the research was reported in the PDH & Orion Health Blog https://orionhealth.com/global/knowledge-hub/blogs/meet-mandy-an-intelligent-and-interactive-medicare-system/.

# References

1. Ahmad, F., Hogg-Johnson, S., Stewart, D.E., Skinner, H.A., Glazier, R.H., Levinson, W.: Computer-assisted screening for intimate partner violence and control a randomized trial. Ann. Intern. Med. **151**(2), 93–102 (2009)
2. Bengio, Y., Ducharme, R., Vincent, P., Jauvin, C.: A neural probabilistic language model. J. Mach. Learn. Res. **3**(Feb), 1137–1155 (2003)
3. Bernstein, S.L., Aronsky, D., Duseja, R., Epstein, S., Handel, D., Hwang, U., McCarthy, M., John McConnell, K., Pines, J.M., Rathlev, N., et al.: The effect of emergency department crowding on clinically oriented outcomes. Acad. Emerg. Med. **16**(1), 1–10 (2009)
4. Bickmore, T., Giorgino, T.: Health dialog systems for patients and consumers. J. Biomed. Inform. **39**(5), 556–571 (2006)
5. Bickmore, T.W., Pfeifer, L.M., Byron, D., Forsythe, S., Henault, L.E., Jack, B.W., Silliman, R., Paasche-Orlow, M.K.: Usability of conversational agents by patients with inadequate health literacy: evidence from two clinical trials. J. Health Commun. **15**(S2), 197–210 (2010)
6. Black, L.-A., McTear, M., Black, N., Harper, R., Lemon, M.: Appraisal of a conversational artefact and its utility in remote patient monitoring. In: Proceedings of 18th IEEE Symposium on Computer-Based Medical Systems, pp. 506–508. IEEE (2005)
7. Caley, M., Sidhu, K.: Estimating the future healthcare costs of an aging population in the UK: expansion of morbidity and the need for preventative care. J. Publ. Health **33**(1), 117–122 (2011)
8. Clancey, W.J., Letsinger, R.: NEOMYCIN: Reconfiguring a rule-based expert system for application to teaching. Stanford University, Department of Computer Science (1982)
9. Delichatsios, H.K., Friedman, R.H., Glanz, K., Tennstedt, S., Smigelski, C., Pinto, B.M., Kelley, H., Gillman, M.W.: Randomized trial of a "talking computer" to improve adults' eating habits. Am. J. Health Promot. **15**(4), 215–224 (2001)
10. Di Somma, S., Paladino, L., Vaughan, L., Lalle, I., Magrini, L., Magnanti, M.: Overcrowding in emergency department: an international issue. Intern. Emerg. Med. **10**(2), 171–175 (2015)
11. Farzanfar, R., Frishkopf, S., Migneault, J., Friedman, R.: Telephone-linked care for physical activity: a qualitative evaluation of the use patterns of an information technology program for patients. J. Biomed. Inform. **38**(3), 220–228 (2005)
12. High, R.: The Era of Cognitive Systems: An Inside Look at IBM Watson and How it Works. IBM Corporation, Redbooks (2012)
13. Hubal, R.C., Day, R.S.: Informed consent procedures: an experimental test using a virtual character in a dialog systems training application. J. Biomed. Inform. **39**(5), 532–540 (2006)
14. Hunt, D.L., Haynes, R.B., Hanna, S.E., Smith, K.: Effects of computer-based clinical decision support systems on physician performance and patient outcomes: a systematic review. JAMA **280**(15), 1339–1346 (1998)
15. Jones, P., Chalmers, L., Wells, S., Ameratunga, S., Carswell, P., Ashton, T., Curtis, E., Reid, P., Stewart, J., Harper, A., et al.: Implementing performance improvement in new zealand emergency departments: the six hour time target policy national research project protocol. BMC Health Serv. Res. **12**(1), 45 (2012)
16. Khoury, M.J., Ioannidis, J.P.A.: Big data meets public health. Science **346**(6213), 1054–1055 (2014)

17. Lipkin, M., Quill, T.E., Napodano, R.J.: The medical interview: a core curriculum for residencies in internal medicine. Ann. Intern. Med. **100**(2), 277–284 (1984)
18. Martínez-Pérez, B., de la Torre-Díez, I., López-Coronado, M., Sainz-De-Abajo, B., Robles, M., García-Gómez, J.M.: Mobile clinical decision support systems and applications: a literature and commercial review. J. Med. Syst. **38**(1), 4 (2014)
19. McFillen, J.M., O'Neil, D.A., Balzer, W.K., Varney, G.H.: Organizational diagnosis: an evidence-based approach. J. Change Manag. **13**(2), 223–246 (2013)
20. Mikolov, T., Chen, K., Corrado, G., Dean, J.: Efficient estimation of word representations in vector space. arXiv preprint arXiv:1301.3781 (2013)
21. Mikolov, T., Sutskever, I., Chen, K., Corrado, G.S., Dean, J.: Distributed representations of words and phrases and their compositionality. In: Advances in neural information processing systems, pp. 3111–3119 (2013)
22. Miller, R.A., Pople Jr., H.E., Myers, J.D.: Internist-I, an experimental computer-based diagnostic consultant for general internal medicine. New Engl. J. Med. **307**(8), 468–476 (1982)
23. Ramelson, H.Z., Friedman, R.H., Ockene, J.K.: An automated telephone-based smoking cessation education and counseling system. Patient Educ. Couns. **36**(2), 131–144 (1999)
24. Realdi, G., Previato, L., Vitturi, N.: Selection of diagnostic tests for clinical decision making and translation to a problem oriented medical record. Clin. Chim. Acta **393**(1), 37–43 (2008)
25. Richardson, D.B.: Increase in patient mortality at 10 days associated with emergency department overcrowding. Med. J. Aust. **184**(5), 213–216 (2006)
26. Scheffler, R.M., Liu, J.X., Kinfu, Y., Dal Poz, M.R.: Forecasting the global shortage of physicians: an economic-and needs-based approach. Bull. World Health Organ. **86**(7), 516–523B (2008)
27. Stern, S., Cifu, A., Altkorn, D.: Symptom to Diagnosis an Evidence Based Guide. McGraw Hill Professional, New York City (2014)
28. Taber, J.M., Leyva, B., Persoskie, A.: Why do people avoid medical care? A qualitative study using national data. J. Gen. Intern. Med. **30**(3), 290–297 (2015)
29. Victor, L.Y., Buchanan, B.G., Shortliffe, E.H., Wraith, S.M., Davis, R., Davis, A.R., Scott, A.C., Cohen, S.N.: Evaluating the performance of a computer-based consultant. Comput. Prog. Biomed. **9**(1), 95–102 (1979)
30. Warren, J.R.: Better, more cost-effective intake interviews. IEEE Intell. Syst. Appl. **13**(1), 40–48 (1998)
31. Weizenbaum, J.: Eliza—a computer program for the study of natural language communication between man and machine. Commun. ACM **9**(1), 36–45 (1966)

# Sequence-Based Measure for Assessing Drug-Side Effect Causal Relation from Electronic Medical Records

Tran-Thai Dang[1]([⊠]) and Tu-Bao Ho[1,2]

[1] Japan Advanced Institute of Science and Technology,
1-1 Asahidai, Nomi, Ishikawa, Japan
{dangtranthai,bao}@jaist.ac.jp
[2] John Von Neumann Institute, VNU-HCM, Ho Chi Minh City, Vietnam

**Abstract.** The recent prevalence of electronic medical records offers a new way to identify likely drug-side effect causal relations. However, it faces with a big challenge due to simultaneously taking multiple drugs by patients then a mixture of side effects of these drugs is observed that forms a huge space of possible relations between the drugs and the effects, and makes a confusion in distinguishing causal relations from non-causal ones. Most of existing methods, which use frequency-based measures to quantify the association strength between the drugs and the effects, perform rather low accuracy. Therefore, we propose a novel measure called sequence-based measure that bases on the assumption about the association between side effects exposing during the treatment period. The experimental results show an effectiveness of using the proposed measure in detecting proper causal relations in comparison with existing methods, as well as reflect the likelihood of the assumption.

**Keywords:** Electronic medical records · Drug-side effect causal relation · Sequence-based measure

## 1 Introduction

Drug side effects are understood as undesirable effects reasonably associated with the use of drugs, which occur as a part of the pharmacological action of the drugs or unpredictable interaction with human body. Detecting side effects caused by a specific drug or a combination of drugs plays an essential role in drug safety. Before being approved and released for using, a drug has to go through a series of clinical trials to evaluate expected indications and possible side effects. However, these trials are often carried out under ideal and controlled circumstances which only test efficacy of the drug but not its effectiveness in practical use.

The effectiveness of a drug is evaluated by pragmatic clinical trials [10,24] that basically analyze textual data coming from patient spontaneous reports, social network, and electronic medical records (EMRs) [12,24,30]. In those, EMR data is well recognized as a precious resource for pragmatic clinical trials due to

© Springer Nature Singapore Pte Ltd. 2017
J. Chen et al. (Eds.): KSS 2017, CCIS 780, pp. 53–65, 2017.
https://doi.org/10.1007/978-981-10-6989-5_5

objectively recording the facts of drug effects under the real condition in a huge patient cohort. The use of EMRs for pragmatic clinical trials, even being still in its infancy, is mentioned under the abbreviation EMRPCT [8,25,29]. In [3], the authors presented and analyzed the theoretical advantages and disadvantages, the ethical and regulatory aspects of EMRPCT, as well as prospects of EMRPCT in drug effectiveness study. Typically, EMR data brings two main properties, one is longitude, and the other is heterogeneity [19] that give a new opportunity in clinical research, as well as pose many challenges in analyzing and mining clinical text [13].

One of important problems in evaluating the drug effectiveness, which we are concentrating on, is to recognize side effects recorded in clinical notes then assess the causal relation between them and the drug [9]. For reliably assessment, reasonable evidences that almost base on statistics on clinical text should be given. Unfortunately, different from the patient reports, the clinical text in EMRs is basically narratives [7], in which the drug-effect causal relations are not explicitly mentioned. Moreover, the use of multiple drugs makes a confusing in determining that which drug or group of drugs side effects are actually related to.

A well-known evidence for drug-side effect causal relation assessment used in most of existing methods is the frequency of the drug-side effect co-occurrence in clinical text. Strength of the causal relation is quantified through several conventional pairwise measures. In [20], Liu *et al.* measured the association between side effects and statin drugs by using log-likelihood ratio which is based on the proportion between the number of statin drug reviews and non-statin drug reviews. Additionally, in [5,21,28], $\chi^2$ statistics was commonly used to confirm the association between drugs and effects. Besides, Wang *et al.* [27] used Pairwise Mutual Information (PMI) for this purpose. Several work view causal drug-effect pairs as the form of association rules, and the strength of the rules is supported by some well-known statistical measures such as Relative Reporting Ration (RR) [11], Support, Confidence, Leverage [2,15,16,31]. Although the frequency-based association measures of drug-side effect pairs are widely used in many studies on detecting drug side effects, they are not sufficient to reflect the real causal relation between the drug and the effect. Moreover, in treatment, patients often take several drugs together then a mixture of their side effects is observed and recorded that leads to an imprecise association measure of non-causal pairs because these pairs can be coincidentally observed in the high frequency. In addition, the proper causal pairs which infrequently co-occur may be left out.

For accurate identification of causal drug-side effect pairs, the supporting evidences should be enriched instead of only measuring the association between drugs and effects. For this, the additional relationship among drugs or among effects needs to be exploited. The fact of the patient's body and health status impacting on the drug effects (pharmacokinetics), a hypothesis about the correlation between adverse events [23], the longitudinal property of EMR data allowing to represent clinical notes as a sequence of medication events [19] inspire us to exploit the longitudinal relation among side effects during the treatment process. Relying on the mechanism of the drug action and the protein-protein interaction, we hypothesize the association between side effects exposing in a hospitalization and use it as an assumption to construct the sequence-based measure for quantifying association strength between drugs and effects. The reason of exploiting the association among side effects is that the treatment in Intensive Care Unit often becomes more complicated due to the presenting of comorbidity which demands using more drugs. Therefore, exploiting the relation between side effects can reduce the uncertainty in determining causal pairs occurring in later stages of the treatment period, as well as enhance an ability to detect infrequent causal pairs.

## 2    Problem Formulation

Side effects can be caused by a single drug or an interaction among multiple drugs, in which identifying the side effects caused by multiple drugs is more complicated than that caused by a single drug but is promising to discover new effects. In our view, the problem of detecting side effects caused by both the single drugs and the interaction between multiple drugs can be decomposed into two sub-problems: (1) detecting likely causal relations between each single drug and its side effects; (2) predicting the drug-drug interaction then inferring side effects caused by this interaction. In the scope of this study, we focus on solving the first sub-problem.

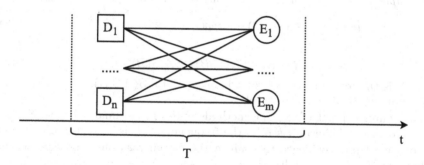

**Fig. 1.** Bipartite graph represents all possible associations between drugs and side effects observed in a time window $T$ of the treatment period, in which function $w(D_i, E_j)$ produces the weight of each connection between a drug $D_i$ and an effect $E_j$ that measures an association strength between $D_i$ and $E_j$.

The treatment period noted in EMRs is often divided into time intervals (time windows) for investigating all possible temporal associations between drugs and side effects within each window $T$ to find likely causal ones [15,16]. The list of all possible associations between drugs and side effects in a time window $T$ can be represented by a bipartite graph illustrated in Fig. 1 with the function $w(D_i, E_j)$

used to measure the association strength between a drug $D_i$ and an effect $E_j$. Obviously, the problem of identifying likely causal drug-side effect pairs can be viewed as the problem of ranking all possible connections in the bipartite graph according to their association strength. As the groundwork of effectively ranking depends on the quality of the associated strength measure, so the objective of our work is to find a measure that well reflects the real causal relation between drugs and their side effects.

## 3    Existing Methods of Measuring Drug-Side Effect Causal Relation Strength

As briefly mentioned in Sect. 1, most of existing work on identifying drug-side effect causal relations from EMRs so far represent the causal relations between drugs and side effects within a window $T$ in form of temporal association rules $D_i \xrightarrow{T} E_j$ with various pairwise statistical association measures for quantifying strength of the rules [15,16,22].

Commonly, association strength of the rules can be estimated through several kinds of measures such as Confidence $(conf)$, Leverage $(lev)$ [15,16], $\chi^2$ test [5], and Relative Reporting Ratio $(RR$, which basically is similar to Pointwise Mutual Information) [11] as follows:

$$conf(A \xrightarrow{T} C) = \frac{supp(A \xrightarrow{T} C)}{supp(A \xrightarrow{T})} \qquad (1)$$

where $supp(A \xrightarrow{T})$ is proportion of $T-$constrained sub-sequences containing $A$.

$$lev = supp(A \xrightarrow{T} C) - supp(A \xrightarrow{T}) \times supp(\xrightarrow{T} C) \qquad (2)$$

$$RR = N \times S(A \cup B)/S(A)S(B) \qquad (3)$$

where $N$ is the total number of records in the data, $S(A \cup B)$, $S(A)$, $S(B)$ are support of $A \cup B$, $A$, $B$, respectively.

The common point of existing methods is to only estimate the association between drugs and effects based on the frequency of their co-occurrence. These methods used poor evidence to confirm the causal relations, and can imprecisely recognize non-causal drug-effect pairs which frequently and coincidentally co-occur.

## 4    Sequence-Based Measure of Drug-Side Effect Causal Relation Strength

### 4.1    Splitting Treatment Period into Time Windows

Our work was conducted on a practical electronic medical record database called MIMIC-III (Medical Information Mart for Intensive Care III)[1] [17]. This database contains prescription records and clinical notes of a large number of patients.

---

[1] https://mimic.physionet.org.

In that, the treatment process for each patient in his/her hospitalization is recorded with timestamps corresponding to when a drug is prescribed or a clinical note is created.

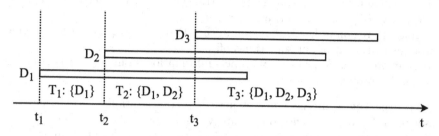

**Fig. 2.** An example of splitting the treatment period into time windows based on the information of starting and ending time of three drugs $D_1$, $D_2$, $D_3$. Each bar indicates the length of period that each drug is used, $t_1, t_2, t_3$ are starting times corresponding to the drugs $D_1$, $D_2$, $D_3$, respectively. The dashed lines are boundaries of time windows.

Splitting the treatment period in a hospitalization aims to form time windows for restricting the scope of possible causal relations between drugs and side effects. It is based on starting time of each drug used during the treatment period. An example of forming time windows is illustrated in Fig. 2. In this figure, a drug is considered to belong to a window if the time interval bounded by the window completely lies on the usage period of this drug. For example, in the time window $T_1$, only the drug $D_1$ is prescribed, then in $T_2$, the drug $D_2$ is started using with $D_1$, and in the last window, all three drugs are prescribed together. After forming time windows, the clinical notes are also mapped to their corresponding window based on the creation time attached with each note.

After mapping the clinical notes, a set of drug effects is determined by extracting words, phrases expressing symptoms, abnormalities from these clinical notes using MetaMap[2] [1]. MetaMap is a well-known Natural Language Processing system for analyzing biomedical text based on Unified Medical Language System (UMLS) Metathesaurus. Two main functions of MetaMap are medical terminology recognition and category (often called as semantic type) identification. In our work, the set of effects is identified using three semantic types including "Acquired Abnormality" and "Finding" and "Sign or Symptom".

## 4.2 Assumption About Association Between Drug Effects

In pharmaceutical science, a drug is essentially a chemical compound, and a drug target is considered as a mass of protein molecules including receptors that receive chemical signal from outside a cell. Due to being protein, the drug target is associated with observed diseases, symptoms which are called phenotype in

---

[2] https://metamap.nlm.nih.gov.

general [14]. To understand about the mechanism of drug action, we briefly introduce some relevant biological concepts as follows:

- Transcription factor: A protein required to bind to regulatory region of DNA (Deoxyribonucleic acid), and helps to translate "genetic message" in DNA into RNA (Ribonucleic acid) and protein.
- Regulatory region of DNA: A region in the DNA sequence that needs specific proteins to turn it on, or sometimes off.
- Gene expression: A process by which information from a gene is used to synthesize protein.

When a drug comes in the body, it activates transcription factors, then the transcription factors can bind to regulatory regions of DNA. The DNA changes its status that leads to the gene expression process taking place to change RNA and protein functions, and exposing diseases and symptoms.

Not only the interaction between drugs and drug targets but also the targets also interact with each other due to the protein-protein interaction. Commonly, the protein-protein interaction can be understood as physical contacts between proteins that occur in cell or in living organism [6]. The physical contacts mean the functional sharing between proteins. Therefore, the change in functions of a protein can lead to the change in functions of the others.

A hypothesis of the association between side effects and therapeutic indications was raised in [26]. In this work, the authors verified this hypothesis using a predictive model. This hypothesis and the inferring from the drug-target interaction and the protein-protein interaction inspire us to propose an assumption about the association between side effects exposing during the treatment period in a hospitalization. In general, we hypothesize two types of this association. The first type is side effects belonging to the same living organ. For example, both respiratory tract infection and rhinitis, which are related to the respiratory system, are side effects of Salbutamol (the drug used to treat asthma)[3] that may co-occur when this drug impacts on the respiratory system. The second type is side effects belonging to organs sharing the function. For example, headache and fever are also side effects of Salbutamol, excluding the directly impact of this drug on brain, another reason can be considered that respiratory side effects cause breathing difficulty that leads to the headache and fever. This example reflects the association between side effects due to the protein-protein interaction.

Since the duration of drug action depends on several factors such as the amount of drug given (doses), the pharmaceutical preparation, the reversibility of drug action, the half-life of the drug, the slope of the concentration-response curve, the activity of metabolites, the influence of disease on drug elimination [4], and different time of taking drug, the associated side effects do not co-occur. That means the observation of these side effects is sequential.

**Assumption.** *Side effects exposing during the treatment period of a hospitalization may have an association to each other.*

---

[3] http://sideeffects.embl.de/drugs/2083/.

## 4.3 Inspiration of Sequence-Based Drug-Side Effect Causal Relation Suspicion

The medication treatment in the Intensive Care Unit (ICU) often becomes more complicated in later stages due to the appearance of additional diseases (comorbidity) requiring the use of more drugs for treatment. The number of drugs increasing pulls the increase in number of side effects. That means a huge number of possible connections between the drugs and the side effects are produced that pushes the uncertainty and difficulty in identifying causal relations up. That is also the key reason to make most of previous work become ineffective [22]. However, the assumption about the association between side effects mentioned in Subsect. 4.2 opens an opportunity to overcome this drawback by using historical treatment information for reducing the uncertainty. Moreover, it also helps to predict infrequent causal drug-effect pairs because although such pairs co-occur in low frequency, they have chance to be recognized thought their association with the side effects observed in the past.

## 4.4 Model

In this subsection, we introduce a novel method for quantifying strength of drug-side effect causal relation using sequence-based measure. This measure is constructed based on the assumption mentioned in Subsect. 4.2. Our proposed method includes two steps as follows:

1. Detecting likely causal drug-side effect pairs in a hospitalization.
2. Pooling likely causal drug-side effect pairs detected in whole hospitalizations.

**Detecting Likely Causal Drug-Side Effect Pairs in a Hospitalization.**
We extract the drug names and side effects in each hospitalization $h$, in which $D$ is the set of all drugs, and $E$ is the set of all side effects observed in $h$. The treatment period in the hospitalization $h$ is divided into time windows using the method mentioned in Subsect. 4.1. For each drug $d \in D$, let's $n$ is the number of time windows belonging to the usage period of this drug. We form all possible connections between $d$ and side effects in $n$ windows for selecting the likely effects caused by $d$, then define the weight of the connection $w_h(d, e_{ik})$ where $e_{ik} \in E$, $1 \leq i \leq |E_k|$ ($E_k$ is the set of side effects in $k^{th}$ window), and $1 \leq k \leq n$ by a recursive function as below:

$$w_h(d, e_{ik}) = \begin{cases} log\Big(P(d|e_{ik})\Big) & \text{if } k = 1 \\ Q & \text{if } 2 \leq k \leq n \end{cases} \qquad (4)$$

where $Q$ is defined as following:

$$Q = \frac{1}{k} \times \left( log\Big(P(d|e_{ik})\Big) + log\Big(P(e_{ik}|e_{j(k-1)})\Big) + w_h(d, e_{j(k-1)}) \right)$$

The probability $P(d|e_{ik})$ is called *emission probability* that measures the association between the side effect and the drug within the $k^{th}$ window, and $P(e_{ik}|e_{j(k-1)})$ is called *transition probability* that measures the association between side effects observed in two consecutive windows. The emission and transition probabilities are estimated, respectively, as follows:

$$P(d|e_{ik}) = \frac{count(d, e_{ik}) + \lambda}{count(e_{ik}) + \lambda \times |E|}$$

$$P(e_{ik}|e_{j(k-1)}) = \frac{count(e_{j(k-1)}, e_{ik}) + \lambda}{count(e_{j(k-1)}) + \lambda \times |E|}$$

where $count(d, e_{ik})$, $count(e_{j(k-1)}, e_{ik})$, $count(e_{j(k-1)})$ are number of patients taking drug $d$ and side effect $e_{ik}$ is observed, number of patients that both $e_{ik}$ and $e_{j(k-1)}$, and only $e_{j(k-1)}$ exposing, respectively. The predefined constant $\lambda$ is Laplacian smoothing coefficient that often takes the value of 0.1.

We make the cumulative process in estimating $w_h(d, e_{ik})$ smooth to avoid the problem of imprecisely estimating transition and emission probability from the data because of the coincident observation of side effects caused by multiple drugs. The smooth means no rapid change in value of the function $w_h$ between two consecutive windows. For this purpose, we add the smoother $\frac{1}{k}$.

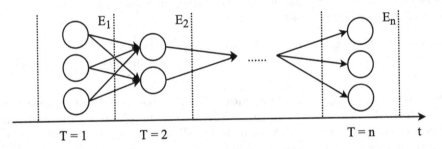

**Fig. 3.** Detecting the sequence of side effects in $n$ windows that maximizes the value of $w_h(d, e_{in})$

Clearly, the recursive function $w_h(d, e_{ik})$ formulates our idea mentioned in Subsect. 4.3 that side effects in the current window may have an association with ones in the previous window, and the association score to the drug for the current effects inherits the score of the previous ones. Additionally, for infrequent causal pairs in $k^{th}$ window, even though their emission probability is low, their scores still have chance to be enhanced through the transition probability and the value of $w_h(d, e_{j(k-1)})$ if the side effects are strongly associated with the effects in $(k-1)^{th}$ window.

As the assumption about the association among side effects of a specific drug, so for each drug $d$, we find an effect sequence $(\hat{e}_1, \hat{e}_2, \ldots, \hat{e}_n)$ where $\hat{e}_k \in E_k$ that maximizes the value of $w_h(d, e_{in})$, which is illustrated in Fig. 3. The value of

$w_h(d, e_{in})$ is the cumulative value of the sequence, which measures how strongly the effects in this sequence are related to the drug. Therefore, the side effects in the selected sequence are identified to have causal relation with the drug in the hospitalization $h$. The most likely sequence is discovered by using Viterbi algorithm.

---

**Algorithm 1.** Viterbi Algorithm

---

$best\_score = \{\}$
$back\_trace = \{\}$
**for** $k := 1$ *to* $n$ **do**
  **if** $k == 1$ **then**
    **for** $x := 1$ *in* $|E_k|$ **do**
      Compute $w(d, e_{xk})$ according to Eq. 4
      $best\_score[d, e_{xk}] = w_h(d, e_{xk})$
      $back\_trace[d, e_{xk}] = None$
  **else**
    **for** $x := 1$ *in* $|E_k|$ **do**
      **for** $y := 1$ *in* $|E_{k-1}|$ **do**
        Compute all values of $w_h(d, e_{xk})$ according to Eq. 4 with different value of $w_h(d, e_{y(k-1)})$ then store the results in an array $W$
        $max\_val = max(W)$
        $ymax = W.index(max\_val)$
      $best\_score[d, e_{xk}] = max\_val$
      $back\_trace[d, e_{xk}] = e_{ymax(k-1)}$
**for** $x := 1$ *in* $|E_n|$ **do**
  Select $\hat{e}_n$ having maximum value of $w_h(d, e_{xn})$
Get the sequence $(\hat{e}_1, \hat{e}_2, \dots, \hat{e}_n)$ using $back\_trace$

---

**Pooling Likely Causal Drug-Side Effect Pairs Detected in Whole Hospitalizations.** The drug-side effect pair $(d, e_{ik})$ can be observed in several hospitalizations $H$ with different values of $w_h(d, e_{ik})$, where $h \in H$, so to get the final score of this pair, we pool all values of $w_h(d, e_{ik})$ by taking their maximum value.

$$w(d, e) = \max_{h \in H} \left( w_h(d, e_{ik}) \right)$$

# 5    Experimental Evaluation

## 5.1    Experimental Design

The data set used for the experiments is MIMIC III (Medical Information Mart for Intensive Care III) briefly mentioned in Subsect. 4.1. This data set is large and freely accessible that contains over 40,000 patients who stayed in the Beth Israel Deaconess Medical Center between 2001 and 2012 [17]. It includes various information of demographics, laboratory test, medication events, clinical notes. For the scope of this study, we used prescriptions and clinical notes for the experiments.

From the MIMIC III database, we exported the prescriptions and clinical notes of 10,000 patients, in which, we detect causal drug-side effect pairs in randomly selected 50 patients by collating with the rest which is used to estimate transition, emission probabilities in our proposed model. The exported raw data was pre-processed by the mechanism mentioned in Subsect. 4.1. We select 49 drugs to detect their side effects using proposed model. The performance of the model is evaluated through checking how many causal drug-side effect pairs that are confirmed by SIDER[4] [18] in the retrieval pairs.

We compare the performance of our proposed model with existing methods mentioned in Sect. 3. The key point is to investigate the quality of association measures used in those methods in reflecting the real drug-side effect causal relation. In previous work, estimating the value of probabilities such as $supp$, $conf$ was carried out in a different way from our method, so for fairly comparison, we make a consensus of probability computing which is based on the proportion between the number of patients presenting the relation or property over the total patients. That means we count the number of patients whom the drug-effect pairs, effect-effect pairs are observed on, instead of counting the frequency of these pairs mentioned in the clinical text.

## 5.2   Evaluation Metrics

In this study, we evaluate the performance of the methods in identifying drug-side effect causal relation by Precision at K ($Prec_K$) which is defined as the fraction of known side effects occurring in the top $K$ ones of the list returned by each method for a specific drug [22].

$$Prec_K = \frac{\sum_{i=1}^{K} y(i)}{K}$$

where $y(i) = 1$ if the $i^{th}$ side effect is proper, and is 0 for otherwise.

## 5.3   Experimental Results and Discussion

Identifying drug-side effect causal relation in electronic health records or electronic medical records is a challenging problem. The solutions for this problem so far are still in early stage that just used conventional statistical measures to directly estimate strength of drug-side effect relation, which mostly produces low performance. For example, in [16], the authors used the leverage to detect causal drug-effect pairs in the Queensland Linked dataset, and got the $Prec_{10}$ is about 0.313.

---

[4] http://sideeffects.embl.de.

In order to investigate the likelihood of the proposed assumption about the association between side effects appearing in the treatment period of a hospitalization as well as the effectiveness of the sequence-based measure utilization for solving this problem, we make a comparison between the proposed method and existing methods with multiple values of $K$ that is showed in Table 1.

**Table 1.** Performance comparison between sequence-based method and existing methods in identifying drug-side effect causal relation

| Method | $Prec_5$ | $Prec_{10}$ | $Prec_{15}$ | $Prec_{20}$ | $Prec_{25}$ | $Prec_{30}$ |
|---|---|---|---|---|---|---|
| $RR$ | 0.331 | 0.33 | 0.33 | 0.337 | 0.333 | 0.339 |
| $conf$ | 0.403 | 0.375 | 0.386 | 0.387 | 0.389 | 0.39 |
| $lev$ | 0.373 | 0.337 | 0.343 | 0.343 | 0.339 | 0.335 |
| $\chi^2$ test | 0.373 | 0.346 | 0.356 | 0.367 | 0.369 | 0.363 |
| Sequence-based measure | **0.437** | **0.447** | **0.439** | **0.439** | **0.433** | **0.427** |

Equation 4 shows that the function $w_h(d, e_{ik})$ when $k = 1$ (without previous windows) is equivalent to pairwise association measure of the drug and effect that is similar to existing methods, and the difference between our method and existing ones is since $k = 2$ when the function $w_h$ incorporates information of causal relation of candidates from historical windows. Table 1 shows a significant improvement of $Prec_K$ over all considering values of $K$ when using sequence-based measure, the precision increases from about 4% (in $Prec_5$ and $Prec_{30}$) to about 7% (in $Prec_{10}$) in comparison with existing methods. These results also show the likelihood of our proposed assumption and the effectiveness of sequence-based measure in identifying causal drug-side effect pairs from EMRs.

Although there is an improvement when using the sequence-based method for this problem, the precision is still low. The reason is that we detect causal drug-effect pairs in the subset of 50 patients based on the information of other patients, however, doing statistic on the other patients does not give precise values of transition and emission probabilities. Therefore, this drawback motivates our future work for effectively solving this problem.

# 6   Conclusion

This paper introduces a novel measure for causal relation quantification, which is called the sequence-based measure built on our assumption about the association between side effects observed in a hospitalization, to overcome the difficulty in identifying drug-side effect causal relation from EMRs in case of multiple drugs usage. The proposed assumption bases on the inference from the mechanism of drug action, and the knowledge of protein-protein interaction. The experimental results show the likelihood of this assumption and the effectiveness of the sequence-based measure to solve this problem in comparison with conventional measures often used in existing methods.

# References

1. Aronson, A.R., Lang, F.-M.: An overview of metamap: historical perspective and recent advances. J. Am. Med. Inform. Assoc. **17**(3), 229–236 (2010)
2. Benton, A., Ungar, L., Hill, S., Hennessy, S., Mao, J., Chung, A., Leonard, C.E., Holmes, J.H.: Identifying potential adverse effects using the web: a new approach to medical hypothesis generation. J. Biomed. Inform. **44**(6), 989–996 (2011)
3. Carcasa, A.J., Santos, F.A., Perrucac, L.S., Dal-Ree, R.: Electronic medical record in clinical trials of effectiveness of drugs integrated in clinical practice. Med. Clin. **145**(10), 452–457 (2015)
4. Carruthers, S.G.: Duration of drug action. Am. Fam. Physician **21**(2), 119–126 (1980)
5. Chen, E.S., Hripcsak, G., Xu, H., Markatou, M., Friedman, C.: Automated acquisition of disease-drug knowledge from biomedical and clinical documents: an initial study. J. Am. Med. Inform. Assoc. **15**(1), 87–98 (2008)
6. De Las Rivas, J., Fontanillo, C.: Protein-protein interactions essentials: key concepts to building and analyzing interactome networks. PLoS Comput. Biol. **6**(6), e1000807 (2010)
7. Deng, Y., Stoehr, M., Denecke, K.: Retrieving attitudes: sentiment analysis from clinical narratives, pp. 12–15 (2014)
8. Elkhenini, H.F., Davis, K.J., Stein, N.D., New, J.P., Delderfield, M.R., Gibson, M., Vestbo, J., Woodcock, A., Bakerly, N.D.: Using an electronic medical record (EMR) to conduct clinical trials: salford lung study feasibility. BMC Med. Inform. Decis. Mak. **15**, 8 (2015)
9. Farcas, A., Bojita, M.: Adverse drug reactions in clinical practice: a causality assessment of a case of drug-induced pancreatitis. J Gastrointest. Liver Dis. **18**(3), 353–358 (2009)
10. Ford, I., Norrie, J.: Pragmatic trials. New Engl. J. Med. **375**(5), 454–463 (2016)
11. Harpaz, R., Haerian, K., Chase, H.S., Friedman, C.: Statistical mining of potential drug interaction adverse effects in FDAs spontaneous reporting system. In: AMIA Annual Symposium Proceedings, pp. 281–285 (2010)
12. Ho, T.B., Le, L., Dang, T.T., Siriwon, T.: Data-driven approach to detect and predict adverse drug reactions. Curr. Pharm. Des. J. **22**(123), 3498–3526 (2016)
13. Hripcsak, G., Albers, D.J.: Next-generation phenotyping of electronic health records. J. Am. Med. Inform. Assoc. **20**(1), 117–121 (2013)
14. Imming, P., Sinning, C., Meyer, A.: Drugs, their targets and the nature and number of drug targets. Nat. Rev. Drug Discov. **5**(10), 821–834 (2006)
15. Ji, Y., Ying, H., Dews, P., Tran, J., Mansour, A., Miller, R.E., Massanari, R.M.: An exclusive causal-leverage measure for detecting adverse drug reactions from electronic medical records. In: 2011 Annual Meeting of the North American Fuzzy Information Processing Society (NAFIPS), pp. 1–6 (2011)
16. Jin, H., Chen, J., He, H., Kelman, C., McAullay, D., O'Keefe, C.M.: Signaling potential adverse drug reactions from administrative health databases. IEEE Trans. Knowl. data Eng. **22**, 839–853 (2010)
17. Johnson, A.E.W., Pollard, T.J., Shen, L., Lehman, L.W.H., Feng, M., Ghassemi, M., Moody, B., Szolovits, P., Celi, L.A., Mark, R.G.: MIMIC-III, a freely accessible critical care database. Sci. data **3**, 160035 (2016)
18. Kuhn, M., Letunic, I., Jensen, L.J., Bork, P.: The sider database of drugs and side effects. Nucleic Acids Res. **44**(D1), D1075–D1079 (2016)

19. Liu, C., Wang, F., Hu, J., Xiong, H.: Temporal phenotyping from longitudinal electronic health records: a graph based framework, pp. 705–714 (2015)
20. Liu, J., Li, A., Seneff, S.: Automatic drug side effect discovery from online patient-submitted reviews: focus on statin drugs. In: Proceedings of First International Conference on Advances in Information Mining and Management (IMMM), Barcelona, Spain, pp. 23–29 (2011)
21. Liu, M., McPeek Hinz, E.R., Matheny, M.E., Denny, J.C., Schildcrout, J.S., Miller, R.A., Xu, H.: Comparative analysis of pharmacovigilance methods in the detection of adverse drug reactions using electronic medical records. J. Am. Med. Inform. Assoc. **20**(3), 420–426 (2013)
22. Reps, J., Garibaldi, J.M., Aickelin, U., Soria, D., Gibson, J.E., Hubbard, R.B.: Comparing data-mining algorithms developed for longitudinal observational databases, pp. 1–8 (2012)
23. Roitmann, E., Eriksson, R., Brunak, S.: Patient stratification and identification of adverse event correlations in the space of 1190 drug related adverse events. Front. Physiol. **5**, 332 (2014)
24. van Staa, T.-P., Goldacre, B., Gulliford, M., Cassell, J., Pirmohamed, M., Taweel, A., Delaney, B., Smeeth, L.: Pragmatic randomised trials using routine electronic health records: putting them to the test. BMJ **344**, e55 (2012)
25. van Staa, T.P., Dyson, L., McCann, G., Padmanabhan, S., Belatri, R., Goldacre, B., Cassell, J., Pirmohamed, M., Torgerson, D., Ronaldson, S., Adamson, J., Taweel, A., Delaney, B., Mahmood, S., Baracaia, S., Round, T., Fox, R., Hunter, T., Gulliford, M., Smeeth, L.: The opportunities and challenges of pragmatic point-of-care randomised trials using routinely collected electronic records: evaluations of two exemplar trials. Health Technol. Assess. **18**(43), 1–141 (2014)
26. Wang, F., Zhang, P., Cao, N., Jianying, H., Sorrentino, R.: Exploring the associations between drug side-effects and therapeutic indications. J. Biomed. Inform. **51**, 15–23 (2014)
27. Wang, X., Hripcsak, G., Friedman, C.: Characterizing environmental and phenotypic associations using information theory and electronic health records. BMC Bioinform. **10**(9), S13 (2009)
28. Wang, X., Hripcsak, G., Markatou, M., Friedman, C.: Active computerized pharmacovigilance using natural language processing, statistics, and electronic health records: a feasibility study. J. Am. Med. Inform. Assoc. **16**(3), 328–337 (2009)
29. Yamamoto, K., Sumi, E., Yamazaki, T., Asai, K., Yamori, M., Teramukai, S., Bessho, K., Yokode, M., Fukushima, M.: A pragmatic method for electronic medical record-based observational studies: developing an electronic medical records retrieval system for clinical research. BMJ open **2**, 1–10 (2012)
30. Yamamoto, K., Sumi, E., Yamazaki, T., Asai, K., Yamori, M., Teramukai, S., Bessho, K., Yokode, M., Fukushima, M.: A pragmatic method for electronic medical record-based observational studies: developing an electronic medical records retrieval system for clinical research. BMJ open **2**(6), e001622 (2012)
31. Yang, C.C., Jiang, L., Yang, H., Tang, X.: Detecting signals of adverse drug reactions from health consumer contributed content in social media. In: Proceedings of ACM SIGKDD Workshop on Health Informatics (2012)

# A Multi-center Physiological Data Repository for SUDEP: Data Curation, Data Conversion and Workflow

Wanchat Theeranaew[1(✉)], Bilal Zonjy[1], James McDonald[1],
Farhad Kaffashi[1], Samden Lhatoo[2,3], and Kenneth Loparo[1,3]

[1] Case Western Reserve University, Cleveland, OH 44106, USA
wanchat.theeranaew@case.edu
[2] University Hospitals of Cleveland, Cleveland, OH 44106, USA
[3] NIH Center for SUDEP Research, Cleveland, OH 44106, USA

**Abstract.** For any rare diseases, patient cohorts from individual medical research centers may not have sufficient statistical power to develop and verify/validate disease biomarkers as results of either small sample size or lack of patient-level predictors of the disease often in the form of recorded biological signals integrated with clinical data. Continuous recording is thus becoming a necessary step in the research to identify these biomarkers. The creation of a biological signals repository on top of a clinical data repository from multiple centers is thus a catalyst for current and future research of rare diseases. In this paper, several issues are considered in order to combine recorded physiological measurements from multiple centers to create a collaborative Big Data repository. Practical challenges including standardization of clinical information as well as physiological data are addressed. A case study of the Big-Data challenges associated with creating a large physiological data repository for the study of SUDEP (Sudden Unexpected Death in Epilepsy) as a part of the CSR (Center for SUDEP Research) study is presented. This includes end-to-end workflow from obtaining the source waveform data to storing standardized data files in the multi-center repository. This workflow has been implemented at Case Western Reserve University in partnership with University Hospitals to standardize data from multiple SUDEP centers that include Nihon Kohden, Micromed, and Nicolet physiological signal formats converted to European Data Format (EDF). A combination of existing third party, proprietary, and in-house-developed software tools used in the workflow are discussed.

**Keywords:** Epilepsy SUDEP · Physiological data repository · Multi-center · EEG

## 1 Introduction

Developing deeper understanding of disease states and discovering biomarkers that provide predictive information, especially for rare diseases, requires sufficient amounts of physiological and clinical data that support basic and translational research (Cimino et al. 2010). This requires collaboration across centers where data recorded at a variety

© Springer Nature Singapore Pte Ltd. 2017
J. Chen et al. (Eds.): KSS 2017, CCIS 780, pp. 66–75, 2017.
https://doi.org/10.1007/978-981-10-6989-5_6

of different sites is integrated into a central data repository. Previous examples of medical repositories are described in (Scully et al. 1997; Brindis et al. 2001; Johnson et al. 1994; Mullins et al. 2006; Gollub et al. 2013). However, these repositories only contain clinical information due to technological limitations, e.g. computer storage at the time of development. It has been shown that having real-time recording of biological signals can propel the research forward (Mark et al. 1982; Moody et al. 2001). As the cost of computer data storage continues to decrease while the size of the storage and computational power progressively increase, it has become feasible to share both clinical data as well as physiological measurements of the patient during admission. Further, with wearable devices and technology continuing to improve, physiological data recording after discharge is also possible. These developments are essential in finding biomarkers for rare diseases, such as SUDEP, as the possible sample size for the number of prospective patients is increased considerably. In order to reliably discover a biomarker for SUDEP, real-time recording of EEG, EKG, Oxygen saturation, and other physiological measurements from multiple centers on prospective patients are critical (Zhang et al. 2014; Gewin 2016).

Sharing data across multiple centers presents multiple technical challenges due to variations of workflow and data collection protocols between various centers participating in the SUDEP study. The protocols at various research centers use different systems for data recoding with the acquired data usually archived in proprietary formats, and include clinical annotations that are not standardized across the sites. These differences make it difficult to integrate the data and thus two data standardization processes are required, one for the physiological data and one for the clinical semantic context of the annotations (Gewin 2016). Thus, developing this standardization requires both medical and technical expertise. On the technical side, physiological signals from different medical centers are acquired through multiple acquisition systems with proprietary data formats. In the Center for SUDEP Research, the acquisition systems include Nihon Kohden, Nicolet, and Micromed. These systems do not provide a reliable means of converting data to an open data format, such as the EDF (Kemp et al. 1992; Kemp and Olivan 2003), that preserves physiological information available while ignoring Protected Health Information (PHI). Normally, each physiological signal is measured through a different sensor (e.g. EEG or ECG electrode) or different device (e.g. Pulse Oximeter or Non-invasive Arterial Blood Pressure) and converted to digital values for archiving. Therefore, we need to know the exact conversion factors to obtain meaningful physiological measurements. Unfortunately, many of the available conversion tools to EDF do not properly extract the conversion coefficients for each measurement rendering the stored signal values meaningless, e.g. for blood pressure and heart rate signals. The variations in clinical data records across centers can manifest in a number of ways that can affect the interpretation of the annotations. For example, the omission of clinical information because it is not part of the clinical workflow at an individual site can lead to misinterpretation of important annotations in the recorded data by analysts who do not have the necessary clinical expertise.

Additional challenges arise from the distribution of this standardization workload. This poses both technical and data-management challenge in implementing a workflow that easily allows simultaneous clinical data review and data conversion to EDF along with complete and accurate clinical annotations while ensuring data from different centers is converted accurately and reliably. Both clinical data review and data conversion are time consuming processes and should occur concurrently, additional information from the review process is required in data conversion. Naturally, these two primary tasks may be split among technical and medical personnel performing the data format standardization as well as the data review for clinical consistency, respectively. This poses the challenging task of creating a workflow that allows both the clinical data review and data conversion to move forward in a seamless process.

In this paper, we describe the challenges and approaches in both the clinical review process and data conversion process for physiological signals. We also propose a workflow that distributes the work across multiple teams while reducing the potential bottle-neck from clinical standardization dependency in clinical data review and data format standardization.

## 2   Clinical Review

Data are collected from different centers each with their own clinical protocol including the classification of seizures. In the SUDEP study, some centers do not annotate post ictal suppression and breathing because those are not part of the regular clinical workflow in their units. Further, some centers annotate motor seizures as one event while others annotate a motor seizure in multiple stages (Lüders et al. 1998). Additionally, different centers may use different terminology to refer to the same type of event. Without standardization, and familiarity with the particular workflow of each unit, the semantic information of these annotations may be lost and may be misinterpreted.

Without a proper clinical review process, non-clinical expert researchers who rely on metadata in these signal files could easily misinterpret the absence of clinical annotation as absence of clinically important events. Analysis of these annotations for cohort building may lead to a poorly representative cohort, resulting in misleading or spurious conclusions. This type of problem decreases the reliability and usability of the data repository and defeats the purpose of creating a centralized data repository and, as such, must be avoided at all cost. Although some centers maybe willing to review and re-annotate their clinical annotations to complete all possible clinical events, reviewers may not be specialized in identifying particular clinical events since they were never included in the workflow in their units. For this reason, even if the clinical annotations are completed with respect to the protocol of the source medical center, an additional clinical curation process is indispensable to ensure the quality and consistency of the signals and clinical data repository. Without such a data review, correct interpretation of the annotations would burden the data analysts to adjust for different clinical workflows across the centers. This restricts semi-automation during the data analysis process and forces researchers to go through laborious processes for each cohort. It is also recommended that the same terminology should be used for the same clinical events to provide the ability to search on the data level (Gewin 2016). The unification of clinical terminology can be integrated into the clinical review process and makes basic and translational research more convenient.

# 3  Data Conversion

In order to plan, manage and select an appropriate open, standard data format, specifications of each source format need to be well understood so that the output files can accurately and efficiently represent the variations of the source data formats. A fundamental aspect of the recorded physiological signals is the resolution of the analog to digital conversion process. Normally, each data point converted to digital values using an analog to digital converter with resolution of 16 or more bits is stored with the same resolution in the data format. Recorded data in proprietary formats includes both 32-bit and 16-bit resolution. If there is a necessity to standardize data with different resolution, the output format should be either 16 bits or 32 bits. Both options have their own benefits and drawbacks. For instance, there will be no loss of information if each sample is stored in 32-bits with the cost of extra storage space for each sample. In contrast, storing each sample of data in 16 bits requires the raw source data be truncated and loss of precision per sample from 32 bits to 16 bits. The problem associated with using a lower number of bytes (bits) to store each sample should be considered carefully when 32-bit data is archived using a 16-bit data format.

It is not uncommon that waveform data file is noncontiguous due to various reasons that require a pause in the recording. In addition, some centers may not send the full recording but rather a clipped version that contains only important clinical events. This makes most of the clipped data noncontiguous. For any noncontiguous signal data, the most important information is the consistency in the number of channels, sampling rate and/or sensitivities of the various channels. If this aforementioned information is not consistent across multiple segments in a data file, the output signal data must be able to cope with this inconsistency. By aggregating the variations of all source data files, a standard file format along with how to structure them for digital storage, retrieval and analysis can be determined.

After a standard file format has been selected, accurately converting different proprietary data formats into that single format is still a challenge. For each data format, direct communication with the vendors of each proprietary data format is necessary. Some companies may agree to provide documentation of the file format under an agreement while other companies may not disclose their file format but provide their protected SDK to extract the signal data from recorded files using their acquisition system. The tools to convert these data have to be developed differently based on available resources and information. This involves development of conversion tools using multiple programming platforms. Thus, development and data conversion for different file formats need to be integrated into the workflow.

It is relatively straightforward to extract EEG signals from a variety of different proprietary recording formats. Also, to the best of our knowledge, digital values can generally be directly mapped into microvolts (uVs) for all of proprietary formats, so that signals measured in microvolt, e.g. EEG and ECG, have the same conversion factor to obtain their corresponding physical values. Additional steps are often required for some proprietary formats to extract the signals with proper physical units from many of the non-EEG channels. For example, to obtain oxygen saturation values, one should know how to map the 0–1 output voltage range of the raw data from the pulse oximeter to 0–100% of oxygen saturation along with the specifications of analog to digital converter of the acquisition system to obtain actual measured voltage from the digital values. The data format provided by commercial systems may not include the necessary information (e.g. conversation factors/calibration coefficients for each connected device) to accurately extract and convert non-EEG signals to their appropriate physical values and units. In this case, extracting non-EEG signals requires utmost care to ensure that the converted data from all centers is converted accurately and reliably.

Additional annotations from the data review process may introduce additional annotations files due to 3rd party software used in the review process. In the CSR workflow, 3rd party software, Persyst, is used in the clinical review stage and output files generated during this review are completely different from any source file format and therefore must be integrated into the standard output files. Understanding how 3rd party software manages and maintains information in its output files for each native file format is critical for proper interpretation of the files. Cross validation of the output data from clinical personnel who perform review and curation is necessary to ensure accuracy of all data in the repository. Custom annotations marking specific events and times should be used instead of standard clinical annotations until the development phase is complete.

## 4    Workflow for a Reliable Multi-center Physiological Data Repository

The first step in creating a data repository is to implement the required tools for clinical review and the data conversion process that cannot be provided by in-house proprietary solutions readily available, e.g. Nihon Kohden EEG viewing software. With these tools developed and implemented, data arriving from each center is routinely reviewed, converted to a standard format and uploaded to the data repository. Because the clinical data is to be uploaded from at regular intervals, it is beneficial that data from each center are managed in a timely manner. In order to produce accurate output signal files, clinical revisions from the clinical review process need to be integrated into standard data format.

To serve the principle of keep the workflow distributed, the data conversion and clinical data review processes should run in parallel. To accomplish this – we integrate annotations into the open data files after they have been converted. However, the annotation files generated from the clinical data review by clinical personnel as well as the unstandardized files from the SUDEP centers are archived. This permits recreation of the transformation of incoming files to standardized output files end-to-end if need be, without repetition of the clinical review process.

In our workflow, the most computational intensive process in data conversion should be able to proceed without the result from the clinical review step. Integration of complete clinical annotations is perform after the clinical review process to standardize data files and generate finalized signal files for the data repository. In addition to data in the repository, both raw data and additional annotations from the clinical review process are archived. Without proper file tracking, it is possible to introduce inaccuracies into the data repository or archive. Therefore, we propose the workflow shown in

Fig. 1 that separates the process into 3 categories; data management, clinical review, and data conversion.

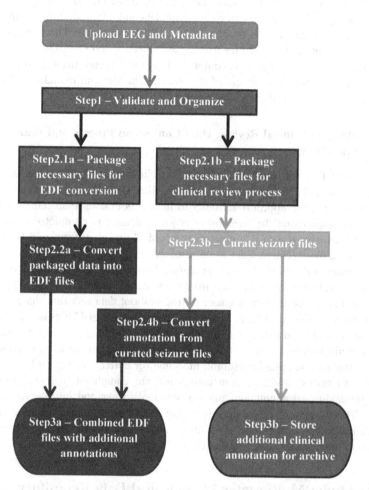

**Fig. 1.** Flowchart shows the workflow for CSR project in data conversion and curation. The green object is the task for each center. The blue objects are the tasks for data management personnel. The red objects are the tasks for data conversion personnel. The purple object is the task for clinical personnel. (Color figure online)

## 4.1 Workflow on Data Management Before Clinical Review and Data Conversion Process

Ideally, each SUDEP center would be able to upload data into a standard file and folder structure. However, due to the varying formats of the raw data, this may not be feasible in practice. In some cases, proprietary readers and viewers of various file formats may assume a specific folder structure of the source data. At the minimum, the source data needs to be organized by a unique patient identifier (e.g. patient ID at the CSR center) so that it enhances the workflow to ensure the reliability of both the archival source data and the EDF data for the data repository. All PHI in the raw data has been removed for HIPAA compliance. Raw data must be de-identified to be used for conversion to the open data format, and only signal data and attached clinical annotations are used in the data conversion process to be HIPAA compliant. The clinical review process requires both the signal files and the associated video recording to accurately review seizure events, and this is accomplished behind a secure firewall to protect all PHI. Two different essential subsets of the source data files and metadata are used in data conversion and clinical curation processes.

### 4.2   Workflow on Clinical Review, Data Conversion Process and Data Archival Process

From the seizure files and associated information, clinical personnel use appropriate $3^{rd}$ party software to verify, re-annotate and unify all clinical terminology. All curated EEG files are placed into the appropriate folder to inform both data management and data conversion personnel that the curation of a given dataset is completed. Any issues related to the raw signal data that obstruct the clinical review process will be reported back to the data management team to be resolved.

Concurrently to the clinical review process, all non-curated signal data is converted into selected standard format. Data sets from the same acquisition system are processed in the same pipeline because it is easier to troubleshoot data sets that share the same source format and folder structure. The results from standard EEG files are reviewed by the data conversion team to make sure that there are no technical issues, Any data that contains technical errors is reported back to the data management team to crosscheck with the center and potentially resubmit the same (or corrected) data to the process.

After both review and raw data conversion are completed, the next step is to integrate the additional annotations into the standard format and into the raw files for archiving. The structure used for data archiving should be similar to the structure used in the data repository, to facilitate any future changes in output file format by providing the ability to trace back to the original data when verification is required.

## 5   Case Study: Multi-center Physiological Data Repository for CSR and Discussion

The data files for CSR project come from 7 different centers that are participating in the SUDEP project. Currently, there are 3 different file formats processed in our workflow, Nihon Koden, Nicolet and Micromed. These three file formats store 16-bit data and support noncontiguous recordings of physiological signals. In addition, the number of channels and sampling rates in different segments may vary. We considered using the EDF+ format because it can store noncontiguous data; however EDF+ requires a consistent number of channels for the entire recording which then requires zero-padding of individual files to include empty channels to store noncontiguous data. As a result, we selected the EDF format and store data from noncontiguous recordings into multiple EDF output files. The data from each visit are stored in their respective folders, and the output EDF files are well organized and it is easy to trace back to their original data files using the date and time of the recording in the EDF file.

Despite five different centers using Nihon Koden acquisition systems, there are minor differences in the file formats from individual centers. Nihon Koden provides most information about their data file format, but cross validation of the output EDF file and the Nihon Koden raw data file is used to provide any missing information. Similar to Nihon Koden, Micromed provides details on the file structure that allowed the development of a Micromed data file reader. Details on the Nicolet format were not available, so the development of the Nicolet data conversion software was based solely on SDK provided in C++. Thus, we developed MATLAB-based data converter tools for Nihon Koden and Micromed, and a C++-based conversion tool for Nicolet. All core data conversion software modules were compiled as platform independent executable files, and a script and GUI front end were developed to unify all three data conversion modules allowing for remapping of important center-specific metadata such as medical annotations and for making sure that the scaling and physical variables associated with all acquired signals converted properly.

We key each patient using the Ontology-driven Patient Information Capture (OPIC) system, and this enables the use of the Multi-modality Epilepsy Data Capture and Integration System (MEDCIS) to query the data repository and develop a desired cohort. For studying SUDEP, dealing with motor seizures is important because frequent generalized tonic-clonic seizure is considered a risk factor for SUDEP (Tomson et al. 2005). For these types of seizures, muscle and movement artifacts frequently obscure the EEG signals so video recording is needed to make sure that annotations marking different stages of motor seizures as well as postictal generalized EEG suppression are consistent and refer to the same semiology and EEG morphology. This information is accessible to clinical personnel who can view patient identifiers under HIPAA.

We use NeuroWorkbench for the clinical review of Nihon Koden raw data files, and the curated data contains the revised clinical annotation from the "source" file. Persyst is used to review Nicolet and Micromed source data files. Persyst does not alter the source file but generates a proprietary metadata file alongside the source file that contains the revised clinical annotations created during the data review process. Integration of the Persyst output file into the EDF output data required some software development effort. The combined annotations are then converted to both external JavaScript Object Notation (JSON) and Tab Separated Values (TSV) files that accompany the output EDF file. The resulting EDF and annotation files are then integrated into our collaborative medical database, MEDCIS.

Although we elaborated on many details regarding the data transformation and workflow, and the data repository on the continuous recording side, considerable additional was required to join the medical data and signal data to make a complete and integrated multi-center repository. The reader may refer to (Zhang et al. 2014) for details on the MEDCIS system.

# 6  Conclusions

The workflow developed allowed for simultaneous clinical review and data conversion processes for the Multi-center SUDEP study. In this paper, we have addressed the work distribution between team members and provided the steps in the process that each member needs to perform to reliably transform raw source data from multiple acquisition systems into standardized, cleaned, OPIC-consistent patient records in the MEDCIS database system. This workflow expedited the data conversion process to minimize bottlenecks between team members at the Center for SUDEP Research (CSR) at Case Western Reserve University and University Hospitals Cleveland Medical Center, sped up the entire process for creating finalized signal files for the multi-center physiological repository, and allowed us to both manage significant backlogs in data curation and conversion and transform the incoming data in a timely manner. The Matlab and C++ tools developed allow for the integration of multi-center data that preserves the accuracy of all EEG and physiological data for three different file formats. Our tools also integrate additional annotations from 3$^{rd}$ party software, Persyst, to generate complete EDF datasets. To date, roughly 1,900 Nihon Kohden recordings from three SUDEP centers, 354 Nicolet recordings from one SUDEP center and 299 Micromed recordings from one SUDEP center, have been clinically reviewed and converted to EDF. We believe the experience we have developed along with the workflow and tools that have been developed will aid in the future development of new multi-center data repository.

**Acknowledgements.** Research reported in this publication was supported by the National Institute of Neurological Disorders and Stroke of the National Institutes of Health under Award Numbers U01NS090407 and U01NS090408. The content is solely the responsibility of the authors and does not necessarily represent the official views of the National Institutes of Health.

# References

Brindis, R.G., Fitzgerald, S., Anderson, H.V., Shaw, R.E., Weintraub, W.S., Williams, J.F.: The American College of Cardiology-National Cardiovascular Data Registry™ (ACC-NCDR™): building a national clinical data repository. J. Am. Coll. Cardiol. **37**(8), 2240–2245 (2001)

Cimino, J.J., Ayres, E.J.: The clinical research data repository of the US National Institutes of Health. Stud. Health Technol. Inform. **160**(Pt 2), 1299 (2010)

Gewin, V.: Data sharing: an open mind on open data. Nature **529**(7584), 117–119 (2016)

Gollub, R.L., Shoemaker, J.M., King, M.D., White, T., Ehrlich, S., Sponheim, S.R., Clark, V.P., Turner, J.A., Mueller, B.A., Magnotta, V., O'Leary, D.: The MCIC collection: a shared repository of multi-modal, multi-site brain image data from a clinical investigation of Schizophrenia. Neuroinformatics **11**(3), 367–388 (2013)

Johnson, S.B., Hripcsak, G., Chen, J., Clayton, P.: Accessing the Columbia clinical repository. In: Proceedings of the Annual Symposium on Computer Application in Medical Care, p. 281. American Medical Informatics Association (1994)

Kemp, B., Värri, A., Rosa, A.C., Nielsen, K.D., Gade, J.: A simple format for exchange of digitized polygraphic recordings. Electroencephalogr. Clin. Neurophysiol. **82**(5), 391–393 (1992)

Kemp, B., Olivan, J.: European data format 'plus'(EDF+), an EDF alike standard format for the exchange of physiological data. Clin. Neurophysiol. **114**(9), 1755–1761 (2003)

Lüders, H., Acharya, J., Baumgartner, C., Benbadis, S., Bleasel, A., Burgess, R., Dinner, D.S., Ebner, A., Foldvary, N., Geller, E., Hamer, H.: Semiological seizure classification. Epilepsia **39**(9), 1006–1013 (1998)

Mark, R.G., Schluter, P.S., Moody, G., Devlin, P., Chernoff, D.: An annotated ECG database for evaluating arrhythmia detectors. In: IEEE Transactions on Biomedical Engineering, vol. 29, no. 8, p. 600, January 1982

Moody, G.B., Mark, R.G.: The impact of the MIT-BIH arrhythmia database. IEEE Eng. Med. Biol. Mag. **20**(3), 45–50 (2001)

Mullins, I.M., Siadaty, M.S., Lyman, J., Scully, K., Garrett, C.T., Miller, W.G., Muller, R., Robson, B., Apte, C., Weiss, S., Rigoutsos, I.: Data mining and clinical data repositories: Insights from a 667,000 patient data set. Comput. Biol. Med. **36**(12), 1351–1377 (2006)

Scully, K.W., Pates, R.D., Desper, G.S., Connors, A.F., Harrell Jr., F.E., Pieper, K.S., Reynolds, R.E.: Development of an enterprise-wide clinical data repository: merging multiple legacy databases. In: Proceedings of the AMIA Annual Fall Symposium, p. 32. American Medical Informatics Association (1997)

Tomson, T., Walczak, T., Sillanpaa, M., Sander, J.W.: Sudden unexpected death in epilepsy: a review of incidence and risk factors. Epilepsia **46**(s11), 54–61 (2005)

Zhang, G.Q., Cui, L., Lhatoo, S., Schuele, S., Sahoo, S.: MEDCIS: multi-modality epilepsy data capture and integration system. In: AMIA Annual Symposium Proceedings, vol. 2014, pp. 1248–1257, November 2014

# Concept Name Similarity Measure
# on SNOMED CT

Htet Htet Htun[✉] and Virach Sornlertlamvanich

School of Information, Computer and Communication Technology,
Sirindhorn International Institute of Technology, Thammasat University,
Bangkok, Thailand
htethtethtun.8910@gmail.com, virach@siit.tu.ac.th

**Abstract.** The semantic similarity measure between biomedical terms or concepts is a crucial task in biomedical information extraction and knowledge discovery. Most of the existing similarity approaches measure the similarity degree based on the path length between concept nodes as well as the depth of the ontology tree or hierarchy. These measures do not work well in case of the "primitive concepts" which are partially defined and have only few relations in the ontology structure. Namely, they cannot give the desired similarity results against human expert judge on the similarity among primitive concepts. In this paper, the existing two ontology-based measures are introduced and analyzed in order to determine their limitations with respect to the considered knowledge base. After that, a new similarity measure based on concept name analysis is proposed to solve the weakness of the existing similarity measures for primitive concepts. Using SNOMED CT as the input ontology, the accuracy of our proposal is evaluated and compared against other approaches with the human expert results based on different types of ontology concepts. Based on the correlation between the results of the evaluated measures and the human expert ratings, this paper analyzes the strength and weakness of each similarity measure for all ontology concepts.

**Keywords:** Concept name similarity measure · Text Similarity · Natural language processing · SNOMED CT · Semantic similarity

## 1 Introduction

Over the years, the determination of semantic similarity between word pairs has been recognized as an important task of text understanding applications such as the proper exploitation, management or classification of textual data [1], information retrieval [2] and decision-support systems that utilize knowledge sources and ontologies [3]. Semantic Similarity measures exploit knowledge sources as the base to perform the estimation. In recent years, knowledge sources and ontologies are widely used for the semantic similarity research area as they offer a structured and unambiguous representation of knowledge in the form of conceptualizations interconnected by means of semantic pointers. At the same

© Springer Nature Singapore Pte Ltd. 2017
J. Chen et al. (Eds.): KSS 2017, CCIS 780, pp. 76–90, 2017.
https://doi.org/10.1007/978-981-10-6989-5_7

time, finding the semantic similarity between concepts based on the medical ontologies becomes crucial for the biomedical domain. As an example, health decision support system retrieves similar treatment cases in the past based on their different similarity levels as guidelines in order to treat the current patient [4]. Therefore, many ontology-based similarity measures have been developed by exploiting the medical ontologies such as SNOMED CT (Systematized Nomenclature of Medicine - Clinical Terms). SNOMED CT is considered as a standard medical terminology [5] that covers all areas of clinical information including body structure, diseases, organisms and clinical findings etc.

The fundamental idea of computing the similarity between words/concepts is based on the taxonomic structure of an ontology by taking the minimum number of shortest path between evaluated two concepts. Leacock and Chodorow [6] proposed a measure that considers both the shortest path length between two concepts and the maximum depth of the taxonomy but performs a logarithmic scaling. Wu and Palmer [7,8] also proposed a path-based measure that takes into account the depth of the two concepts in the hierarchy and also the depth from their least common subsumer (LCS) to the root of the ontology. Choi and Kim [9] also proposed an taxonomic-based measure according to the difference in the depth levels of two concepts and the distance of the shortest path between them. Al-Mubaid and Nguyen [10] proposed a cluster-based measure that combines path length and common specificity by subtracting the depth of their LCS from the depth of the ontology. These previous measures use an ontology as background knowledge and determine the similarity based on taxonomic structure of an ontology by taking shortest path length and depth of evaluated concepts. But the new ontology taxonomic-based measure [11] analyzed previous measures and presented as taking only the minimum path length between evaluated concepts omits a large amount of taxonomic knowledge of the ontology for the evaluated concepts and waste a lot of explicitly available knowledge. Therefore, they proposed a new taxonomic-based measure by taking all possible number of parent concepts and they got the highest correlation results among previous measures against human expert ratings. In the literature, SNOMED CT is chosen as the input ontology to find the semantic similarity. When we analyze the SNOMED CT ontology, there are two kinds of concepts "defined concepts" and "primitive concepts" according to the available hierarchical information of the concepts in the ontology [12]. For example, definition of primitive concept "Tumor of dermis" is as follows.

```
(Tumor of dermis ⊑ special concept) ⊓ (special concept ⊑ SNOMED
CT concept)
```

It means "Tumor of dermis" has "is-a" relation with "special concept" and "special concept" has "is-a" relation with the top "SNOMED CT concept". The definition of another primitive concept "Vibrio species n-z" is the same with "tumor of dermis".

```
(Vibrio species n-z ⊑ special concept) ⊓ (special concept ⊑
SNOMED CT concept)
```

Therefore, their definitions are the same and not sufficient to distinguish from each other and they needed to be additionally defined with the specific information in the ontology. For the "defined concepts", they are sufficiently defined in the ontology as follows.

(Hypoxia of brain ≡ Hypoxia ⊓ ∃FindingSite. Brain Structure)

"Hypoxia of brain" has "is-a" relation with "Hypoxia" and it has "attribute-value" relation type "findingSite" with another concept "brain structure". Therefore, "defined concepts" have specific and complete information in the ontology but "primitive concepts" have partially hierarchical information and they are actually needed to define with complete information [12]. Therefore, released ontology versions include different amounts of concepts and relations in every year because ontology builders always redefine the concepts with more complete and specific information from the actual medical records. For this reason, ontology builders call the concepts that have been completely defined in the ontology as "defined concepts" and other concepts that are needed to add specific hierarchical information and relations in the ontology as "primitive concepts". In the literature, there is no experiments based on the different types of concepts for the SNOMED CT ontology. These facts push us to measure the similarity between different types of concepts. Therefore, we consider three cases of experiment for measuring the degree of similarity (1) between the primitive concepts, (2) between primitive concepts and defined concepts, and (3) between the defined concepts.

In this paper, we first review taxonomic-based measure [11] that got the highest correlation with the human expert result among most of the previous ontology-based measures in the literature. From the logic point of view, SNOMED CT definitions are written by Description Logic EL family, therefore, we review the description logic ELH similarity measure [13]. Then, we make the detail analysis between three cases of experiment by identifying limitations of previous measures according to dimensions of expected accuracy. In order to overcome the limitations of previous measures, we propose a new similarity measure based on concept name for all types of concepts. To compare all measures in a practical setting, we evaluate them against human expert results. The results show that our proposal solves the limitation of existing measures.

The rest of the paper is organized as follows. Section 2 presents and analyzes previous ontology-based similarity measures. Section 3 presents our similarity measure and its main benefits. Section 4 evaluates all measures based on three cases of experiments using SNOMED CT as the domain ontology and makes the comparison of evaluated measures with human expert ratings. Section 5 presents the conclusions.

## 2  Ontology-Based Similarity Measures

### 2.1  Taxonomic-Based (Path-Based) Measure

This measure computes the similarity based on the taxonomic paths connecting the two concepts [11]. It considers all of the superconcepts belonging to all the possible taxonomic paths between concepts. This relation is based on the idea that pairs of concepts belonging to an upper level of the taxonomy (i.e. they share few superconcepts) should be less similar than those in a lower level (i.e. they have more superconcepts in common). It defines the similarity between concept $c_1$ and $c_2$ as the ratio between the amount of non-shared knowledge and the sum of shared and non-shared knowledge, and then it takes the inverted logarithm function as shown in Eq. (1).

$$sim(c_1, c_2) = -log_2 \frac{|T(c_1) \cup T(c_2)| - |T(c_1) \cap T(c_2)|}{|T(c_1) \cup T(c_2)|} \qquad (1)$$

In the full concept hierarchy $H^c$ of concepts (C) of an ontology, $T(c_i) = \{ c_j \in C - c_j$ is superconcept of $c_i \} \cup \{ c_i \}$ is defined as the union of the ancestors of the concept $c_i$ and $c_i$ itself.

This measure takes into account all the superconcepts regarding the evaluated concepts and it relies on the taxonomic paths. According to the SNOMED CT ontology, all of the ontology concepts are not well defined and completely defined especially for all primitive concepts. So similarity measure based on taxonomic structure of an ontology will be the main problem for the primitive concepts because they are actually needed to define with full hierarchical information from all actual medical records. As a consequence, similarity degree between primitive and defined concepts may not get the correct similarity value because primitive concepts have few inter-links and defined concepts have full inter-links between them so their similarity may be low but their actual similarity according to the judgment of human expert will be high.

## 2.2   ELSIM Similarity Measure

The ELSIM semantic similarity measure computes the similarity between $\mathcal{ELH}$ concepts based on homomorphism tree function. This function provides a numerical value that represents structural similarity of one concept description against the another concept description. This measure is used for description logic $\mathcal{ELH}$ definitions. It first constructs description tree for each concept from Top to evaluated concept using Algorithm 1 (see in detail [13]). Secondly, it maps between two description trees using homomorphism degree function as the following.

**Definition 3.1** (Homomorphism Degree). Let $T^{\mathcal{ELH}}$ be a set of all $\mathcal{ELH}$ description trees and $\mathcal{T}_C, \mathcal{T}_D \in T^{\mathcal{ELH}}$ corresponds to two $\mathcal{ELH}$ concept names C and D, respectively [13]. The hd: $T^{\mathcal{ELH}} \times T^{\mathcal{ELH}} \rightarrow [0, 1]$ is inductively defined as follows:

$$hd(\mathcal{T}_C, \mathcal{T}_D) = \mu \cdot p - hd(\mathcal{P}_C, \mathcal{P}_D) + (1 - \mu) \cdot \text{e-set-hd}(\mathcal{E}_C, \mathcal{E}_D), \qquad (2)$$

where $0 \leq \mu \leq 1$;

$$p - hd(\mathcal{P}_C, \mathcal{P}_D) := \begin{cases} 1 & if \ \mathcal{P}_C = \emptyset \\ \frac{|\mathcal{P}_C \cap \mathcal{P}_D|}{|\mathcal{P}_C|} & otherwise, \end{cases} \qquad (3)$$

e-set-hd$(\mathcal{E}_C, \mathcal{E}_D) :=$

$$
\begin{cases}
1 & \text{if } \mathcal{E}_C = \emptyset \\
0 & \text{if } \mathcal{E}_C \neq \emptyset \text{ and } \mathcal{E}_D = \emptyset \\
\sum_{\epsilon_i \in \mathcal{E}_C} \frac{max\{e-hd(\epsilon_i, \epsilon_j) : \epsilon_j \in \mathcal{E}_D\}}{|\mathcal{E}_C|} & \text{otherwise},
\end{cases}
\tag{4}
$$

$$
e\text{-}hd(\exists r.X, \exists s.Y) := \gamma(\nu + (1 - \nu).hd(\mathcal{T}_X, \mathcal{T}_Y)) \tag{5}
$$

**Definition 3.2** ($\mathcal{ELH}$ Similarity Degree). The final similarity degree between concept C and D is defined by taking the average of homomorphism degree from C to D and D to C as follows:

$$
\text{sim}(C, D) = \frac{hd(\mathcal{T}_C, \mathcal{T}_D) + hd(\mathcal{T}_D, \mathcal{T}_C)}{2} \tag{6}
$$

We can use the implementation of this measure on the website (http://ict. siit.tu.ac.th/sun.html). This measure also calculates the similarity based on two structural trees of concepts using the description tree algorithm. This measure is constructed using a specific language Description logic $\mathcal{ELH}$. But there are many primitive concepts in an ontology which their definitions are not sufficiently distinguish from each other. For this reason, ELSIM also cannot give the correct similarity degrees between all types of concepts in an ontology. Similarity degree between two concepts will be low if there has few inter-links between them but it will be high when ontology builder can add complete hierarchical information or more related links for these concepts. Another important thing is similarity values will be changed when we use different ontologies as ontologies have different structures. As a result, estimation of semantic similarity between ontology concepts based on taxonomic structure has the weakness for the concepts that have few hierarchical links in the ontology and we perform some experiments by applying existing two similarity measures mainly for the partially defined concepts or primitive concepts. Some evidence are shown in Tables 1 and 2.

**Table 1.** Incomparable similarity values between primitive concepts using path-based measure and ELSIM with human expert results

| Primitive concept $P_1$ | Primitive concept $P_2$ | Path-based | ELSIM | Human expert |
|---|---|---|---|---|
| Infiltrative lung tuberculosis | Nodular lung tuberculosis | 0.2 | 0.0 | 0.7 |
| Maternal autoimmune hemolytic anemia | Autoimmune hemolytic anemia | 0.2 | 0.0 | 0.8 |
| Phakic corneal edema | Corneal epithelial edema | 0.2 | 0.0 | 0.5 |

**Table 2.** Incomparable similarity values between primitive and defined concepts using path-based measure and ELSIM with human expert results

| Primitive concept $P_1$ | Defined concept $P_2$ | Path-based | ELSIM | Human expert |
|---|---|---|---|---|
| Coronary artery thrombosis | Vertebral artery thrombosis | 0.2 | 0.0 | 0.6 |
| Corneal epithelial edema | Idiopathic corneal edema | 0.1 | 0.0 | 0.6 |
| Infectious mononucleosis hepatitis | Chronic alcoholic hepatitis | 0.2 | 0.0 | 0.5 |

## 3  Proposed Similarity Measure Based on Concept Name

From the study of previous ontology-based similarity measures, they do not give the desired similarity degrees with the human expert result. If the ontology builders redefine the concepts with full relations, they will get higher similarity degree for path-based measures. Therefore, we want to fill the gap of ontology-based similarity measure and consider semantic similarity according to textual annotations (concept names) because each ontology concept is uniquely identified by a concept ID (e.g. id = 10365005), annotated with a short textual description (e.g. "right main coronary artery thrombosis") and equipped with a definition in description logic. Moreover, ontology concept names are taken from the actual patient medical treatment records so they are very informative and can illustrate the complete meaning of the concept.

From this point of view, we propose a new similarity measure based on concept label from the natural language processing views. We modify concept name similarity by using following features.

1. Put different weights based on the headword of noun phrase to obtain a better similarity value.
2. Use context-free grammar to compute the syntactic similarity based on the noun phrase structure of concept name.

### 3.1  Linguistic Headword Structure (Semantic Similarity)

All of the text labels of concept name are expressed in the form of noun phrase, in which the "headword" holds the core meaning of the phrase [14]. We cannot omit the headword in noun phrase. Therefore we should consider the highest weight for the headword when comparing the similarity of two concept names. In English, the structure of noun phrase can be defined as in the following cases.

1. Det + Pre-modifiers + noun (headword)
2. noun (headword) + Post-modifier/complement
3. noun + noun

All of the SNOMED CT concept names appear as the first case. Therefore, the rightmost noun is the headword of the concept name. After some experiments, we can conclude that the suitable weight for the headword is 0.6, and 0.4 is for the remaining components.

Let's consider concept $P_1$ = "right main coronary artery thrombosis" and concept $P_2$ = "superior mesenteric vein thrombosis" For concept $P_1$,

- Weight for headword "thrombosis" is 0.6.
- Weight for remaining components is 0.4 (0.1 for each remaining component).
- To assign different weights for each component, we consider positions of the component because the nearer component to the headword should get higher weight and it has higher semantic influence on the headword than other words [14]. Therefore, we give the weight for each component based on the distance from the headword. And then each component is divided by the distance value. For the component nearest from the headword, we subtract the sum of all other remaining components from 0.4. So, the sum of all weights of concept name is 1. As a result, the weight can be distributively estimated as shown in Tables 3 and 4.

**Table 3.** Different weights of concept $P_1$

| 4 | 3 | 2 | 1 | | 0 |
|---|---|---|---|---|---|
| right | main | coronary | artery | | thrombosis |
| 0.1 | 0.1 | 0.1 | 0.1 | | 0.6 |
| $0.1/4 =$ 0.025 | $0.1/3 =$ 0.033 | $0.1/2 =$ 0.05 | $0.4 - (0.025 + 0.033 + 0.05)$ $= 0.292$ | | 0.6 |

**Table 4.** Different weights of concept $P_2$

| 3 | 2 | 1 | 0 |
|---|---|---|---|
| superior | mesenteric | vein | thrombosis |
| 0.133 | 0.133 | 0.133 | 0.6 |
| $0.133/3 = 0.044$ | $0.133/2 = 0.067$ | $0.4 - (0.044 + 0.067) = 0.289$ | 0.6 |

We define the headword noun phrase structure denoted by $\mathbf{sim}_{Headword}$ based on the Jaccard similarity [15] (the number of shared terms over the number of all unique terms). Therefore,

$|\text{wset}(P_1) \cap \text{wset}(P_2)|$ is the sum of the weights of shared terms and $|\text{wset}(P_1) \cup \text{wset}(P_2)|$ is the sum of the weights of all unique terms.

$\mathbf{sim}_{Headword}(P_1, P_2)$

$$= \frac{|wset(P_1) \cap wset(P_2)|}{|wset(P_1) \cup wset(P_2)|}$$

$$= \frac{0.6}{(0.025 + 0.033 + 0.05 + 0.292 + 0.6 + 0.044 + 0.067 + 0.289)}$$

$$= 0.43$$

There are two points that we need to consider for this surface-matching similarity.

1. Some words are lexically similar but they have different meanings
   - For example, "kidney parenchyma" and "kidney beans".
   - "kidney parenchyma" is about human tissue of kidney and "kidney beans" is about a kind of bean.
   - In this case, it cannot occur because we compute the similarity based on the same category e.g.: for the disease category, all the concepts are about health such as illness, sickness and unwellness.
2. Some words are lexically different but they have similar meaning
   - For example, illness and sickness.
   - To fulfill this requirement, we used WordNet ontology to calculate the synsets similarity $S_{synset}$ because two terms are similar if their synsets of these terms are lexically similar [16].

$$S_{synset}(h_1, h_2) = \frac{|A \cap B|}{|A \cup B|} \tag{7}$$

   - A is the synset of headword $h_1$ and B is the synset of headword $h_2$.
   - The main idea of our proposed method is based on the importance of two headwords terms. Therefore, we apply the synset similarity calculation to only the two important headwords. If the degree of similarity of synsets is greater 0, then the two words are considered to be the same. Otherwise, they are different.

$$Sim(h_1, h_2) = \begin{cases} 1, & \text{if } S_{synset}(h_1, h_2) > 0 \\ 0, & \text{if } S_{synset}(h_1, h_2) = 0 \end{cases} \tag{8}$$

## 3.2 Syntactic Structure Similarity

In order to know the syntactic structure of noun phrases for estimating the syntactic of the two noun phrases, we apply the context-free grammar (CFG) [17]. The grammar G= ⟨T, N, S, R⟩

- T is set of terminals
- N is set of non-terminals (NP in this case)
- S is the starting symbol
- R is rules or productions of the form

We construct noun phrase rules that cover all types of noun phrases in SNOMED CT concepts as listed in the following.

1. NP → N
2. NP → N NP
3. NP → Adj NP
4. NP → Det NP
5. NP → Adv NP

After applying CFG rule, the parsing orders of $P_1$ and $P_2$ are shown in the following list.

- Parsing order of $P_1$ : 3-3-3-2-1
- Parsing order of $P_2$ : 3-3-2-1

Syntactic similarity measure is estimated by the similarity of the applied CFG parsing rule. For the similarity calculation, numerator is the intersection of rules and denominator is the maximum number of rules.

$$\mathbf{sim}_{CFG}(P_1, P_2) = \frac{4}{5}$$
$$= 0.8$$

### 3.3   Proposed Measure

After getting similarity values from two dimensions: headword structure and syntactic structure, we consider finalize similarity values by giving different weights based on their generalization. If two concepts are exactly same syntactic structure, but different headword terms, they have so much different meanings. But for the headword structure, it gives the accurate similarity value according to their headword position. This means that headword structure can decide the similarity more effective than syntactic structure. Accordingly, we make experiments by setting various weights as in Table 5.

**Table 5.** Different weights for headword and syntactic structure (CFG)

| Concept $P_1$ | Concept $P_2$ | 0.7 and 0.3 | 0.8 and 0.2 | 0.9 and 0.1 | 0.6 and 0.4 | Human result |
|---|---|---|---|---|---|---|
| Mosquito-borne hemorrhagic fever | Glandular fever pharyngitis | 0.5 | 0.4 | 0.3 | 0.6 | 0.5 |
| Gangrenous paraesophageal hernia | Congenital bladder hernia | 0.6 | 0.5 | 0.5 | 0.7 | 0.6 |

For the overall experiments, 0.7 and 0.3 get the highest correlation values. Therefore, we decide to set different weights as 0.7 for headword structure and 0.3 for syntactic structure.

$$
\begin{aligned}
\mathbf{Wsim}(P_1, P_2) \quad &= a * \mathbf{sim}_{Headword}(P_1, P_2) + b * \mathbf{sim}_{CFG}(P_1, P_2) \\
&= 0.7 * 0.43 + 0.3 * 0.8 \\
&= 0.54
\end{aligned}
$$

# 4   Experimental Results on SNOMED CT

In the experiments, we use SNOMED CT which is the DL version released in January 2005 which contains 364,461 concept names. From the SNOMED CT disorder category, 90 disease concept pairs are selected for evaluation of three cases so 30 concepts pairs for each experiment using path-based measure, ELSIM and our proposed measure. Similarity values between only primitive concepts are shown in Table 6. Similarity values between primitive and defined concepts are

**Table 6.** Results of degree of similarity on 30 pairs between primitive concepts estimated by path-based method, ELSIM, our proposed method, and human expert

| Primitive concept $P_1$ | Primitive concept $P_2$ | Path-based | ELSIM | Proposed method | Human expert |
|---|---|---|---|---|---|
| Hormonal tumor | Malignant mast cell tumor | 0.2 | 0.0 | 0.5 | 0.6 |
| Maternal autoimmune hemolytic anemia | Autoimmune hemolytic anemia | 0.2 | 0.0 | 0.8 | 0.8 |
| Hypertensive leg ulcer | Solitary anal ulcer | 0.3 | 0.7 | 0.5 | 0.4 |
| Bovine viral diarrhea | Bovine coronoviral diarrhea | 0.6 | 0.6 | 0.7 | 0.7 |
| Acute uterine inflammatory disease | Mycoplasmal pelvic inflammatory disease | 0.4 | 0.2 | 0.9 | 0.9 |
| Primary cutaneous blastomycosis | Primary pulmonary blastomycosis | 0.7 | 0.9 | 0.7 | 0.6 |
| Iodine-deficiency-related multinodular endemic goiter | Non-toxic multinodular goiter | 0.8 | 0.7 | 0.8 | 0.8 |
| Congenital pharyngeal polyp | Uterine cornual polyp | 0.4 | 0.6 | 0.5 | 0.5 |
| Phakic corneal edema | Corneal epithelial edema | 0.2 | 0.0 | 0.5 | 0.5 |
| Knee pyogenic arthritis | Gonococcal arthritis dermatitis syndrome | 0.9 | 0.8 | 0.4 | 0.4 |
| Hereditary canine spinal muscular atrophy | Spinal cord concussion | 0.5 | 0.7 | 0.3 | 0.5 |
| Mite-borne hemorrhagic fever | Meningococcal cerebrospinal fever | 0.4 | 0.5 | 0.6 | 0.5 |
| Congenital cleft larynx | Congenital spastic foot | 0.6 | 0.8 | 0.3 | 0.3 |
| Congenital acetabular dysplasia | Short rib dysplasia | 0.5 | 0.9 | 0.5 | 0.5 |
| Intestinal polyposis syndrome | Ovarian vein syndrome | 0.6 | 0.8 | 0.6 | 0.5 |
| Extrapulmonary subpleural pulmonary sequestration | Pulmonary alveolar proteinosis | 0.7 | 0.6 | 0.4 | 0.4 |
| Atypical chest pain | Psychogenic back pain | 0.3 | 0.1 | 0.5 | 0.5 |
| Puerperal pelvic cellulitis | Chronic female pelvic cellulitis | 0.9 | 0.7 | 0.8 | 0.7 |
| Spinal cord hypoplasia | Spinal cord rupture | 0.5 | 0.7 | 0.6 | 0.6 |
| Infiltrative lung tuberculosis | Nodular lung tuberculosis | 0.2 | 0.0 | 0.9 | 0.7 |
| Early gastric cancer | Primary vulval cancer | 0.4 | 0.8 | 0.4 | 0.4 |
| Congenital mesocolic hernia | Gangrenous epigastric hernia | 0.2 | 0.0 | 0.5 | 0.4 |
| Congenital nonspherocytic hemolytic anemia | Congenital macular corneal dystrophy | 0.2 | 0.0 | 0.3 | 0.2 |
| Congenital cerebellar cortical atrophy | Congenital renal atrophy | 0.6 | 0.9 | 0.7 | 0.2 |
| Puerperal pyrexia | Heat pyrexia | 0.3 | 0.0 | 0.5 | 0.6 |
| Methylmalonyl-CoA mutase deficiency | Muscle phosphoglycerate mutase deficiency | 0.2 | 0.0 | 0.7 | 0.5 |
| Recurrent mouth ulcers | Multiple gastric ulcers | 0.5 | 0.8 | 0.4 | 0.4 |
| Infantile breast hypertrophy | Sebaceous gland hypertrophy | 0.6 | 0.7 | 0.6 | 0.4 |
| Congenital pyloric hypertrophy | Synovial hypertrophy | 0.3 | 0.0 | 0.5 | 0.3 |
| Inflammatory testicular mass | Inflammatory epidermal nevus | 0.5 | 0.7 | 0.3 | 0.3 |

**Table 7.** Results of degree of similarity on 30 pairs between primitive and defined concepts estimated by path-based method, ELSIM, our proposed method, and human expert

| Primitive concept $P_1$ | Defined concept $P_2$ | Path-based | ELSIM | Proposed method | Human expert |
|---|---|---|---|---|---|
| Mosquito-borne hemorrhagic fever | Glandular fever pharyngitis | 0.4 | 0.7 | 0.5 | 0.5 |
| Right main coronary artery thrombosis | Coronary artery rupture | 0.9 | 0.9 | 0.5 | 0.4 |
| Right main coronary artery thrombosis | Superior mesenteric vein thrombosis | 0.7 | 0.9 | 0.5 | 0.6 |
| Infectious mononucleosis hepatitis | Chronic alcoholic hepatitis | 0.2 | 0.0 | 0.5 | 0.5 |
| Cerebral venous sinus thrombosis | Phlebitis cavernous sinus | 1.0 | 0.9 | 0.6 | 0.6 |
| Third degree perineal laceration | Complex periorbital laceration | 0.3 | 0.7 | 0.5 | 0.5 |
| Congenital subaortic stenosis | Rheumatic aortic stenosis | 0.9 | 0.7 | 0.6 | 0.7 |
| Congenital acetabular dysplasia | Aortic valve dysplasia | 0.5 | 0.6 | 0.5 | 0.3 |
| Intestinal polyposis syndrome | Fetal cytomegalovirus syndrome | 0.4 | 0.4 | 0.6 | 0.3 |
| Anterior choroidal artery syndrome | Juvenile polyposis syndrome | 0.4 | 0.7 | 0.5 | 0.3 |
| Puerperal pelvic cellulitis | Streptococcal cellulitis | 0.3 | 0.5 | 0.5 | 0.3 |
| Benign hypertensive renal disease | Pulmonary hypertensive venous disease | 0.7 | 0.8 | 0.6 | 0.4 |
| Corneal epithelial edema | Idiopathic corneal edema | 0.1 | 0.0 | 0.8 | 0.6 |
| Chronic sarcoid myopathy | Hereditary hollow viscus myopathy | 0.3 | 0.6 | 0.5 | 0.5 |
| Primary cutaneous blastomycosis | Chronic pulmonary blastomycosis | 0.7 | 0.9 | 0.6 | 0.6 |
| Gingival pregnancy tumor | Granular cell tumor | 0.4 | 0.5 | 0.6 | 0.4 |
| Borderline epithelial tumor | Melanotic malignant nerve sheath tumor | 0.4 | 0.6 | 0.4 | 0.4 |
| Congenital sternomastoid tumor | Malignant mast cell tumor | 0.4 | 0.5 | 0.4 | 0.4 |
| Congenital pharyngeal polyp | Rhinosporidial mucosal polyp | 0.4 | 0.4 | 0.6 | 0.5 |
| Mercurial diuretic poisoning | Lobelia species poisoning | 0.4 | 0.4 | 0.4 | 0.5 |
| Branch macular artery occlusion | Acute mesenteric arterial occlusion | 0.5 | 0.9 | 0.5 | 0.6 |
| Intrarenal hematoma | Stomach hematoma | 0.5 | 0.9 | 0.5 | 0.6 |
| Spinal cord hypoplasia | Spinal cord dysplasia | 0.9 | 0.9 | 0.6 | 0.6 |
| Coronary artery thrombosis | Vertebral artery thrombosis | 0.2 | 0.0 | 0.9 | 0.6 |
| Duodenal papillary stenosis | Congenital bronchial stenosis | 0.5 | 0.6 | 0.6 | 0.4 |
| Arteriovenous fistula stenosis | Subclavian vein stenosis | 0.4 | 0.2 | 0.6 | 0.5 |
| Mechanical hemolytic anemia | Hereditary sideroblastic anemia | 0.7 | 0.7 | 0.6 | 0.5 |
| Malignant catarrhal fever | Malignant lipomatous tumor | 0.7 | 0.2 | 0.3 | 0.3 |
| Bolivian hemorrhagic fever | Dengue hemorrhagic fever | 0.6 | 0.9 | 0.8 | 0.6 |
| Benign brain tumor | Benign neuroendocrine tumor | 0.4 | 0.0 | 0.5 | 0.5 |

shown in Tables 8 and 9 is shown the similarity results between only defined concepts. To examine the validity of all approaches, we evaluate the results of other two measures and our proposed method against human expert judgment. Five medical doctors make a consensus on the degree of similarity of the concepts as shown in each Tables 6, 7 and 8.

**Table 8.** Results of degree of similarity on 30 Defined disease concepts estimated by path-based method, ELSIM, our proposed method, and human expert

| Defined concept $P_1$ | Defined concept $P_2$ | Path-based | ELSIM | Proposed method | Human expert |
|---|---|---|---|---|---|
| Rheumatic heart valve stenosis | Coronary artery stenosis | 0.6 | 0.8 | 0.5 | 0.5 |
| Nasal septal hematoma | Vocal cord hematoma | 0.3 | 0.9 | 0.5 | 0.8 |
| Simple periorbital laceration | Brain stem laceration | 0.5 | 0.9 | 0.4 | 0.8 |
| Peritonsillar cellulitis | Dentoalveolar cellulitis | 0.5 | 0.9 | 0.6 | 0.6 |
| Parainfluenza virus laryngotracheitis | Acute viral laryngotracheitis | 1.0 | 0.9 | 0.4 | 0.9 |
| Bone marrow hyperplasia | Retromolar gingival hyperplasia | 0.8 | 0.8 | 0.5 | 0.8 |
| Chronic proctocolitis | Chronic viral hepatitis | 0.5 | 0.8 | 0.4 | 0.5 |
| Obstructive biliary cirrhosis | Syphilitic portal cirrhosis | 0.6 | 0.8 | 0.5 | 0.6 |
| Peripheral T-cell lymphoma | Primary cerebral lymphoma | 0.5 | 0.8 | 0.5 | 0.6 |
| Mast cell leukemia | Prolymphocytic leukemia | 0.9 | 0.9 | 0.4 | 0.9 |
| Tricuspid valve regurgitation | Rheumatic mitral regurgitation | 0.9 | 0.9 | 0.5 | 0.8 |
| Gangrenous paraesophageal hernia | Congenital bladder hernia | 0.5 | 0.8 | 0.6 | 0.6 |
| Congenital mandibular hyperplasia | Atypical endometrial hyperplasia | 0.3 | 0.6 | 0.5 | 0.7 |
| Tuberculous adenitis | Acute mesenteric adenitis | 0.7 | 0.6 | 0.5 | 0.6 |
| Congenital skeletal dysplasia | Aortic valve dysplasia | 0.4 | 0.7 | 0.5 | 0.7 |
| Histiocytic sarcoma | Alveolar soft part sarcoma | 1.0 | 0.9 | 0.5 | 0.8 |
| Drug-induced ulceration | Amebic perianal ulceration | 0.9 | 0.6 | 0.5 | 0.7 |
| Cervical radiculitis | Cervical lymphadenitis | 0.4 | 0.8 | 0.7 | 0.7 |
| Basilar artery embolism | Obstetric pulmonary embolism | 0.6 | 0.7 | 0.5 | 0.6 |
| Acute apical abscess | Chronic apical abscess | 0.5 | 1.0 | 0.9 | 0.8 |
| Acute glossitis | Chronic glossitis | 0.4 | 0.4 | 0.5 | 0.6 |
| Acute bronchitis | Acute purulent meningitis | 0.3 | 0.6 | 0.4 | 0.4 |
| Acute lower gastrointestinal hemorrhage | Stromal corneal hemorrhage | 0.4 | 0.8 | 0.4 | 0.4 |
| Epidural hemorrhage | Tracheostomy hemorrhage | 0.6 | 0.7 | 0.5 | 0.6 |
| Thallium sulfate toxicity | Ammonium sulfamate toxicity | 1.0 | 0.6 | 0.6 | 0.6 |
| Simple periorbital laceration | Complex periorbital laceration | 0.8 | 1.0 | 0.8 | 0.9 |
| Biceps femoris tendinitis | Profunda femoris artery thrombosis | 0.5 | 0.8 | 0.2 | 0.6 |
| Hyperplastic thrush | Hyperplastic gingivitis | 0.3 | 0.5 | 0.7 | 0.5 |
| Acer rubrum poisoning | Penicillium rubrum toxicosis | 0.5 | 0.6 | 0.6 | 0.6 |
| Acute vesicular dermatitis | Herpesviral vesicular dermatitis | 0.4 | 0.5 | 0.9 | 0.4 |

## 4.1 Discussion

In order to evaluate the validaty of all measures against human expert result, we compute the correlation values and error values based on the results in Tables 6, 7 and 8 and present in Tables 9 and 10. As all of the ontology concepts are not completely structured with full relations and some concepts do not have enough hierarchical information in the ontology. Moreover, there is no useful hierarchical information for the primitive concepts. According to these facts, path-based and ElSIM offer very small correlation values for primitive concept similarity (0.04 and −0.19). ELSIM got the negative correlation so it means that these two results are totally different from each other. Based on the hard evidence,

**Table 9.** Correlation values between similarity measures and human experts for each case

| Method | Type | Primitive concepts | Primitive and defined | Defined concepts |
|---|---|---|---|---|
| Path-based | Ontology-based | 0.04 | 0.2 | 0.61 |
| ELSIM | Ontology-based | −0.19 | 0.18 | 0.6 |
| Proposed method | Concept name | 0.74 | 0.5 | 0.02 |

**Table 10.** Error values between similarity measures and human experts for each case

| Method | Type | Primitive concepts | Primitive and defined | Defined concepts |
|---|---|---|---|---|
| Path-based | Ontology-based | 0.1 | 0.1 | 0.04 |
| ELSIM | Ontology-based | 0.2 | 0.1 | 0.03 |
| Proposed method | Concept name | 0.02 | 0.02 | 0.06 |

we assume to measure the similarity based on the concept name for the primitive concepts and our proposed method gets the highest correlation degree (0.74) and smallest error value (0.02) for the primitive concepts. Therefore, it points out ontology concept names are also essential feature for the similarity measure between the concepts who do not have complete hierarchical information in the ontology. Concept names are taken from the actual patient medical treatment records, therefore, they are systematically constructed and very informative for each concept. Moreover, our proposed similarity measure calculates the similarity based on the linguistic headword structure by applying different weights and including Wordnet synsets similarity for headwords to include semantic similarity and also considers the similarity based on the syntactic structure. Therefore, our proposed measure gains benefit from both semantic and syntactic similarity of concept names.

In a consequence, our proposed method also gives a better accuracy (0.5) than other existing two approaches (0.2 and 0.18) for evaluating the similarity between primitive and defined concepts. For the case of defined concept, path-based and ELSIM get the highest correlation (0.61 and 0.6) when our proposed method gets small correlation (0.02) so it means ontology-based measures are the best similarity measure for the fully defined concepts in the ontology. But they cannot give the correct similarity degrees for the partially defined or primitive concepts who do not have the useful hierarchical information. But, our proposed measure overcomes the weakness of ontology-based measures for the primitive concepts. The most important merit of our proposed method is that it does not rely on the ontology structure, but can effectively capture the syntactic and semantic information of the concept names for the similarity measurement.

# 5    Conclusions

This research reviews the existing ontology-based semantic similarity measures and points out the limitations of these measures based on the three cases of experiment. To overcome the weakness, a new similarity measure has been proposed based on the similarity measure of the semantic and syntactic structure of concept name. This paper also shows the strength and weakness of different measures (based on hierarchical information and concept names) according to the evaluation results against human experts.

This research has many continuous works. Firstly, we are planning to combine our proposed method and existing ontology-based measures to fill the weakness of each other so we will get the perfect and desired similarity results for all different types of concepts. As the next step, we will evaluate the new combined similarity method against more number of different concepts. Finally, we want to apply the new combined similarity method to other medical ontologies such as MeSH ontology, to evaluate the effectiveness of our proposed method.

**Acknowledgments.** The first author is supported by Sirindhorn International Institute of Technology (SIIT) under an Excellent Foreign Student (EFS) scholarship. We are thankful to all those anonymous experts who spent their valuable time to evaluate the similarity between diseases.

# References

1. Resnik, P.: Semantic similarity in a taxonomy. An Information-based measure and its application to problems of ambiguity in natural language. J. Artif. Intell. Res. **11**, 95–130 (1999)
2. Hliaoutakis, A., Varelas, G., Voutsakis, E., Petrakis, E.G.M., Milios, E.: Information retrieval by semantic similarity. Int. J. Semant. Web Inf. Syst. **2**, 55–73 (2006)
3. Pedersen, T., Pakhomov, S., Patwardhan, S., Chute, C.: Measures of semantic similarity and relatedness in the biomedical domain. J. Biomed. Inf. **40**, 288–299 (2007)
4. Tongphu, S., Suntisrivaraporn, B., Uyyanonvara, B., Dailey, M.N.: Ontology-based object recognition of car sides. In: Proceedings of 9th International Conference on Electrical Engineering/Electronics Computer Telecommunications and Information Technology (ECTI) (2012)
5. Snomed, C.T.: Systematized Nomenclature of Medicine - Clinical Terminology. http://www.snomed.org/snomedct/index.html
6. Garla, V.N., Brandt, C.: Semantic similarity in the biomedical domain: an evaluation across knowledge sources. J. BMC Bioinf. **13**, 261 (2012)
7. Abdelrahman, A., Kayed, A.: A survey on semantic similarity measures between concepts in health domain. Am. J. Comput. Math. **5**, 204–214 (2015)
8. Zare, M., Pahl, C., Nilashi, M., Salim, N., Ibrahim, O.: A review of semantic similarity measures in biomedical domain using SNOMED CT. J. Soft Comput. Decis. Support Syst. **2**(6), 1–13 (2015)

9. Choi, I., Kim, M.: Topic distillation using hierarchy concept tree. In: Proceedings of the 26th Annual International ACM SIGIR Conference on Research and Development in Information Retrieval, Toronto, Canada, pp. 371–372 (2003)
10. AI-Mubaid, H., Nguyen, H.: A cluster-based approach for semantic similarity in the biomedical domain. In: Proceedings of the 28th IEEE EMBS Annual International Conference. New York City, USA, 3 September 2006
11. Batet, M., Sanchez, D., Valls, A.: An ontology-based measure to compute semantic similarity in biomedicine. J. Biomed. Inf. **44**, 118–125 (2011)
12. SNOMED CT Basics. https://www.scribd.com/Snomed-Ct/Basics/Nehta/Taping/Aug08
13. Tongphu, S., Suntisrivaraporn, B.: Algorithms for measuring similarity between ELH concept descriptions: a case study on SNOMED CT. J. Comput. Inf. **36**(4), 733–764 (2017)
14. Liberman, M., Sproat, R.: The Stress and Structure of Modified Noun Phrases in English. Stanford University (1992)
15. Gomaa, W.H., Fahmy, A.A.: A survey of text similarity approaches. Int. J. Comput. Appl. **68**(13), 13–18 (2013)
16. Petrakis, E.G.M., Hliaoutakis, G.V.A., Raftopoulou, P.: X-Similarity: computing semantic similarity between concepts from different ontologies. J. Digital Inf. Manag. **4**, 233–237 (2006)
17. Ko, S., Han, Y., Salomma, K.: approximate matching between a context-free grammar and a finite-state automaton. Information and Computation, pp. 278–289, February 2016

# Comparative Study of Using Word Co-occurrence to Extract Disease Symptoms from Web Documents

Chaveevan Pechsiri[1(✉)] and Renu Sukharomana[2]

[1] College of Innovative of Technology and Engineering,
Dhurakij Pundit University, Bangkok, Thailand
`chaveevan.pec@dpu.ac.th`
[2] Institute of Social and Economic Science, TRF, Bangkok, Thailand
`sukharenu@gmail.com`

**Abstract.** The research aim is a comparative study of using different word co-occurrence sizes as the two word co-occurrence and the N word co-occurrence on verb phrases to extract disease symptom explanations from downloaded hospital documents. The research results are applied to construct the semantic relations between disease-topic names and symptom explanations for enhancing the automatic problem-solving system. The machine learning technique, Support Vector Machine, and the similarity score determination are proposed to solve the boundary of simple sentences explaining the symptoms for the two word co-occurrence and the N word co-occurrence respectively. The symptom extraction result by the N word co-occurrence provides the higher precision than the two word co-occurrence from the documents.

**Keywords:** Word co-occurrence · Event boundary · Symptom explanation

## 1 Introduction

The research objective is the comparative study of using different word co-occurrence sizes as the two word co-occurrence and the N word co-occurrence on verb phrases to extract the disease symptom explanations from the downloaded health-care documents on the hospital web-boards. The research results are beneficial to the automatic problem-solving system after each semantic relation is constructed between a disease name from a document topic name and the extracted disease-symptom explanation from the document. Moreover, the disease symptom explanation mostly consists of event expressions on several EDUs (where EDU is an Elementary Discourse Unit expression defined as a simple sentence or a clause, [1]) as the symptom-concept explanation on a document of a certain disease as follow.

J. Chen et al. (Eds.): KSS 2017, CCIS 780, pp. 91–100, 2017.
https://doi.org/10.1007/978-981-10-6989-5_8

Example 1

Topic name: โรคหลอดลมอักเสบ/ *Bronchitis Disease*
EDU1 (symptom) : " เมื่อผมไอออกมา/*When I cough,* "
     (เมื่อ/*When* ( ผม/*I*)/NP (ไอออกมา/*cough*)/VP)
EDU2 (symptom): " [มัน]จะมีเสมหะเป็นเลือด/[*It*]*will have phlegm containing blood.* "
     ([มัน/*It*] (จะมี/*will have* เสมหะ/*phlegm* (*sputum*) เป็นเลือด/*as blood*)/VP)
EDU3 (symptom): " แต่ว่า[ผม]ไม่เป็นไข้/*But* [*I*] *have no fever.*"
     ( แต่ว่า/*But* [ผม/*I*] (ไม่เป็น/*have no* ไข้/*fever*)/VP)
EDU4 (symptom): " [ผม]เป็นมาประมาณ 2 วันครับ/[*I*] *have the symptoms about 2 days.*"
     ( [ผม/*I*] (เป็นมา/*get* [*symptom*] ประมาณ 2 วันครับ/*about 2 days* )/VP)
EDU5: "[ผม]สงสัยเป็นโรคหลอดลมอักเสบ/ [*I*] *doubts to get bronchitis?* "
     ( [ผม/*I*] (สงสัย/*doubt* เป็น/*get* โรคหลอดลมอักเสบ/*bronchitis*)/VP)

where the [..] symbol means ellipsis, NP is a noun phrase, and VP is a verb phrase.
  A symptom-concept EDU boundary occurs on EDU1, EDU2, EDU3 and EDU4. According to Example 1, the research emphasizes on the event expressions by verb phrases because of most symptom-concept expressions on the verb phrases of EDUs. Each EDU is based on the following Thai linguistic pattern after stemming words and eliminating stop words.

EDU → NP1 VP
VP → V1 | V1 NP2 |V1 Adverb | V2 NP3 | V2 NP3 VP | V2 Adj
V1 → $V_{strong}$ | Preverb $V_{strong}$
V2 → $V_{weak}$ | Preverb $V_{weak}$
NP1→ Noun1 | Noun2 | Noun3
NP2 → Noun2 | Noun2 NP2 | Noun2 AdjectivePhrase
NP3→ Noun3 | Noun3 Adj prep NP2
Noun1→{'ผู้ป่วย/*patient*' 'โรค/*disease*'...} ;
Noun2→{'อวัยวะ/*organ*''บริเวณ/*area*' 'อุจจาระ/*stool* ' ...}
Noun3→{'อาการ/*symptom*' 'แผล/*scar*' 'รอย/*mark*' 'ไข้/*fever*' 'ผื่น/*rash*' 'หนอง/*pus*' ...}
$V_{strong}$→{'คลื่นไส้/*nauseate*' 'อาเจียน/*vomit*' 'ปวด/*pain*' 'เจ็บ/*pain*' 'แน่น/*constrict*' 'คัน/*itchy*'...}
$V_{weak}$ → {'เป็น/*be*' 'มี/*have*''รู้สึก/*feel*'}
Adv→ {'ยาก/*difficultly*' .... } ; Adj →{'สี.../...*color*' 'เหลว/*watery*' ....} ;
Preverb → {'ไม่/*not*' ... }

where NP1, NP2, and NP3 are noun phrases. $V_{strong}$ is a strong verb set with the symptom concept. $V_{weak}$ is a weak verb set which need more information to have the symptom concept. Noun3 is a noun set with a symptom concept. Adv is an adverb set with the symptom concept. Adj is the adjective set with the symptom concept. Prep is a preposition.
  There are several techniques [2–5] having been used for event extraction from text (see Sect. 2). However, the Thai documents have several specific characteristics, such as zero anaphora or the implicit noun phrase, without word and sentence delimiters, and

etc. All of these characteristics are involved in two main problems of extracting the explanation of the symptom-concept EDUs. The first problem is how to identify an EDU verb phrase having symptom concept whilst some verb phrases contain $V_{weak}$ which needs some following words to provide the symptom concepts. Thus, we apply two different word co-occurrence types, as a fixed size and a varied size, on an EDU verb phrase with a $V_{strong}/V_{weak}$ element as the first word of the word co-occurrence for the comparative study of using the fixed size word co-occurrence and the varied size N-Word-Co to determine the symptom-concept EDU. The fixed size word co-occurrence is the co-occurrence between two words (called Word-Co acquiring the symptom/problem concept from the fixed size two words or one word having the null second word). The varied size word co-occurrence is the co-occurrence between N words (called N-Word-Co acquiring the symptom/problem concept from N words where N = 1, 2, ..., N), where N-Word-Co size (or the N value) is solved by Support Vector Machine (SVM) learning with the linear kernel function [6]. The second problem is how to determine the symptom explanation as the symptom-concept EDU boundary, i.e. EDU1-EDU4 of Example 1 (see Sect. 3.2). With regard to the second problem, we need to develop a framework which combines the machine learning technique and the linguistic phenomena to learn the several EDU expressions of the disease-symptom explanation on the health-care hospital web boards. Therefore, we propose the linear SVM learning to solve the boundary having Word-Co as input features and the similarity score [7] to solve the boundary having N-Word-Co as input features. The reason of applying SVM learning with the linear kernel function for our research is that this machine can provide results sufficiently good at very low cost and high speed [8].

Our research is organized into 5 sections. In Sect. 2, related work is summarized. Problems in extracting the disease symptom explanation from the documents are described in Sects. 3 and 4 shows our framework for extracting the disease symptom explanation from the documents. In Sect. 5, we evaluate and conclude our proposed model.

## 2  Related Work

Several strategies [2–5] have been proposed to solve the event extraction from text.

In 2011, [2] introduced syntactic constraints and lexical constraints. The syntactic constraints, such as "every multi-word relation phrase must begin with a verb, end with a preposition, and be a contiguous sequence of words in the sentence", i.e. 'has a cameo in', 'made a deal with', etc., can eliminate the problems of the uninformative and incoherent extractions. If the relation phrase has too many words, a lexical constraint is used to separate valid relation phrases with a confidence score using a logistic regression classifier. Their precision and recall were 0.8 and 0.62 respectively.

Ando et al. [3] proposed methods for filtering harmful sentences based on multiple word co-occurrences. They compare harmless rate between two-word co-occurrence and three-word co-occurrence. The precision of identify and filtering the harmful sentences through three-word co-occurrence method exceeds 90% whereas the precision of the two-word co-occurrence is under 50%.

In 2014, [4] worked on a model for identifying causality in verb-noun pairs to encode cause or non-cause relation. The result of this research achieves 14.74% and

41.9% F-scores for the basic supervised classifier and the knowledge of semantic classes of verbs respectively.

In 2016, [5] studied the temporal variation in word co-occurrence (i.e. Noun-Noun, Verb-Noun) statistics, with application to event detection. [5] developed an efficient unsupervised spectral clustering algorithm that uncovers clusters of co occurring words which can be related to events in the dataset. The performance of [5] methods for event detection on F-score, obtaining higher recall at the expense of precision informative terms occurring in discrete time frames.

However, most of previous researches identify an event by two-word/three-word co-occurrence without the EDU/simple-sentence boundary consideration as our research. Whilst the symptom-concept expression on each EDU of our research mostly consists of several words, i.e. EDU2 of Example 1.

## 3 Problems of Extracting Disease-Symptom Concepts

Our research contains two problems of determining the symptom-concept explanation; how to identify an EDU verb phrase having symptom concept and how to determine the symptom-concept EDU boundary.

### 3.1 How to Identify Verb Phrase Having Symptom Concept

According to the hospital's health-care web-boards, there are several verb phrases with/without the symptom concepts as shown in the following Example 2:

Example 2

EDU1: "หลังจาก/*After* คนไข้/ *patient* ((ทาน/*consume*)/verb (อาหาร/ *meal*)/noun)/VP"
(*After a patient has consumed a meal*,)

EDU2: " [คนไข้/*he*] ((มี/*have*)/weak-verb (ไข้/*fever*)/noun)/VP"
([*he*] *has a fever*.)

EDU3: "และ/*and* [คนไข้/*he*] ((มี/*have*)/weak-verb (อาการ/*symptom*)/noun (อุจจาระ/ *stools*)/noun (เหลว/*watery*)/ Adj (หลาย/*several*)/Adj (ครั้ง/ *times*)/noun)/VP"
(*and* [ *he*] *has a symptom of watery stools within several times*.)

According to Example 2, the verb phrases (VP) of EDU2 and EDU3 have the weak verbs with the symptom concepts whereas EDU1 having VP without the symptom concept. Thus, the research applies Word-Co and N-Word-Co on the verb phrases (which contain $w_1$ as either $v_{stromg} \in V_{strong}$ or $v_{weak} \in V_{weak}$ and the co-occurred word as $w_2 \in$ Noun3; Noun3 exists in either NP3 or NP1) to identify the verb phrase having the symptom concepts. Using the N-Word-Co to identify the verb phrase with the symptom concept has another problem of how to determine the size of N-Word-Co or the N value, i.e. in Example 2 having the EDU2 verb phrase with N-Word-Co as "มี/*have* ไข้/*fever*" (N = 2) and the EDU3 verb phrase with N-Word-Co as ' 'มี/*have* อาการ/ *symptom* อุจจาระ/*stools* เหลว/*watery*" (N = 4). Thus, we apply the SVM learning to solve the N value (by sliding the window size of two consecutive words with one sliding word distance after stemming words and the stop word removal).

## 3.2    How to Determine Symptom-Concept EDU Boundary

In regard to Example 1 and the following Example 3, how to determine the symptom-concept EDU boundary is challenge.

Example 3

> EDU1: "[ผม] ไม่มีน้ำมูก/ [I] do not have mucus."
> EDU2: "[ผม] ไม่ไอ /[I] do not cough."
> EDU3: "แต่[ผม] คัดจมูกบ้าง /But [I] have a congested nose."
> EDU4: "[ผม] เจ็บคอ/[I] have sore throat ."
> EDU5: "[ผม] ทานยาแก้อักเสบ/[I] take the antibiotic medicine."
> EDU6: "แต่มันก็ไม่หาย/ but it does not work."

where Example 3 has the symptom-concept EDU boundary occurrence on EDU1 through EDU4. Therefore, we propose the SVM learning to solve the boundary having Word-Co as input features (by sliding the window size of two consecutive EDUs with one sliding EDU distance) and the similarity score <0.9 to solve the boundary having N-Word-Co as input features.

# 4    A Framework for Extracting Disease-Symptoms

There are three main steps in determining the disease-symptom explanation for each document topic name by using the Word-Co or N-Word-Co technique, Corpus Preparation step, Learning Step and Symptom-Concept EDU Boundary Determination Step, as shown in Fig. 1.

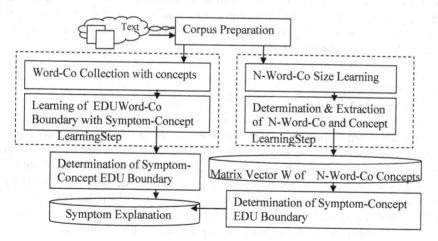

**Fig. 1.**  System overview

---

**Disease Topic :** โรคเกี่ยวกับทางเดินอาหาร / **Gastrointestinal tract disease**

EDU:  [ผู้ป่วย]รู้สึกแน่นที่หน้าอกด้านขวาเป็นบางครั้ง

[ผู้ป่วย/*A patient*]  รู้สึก/*feel*  แน่น/*press against*  ที่/*at* หน้าอก/*chest* ด้านขวา/*right side* เป็นบางครั้ง/*sometime*

<EDU1> ( [((ผู้ป่วย/*A patient*)/ncn])/NP1

(<**Word-CoExpression**   location = chest from Noun2>

   < w₁: setType='weak-verb' ; concept= 'feel' boundary = 'y'>รู้สึก</ w₁>

    < w₂: setType='strong-verb' ; concept='oppress/press against' boundary ='y'>แน่น</ w₂>

     < w₃: setType='Noun2' ; concept= 'chest/organ' boundary= 'y'>หน้าอก</w₃>

      < w₄: setType='Adj' ; concept= 'right side' boundary= 'y'>ด้านขวา</w₄>

       < w₅: setType='Adv' ; concept= 'sometime' boundary= 'n'>เป็นบางครั้ง</w₅>

</ **Word-CoExpression**>)/VP </EDU1>.................

The Word-CoExpression tag is the word boundary tag including 2-Word-Co expression (w1 and w2) and N-Word-Co expression (w1 through w4 with boundary property= 'y'). The w1 tag is the verb tag and the wᵢ tag is the co-occurred wordᵢ tag where i=2,3,..,*num.*

The [..] symbol means ellipsis (Zero Anaphora)

---

**Fig. 2.** Word co-occurrence annotation

## 4.1   Corpus Preparation

This step is the corpus preparation in the form of EDUs from the medical-care documents on the hospital's web-board (http://haamor.com/). The step involves using Thai word segmentation tools [9] including Name entity [10] followed by EDU segmentation [11]. These annotated EDUs are used as an EDU corpus which contains 3000 EDUs of gastrointestinal tract diseases and childhood diseases and is separated into 2 parts; the first part of 2000 EDUs for the learning step of both Word-Co and N-Word-Co; and the second part of 1000 EDUs for determining the symptom-concept EDU boundary. We then semi-automatically annotate the Word-Co expressions with symptom concepts for the w₁ and w₂ tags as Word-Co and for the w₁ through wᵢ as N-Word-Co after stemming words and the stop word removal as shown in Fig. 2. All symptom concepts are referred to WordNet (http://word-net.princeton.edu/obtain) and MeSH (https://www.nlm.nih.gov/mesh/) after translating from Thai to English, by Lexitron (http://lexitron.nectec.or.th/).

## 4.2   Learning

### 4.2.1   Word-Co Learning

We collect each Word-Co feature, $w_1$ $w_2$ or $v_{co}$ $w_{co}$, with the symptom concept into VW from annotated corpus where VW is a Word-Co set with the symptom concepts; $w_1$ is a verb represented by $v_{co}$; and $w_2$ is a co-occurred word represented by $w_{co}$. VW is used for identifying and extracting the consecutive symptom-concept Word-Co occurrences for learning the EDU's Word-Co boundary with the symptom concept by SVM (using Weka, http://www.cs.wakato.ac.nz/ml/weka/). SVM is the linear kernel: the linear function, $f(x)$, of the input x = $(x_1...x_n)$ assigned to the positive class if $f(x) \geq 0$, and otherwise to the negative class if $f(x) < 0$, can be written as

$$f(x) = \langle \mathrm{wt} \cdot \mathrm{x} \rangle + b$$

$$= \sum_{j=1}^{n} wt_j x_j + b \qquad (1)$$

where x is a dichotomous vector number, wt is a weight vector, $b$ is a bias, and $(w, b) \in R^n \times R$ are the parameters that control the function. The SVM learning is to determine $wt_j$ and $b$ for each Word-Co feature, $v_{co-j}\, w_{co-j}\,(x_j)$ in each Word-Co pair, $v_{co-j}\, w_{co-j}\, v_{co-j+1}\, w_{co-j+1}$, from the supervised learning of SVM by sliding the window size of two consecutive EDUs with one sliding EDU distance where $j = 1, 2, \ldots, n$ and $n$ is End-of-Boundary as shown in the following.

| $vw_1\ vw_2.. \ vw_t$ | $vw_1\ vw_2.. \ vw_t$ | $vw_1\ vw_2.. \ vw_t$ | | $vw_1\ vw_2.. \ vw_t$ | $vw_1\ vw_2.. \ vw_t$ |
|---|---|---|---|---|---|
| 0  1... 0 | 0 ..1.. 0 | 1  0... 0 | ·········· | 0  1... 0 | 0  0 ... 1 |
| $v_{co-1}\ w_{co-1}$ | $v_{co-2}\ w_{co-2}$ | $v_{co-3}\ w_{co-3}$ | | $v_{co-n}\ w_{co-n}$ | |

where $vw_k \in VW$; $k = 1, 2, \ldots, t$; and $t$ is the VW cardinality.

### 4.2.2 N-Word-Co Learning

In regard to Eq. 1, the features used for learning N-Word-Co size by SVM are obtained by the following concept sets: $Verb_{strong}$, $Verb_{weak}$, Noun2, Noun3, Adj, and Adv. The SVM learning is to determine $wt_j$ and $b$ for each word feature, $w_j$ (or $x_j$) in each word-concept pair $(w_j\ w_{j+1,})$ with a symptom concept. The N-Word-Co size/boundary learning from $w_j\ w_{j+1}$ (where $w_1 \in V_{strong} \cup V_{weak}$; $w_j\ w_{j+1} \in$ Noun2 $\cup$ Noun3 $\cup$ Verb$_{strong}$ $\cup$ Verb$_{weak}$ $\cup$ Adj $\cup$ Adv; $j = 2, 3, \ldots, n$) of VP is the supervised learning of SVM by sliding the window size of two consecutive words with one sliding word distance after stemming words and the stop word removal. Where $j = 1, 2, \ldots, n$ and $n$ is End-of-Boundary and is equivalent to the N value of N-Word-Co size as shown in the following.

| $w_{e1}\ w_{e2}.. \ w_{et}$ | $w_{e1}\ w_{e2}.. \ w_{et}$ | $w_{e1}\ w_{e2}.. \ w_{et}$ | | $w_{e1}\ w_{e2}.. \ w_{et}$ | $w_{e1}\ w_{e2}.. \ w_{et}$ |
|---|---|---|---|---|---|
| 0  1... 0 | 0  0 ..1.. 0 | 1  0... 0 | ·········· | 0  1... 0 | 0  0 ... 1 |
| $w_2$ | $w_3$ | $w_4$ | | $w_N$ | |

where $w_{ek}$ is a word element $(w_{ek} \in$ Noun2 $\cup$ Noun3 $\cup$ Verb$_{strong}$ $\cup$ Verb$_{weak}$ $\cup$ Adj $\cup$ Adv); $W_1 =$ Noun2 $\cup$ Noun3 $\cup$ Verb$_{strong}$ $\cup$ Verb$_{weak}$ $\cup$ Adj $\cup$ Adv; $k = 1, 2, \ldots, t$; and $t$ is the $W_1$ cardinality.

## 4.3 Symptom-Concept EDU Boundary Determination

### 4.3.1 Symptom-Concept EDU Boundary Determination by Using Word-Co

After using VW to identify a symptom concept EDU from the testing corpus, the wt vector of all $v_{co-j}\ w_{co-j}$ from the SVM learning are used to determine the boundary of the symptom-concept EDUs with Eq. 1 by sliding the window size of two consecutive EDUs with one sliding EDU distance. If $f(x) < 0$ then the boundary is ended as the symptom-concept EDU boundary; otherwise continuing.

### 4.3.2   Symptom-Concept EDU Boundary Determination by Using N-Word-Co

The symptom- concept EDU boundary is determined after the N-Word-Co size determination and extraction. After $w_1 \in V_{strong} \cup V_{weak}$ and $w_1$ is the first word of VP on the testing corpus, the wt vector of all $w_j$ from the SVM learning in Sect. 4.2.2 which are used to determine and extract the N-WordCo size/boundary with symptom-concept collected into the matrix vector (W) of symptom concepts with Eq. 1 by sliding the window size of two consecutive words with one sliding word distance.

If $f(x) < 0$ then the boundary is ended as a word vector of N-Word-Co; otherwise continuing. All extracted N-WordCo expressions are collected into W of symptom concepts as shown in Table 1. The symptom-concept EDU boundary is then determined by the similarity score determination as Max Similarity Score (MaxSimScore) [6] in Eq. 2. MaxSimScore is determined between the N-Word-Co of the testing corpus's EDU and the candidate N-Word-Co expressions from W. The N-Word-Co concept of each consecutive EDU verb phrase is the symptom concept if MaxSimScore $\geq 0.9$ to W; otherwise the symptom vector is ended.

**Table 1.** The N-word-co expression on the health care documents

| N-Word-Co Occurrence on VP | Symptom concept |
|---|---|
| 'เป็น/*be* ผื่น/*rash* แดง/*red* ' | To occur red rash |
| 'เป็น/*be* แผล/*scar* พุพอง/*blister* ' | To occur blister mark |
| 'เป็น/*be* แผล/*scar* อักเสบ/*inflame* ' | To occur inflamed mark |
| ········ | ······ |
| 'แน่น/*constrict* หน้าอก/*chest*' | To constrict chest pain |
| 'แน่น/*constrict* ท้อง/*abdominal*' | To constrict abdominal pain |
| ········ | ······ |
| 'รู้สึก/*feel* แน่น/*constrict* หน้าอก/*chest*' | To constrict chest pain |
| 'รู้สึก/*feel* คลื่นไส้/*be nauseate* ' | To be nauseate |
| 'รู้สึก/*feel* เวียนศีรษะ/*dizzy*' | To be dizzy |
| 'รู้สึก/*feel* ปวด/*pain* ศีรษะ/*head*' | To have an headache |
| 'รู้สึก/*feel* ปวด/*pain* ท้อง/*abdominal*' | To have an abdominal pain |
| ····· | ······ |
| 'มี/*have* ไข้/*fever*' | To have a fever |
| 'มี/*have* รอย/*lesion* ช้ำ/*blue*' | To occur blue lesion |
| ····· | ······· |
| 'มี/*have* อาการ/*symptom*   คลื่นไส้/*nauseate*' | To occur nauseated symptom |
| 'มี/*have* อาการ/*symptom*   ปวด/*pain*' | To occur pain |
| 'มี/*have* อาการ/*symptom* ปวด/*pain* ศีรษะ/*head*' | To have an headache |
| 'มี/*have* อาการ/*symptom* ปวด/*pain* ท้อง/*abdominal*' | To have an abdominal pain |
| ······ | ······ |

$$MaxSimScore = ArgMaxSimilarity_{t=1}^{Cardinality} \left( \frac{|NWCcorpus \cap NWCcandidate_t|}{\sqrt{|NWCcorpus| \times |NWCcandidate_t|}} \right)$$

(2)

*where Cardinality* is the number of N - Word - Co elements of W.

W is the Matrix vector of N - Word - Co (the N - Word - Co set) with the symptom concept.

*NWCcandidate* is a candidate N - Word - Co element of the N - Word - Co set with the symptom concept.

*NWCcorpus* is an N - Word - Co of EDU from the testing corpus.

# 5  Evaluation and Conclusions

The testing corpus of 500 EDUs of gastrointestinal tract diseases and 500 EDUs of childhood diseases collected from the hospital web sites is used for evaluating the symptom-concept EDU explanation/boundary determination from texts. Both evaluations of the symptom-concept EDU explanation determinations by using Word-Co and by using N-Word-Co from the testing corpus are based on the precision and the recall which are evaluated by three expert judgments with max win voting. The average of precisions of determining the symptom-concept EDU explanation are 91.3% and 84.8% with average recalls of 69.2% and 73.1% by using N-Word-Co and Word-Co respectively, as shown in Table 2. The reason of low recall is the anaphora problem, especially with Noun3. For example: there are some pronoun words, i.e. '(บางสิ่ง/*something*)/*pronoun*' '(อะไร/*something*)/*pronoun*', appearing among the consequence words of some verb phrases with the symptom concept, which result in the low recall as shown in the following

VP="(รู้สึก/**feel**)/*serialverb*    (มี/**have**)/*weak-verb*    (บางสิ่ง/**something**)/*pronoun*
(ข้างใน/**inside**)/*prep*  (จมูก/**nose**)/*noun* (ระหว่าง/**during**)/*prep* (เวลาเช้า/
**morning**)/*noun*"
("*feel to have something inside the nose during the morning*")

**Table 2.** Evaluation of symptom vector determination from web documents

| Health-care-symptom corpus | Correctness of symptom vector determination | | | |
|---|---|---|---|---|
| | Using N-word-co | | Using 2-word-co | |
| | Precision | Recall | Precision | Recall |
| Gastrointestinal tract diseases 500EDUs | 92.4% | 63.05% | 84.2% | 70% |
| Childhood diseases 500EDUs | 90.2% | 75.4% | 85.4% | 76.2% |

However, the research results provide the higher precision by using N-Word-Co to determine the symptom-concept EDU explanation from the documents because N-Word-Co contains more information. However, the results also provide the higher recall by using Word-Co to determine the symptom-concept EDU explanation from the documents because Word-Co is more general than N-Word-Co. Thus, the symptom-concept EDU explanation are determined and extracted to construct the semantic relation as the disease Name-symptom Explanation relation where the disease-names are obtained by the document topics. The disease Name-symptom Explanation relation is beneficial to the automatic diagnosis of the problem solution.

Moreover, the proposed method of using either N-Word-Co or Word-Co to determine the information or knowledge can also be applied to the other areas such as the industrial finance problems.

**Acknowledgements.** This work has been supported by The Thailand Research Fund MU-SSIRB:2014/092(B1). The medical-care knowledge and the pharmacology knowledge applied in this research are provided by Puangthong Kraipiboon, a clinician of Division of Medical Oncology, Department of Medicine, Ramathibodi Hospital, and Uraiwan Janviriyasopak, a pharmacist of RexPharmcy, respectively.

# References

1. Carlson, L., Marcu, D., Okurowski, M.E.: Building a discourse-tagged corpus in the framework of rhetorical structure theory. In: van Kuppevelt, J., Smith, R.W. (eds.) Current and New Directions in Discourse and Dialogue. Text, Speech and Language Technology, vol. 22. Springer, Dordrecht (2003). doi:10.1007/978-94-010-0019-2_5
2. Fader, A., Soderland, S., Etzioni, O.: Identifying relations for open information extraction. In: Proceedings of the Conference on Empirical Methods in Natural Language Processing, pp. 1535–1545 (2011)
3. Ando, S., Fujii, Y., Ito, T.: Filtering harmful sentences based on multiple word co-occurrence. In: IEEE/ACIS 9th International Conference on Computer and Information Science(ICIS) (2010)
4. Riaz, M., Girju, R.: Recognizing causality in verb- noun pairs via noun and verb semantics. In Proceedings of the EACL 2014 Workshop on Computational Approaches to Causality in Language, pp. 48–57 (2014)
5. Preoţiuc-Pietro, D., Srijith, P.K., Mark, H., Trevor, C.: Studying the temporal dynamics of word co-occurrences: an application to event detection. In: LREC (2016)
6. Mitchell, T.M.: Machine Learning. The McGraw-Hill Companies Inc. and MIT Press, Singapore (1997)
7. Biggins, S., Mohammed, S., Oakley, S.: University of sheffield: two approaches to semantic text similarity. In: Proceedings of First Joint Conference on Lexical and Computational Semantics, Montreal, Canada, pp. 655–661 (2012)
8. Campbell, C.: An introduction to kernel methods. In: Howlett, R.J., Jain, L.C., Kacprzyk, J. (eds.) Radial Basis Function Networks 1: Recent Developments in Theory and Applications, pp. 155–192. Springer Physica Verlag Rudolf Liebing KG, Vienna (2001)
9. Sudprasert, S., Kawtrakul, A.: Thai word segmentation based on global and local unsupervised learning. In: Proceedings of the 7th National Computer Science and Engineering Conference (2003)
10. Chanlekha, H., Kawtrakul, A.: Thai named entity extraction by incorporating maximum entropy model with simple heuristic information. In: First International Joint Conference, Hainan Island, China (2004)
11. Chareonsuk, J., Sukvakree, T., Kawtrakul, A.: Elementary discourse unit segmentation for thai using discourse cue and syntactic information. In: Proceedings of the 9th National Computer Science and Engineering Conference (2005)

# Forecasting the Duration of Network Public Opinions Caused by the Failure of Public Policies: The Case of China

Ying Lian[1,2], Xuefan Dong[1,2], Ding Li[3], and Yijun Liu[1(✉)]

[1] Institutes of Science and Development, Chinese Academy of Sciences,
Beijing 100090, People's Republic of China
yijunliu@casipm.ac.cn
[2] University of Chinese Academy of Sciences,
Beijing 100190, People's Republic of China
[3] Institute of Development, Southwest University of Finance and Economics,
Chengdu 610074, Sichuan Province, People's Republic of China

**Abstract.** This paper is an effort to identify the factors that may affect the duration of network public opinions caused by failed public policies, and accordingly to propose a model that could predict the duration before the release of policies, in order to provide some rational suggestions to decision makers to reduce the risk of publishing public policies. This paper argues that these factors involve four dimensions: audience, environment, reality, and the Internet. 23 typical Chinese failed public policies happened in recent years and their caused network public opinions are taken as the dataset, and the multivariate regression model and the Cobb-Douglas production function are applied to form the models. Results show that the Cobb-Douglas production form based models could accurately predict the duration of network public opinions aroused by the failure of public policies.

**Keywords:** Public policy · Network public opinion · Forecasting practice · Multivariate regression model · Cobb–Douglas production function

## 1 Introduction

Public policy has been cited as a focus of the public management research area all around the world (Burstein 2010; Marchi et al. 2016; Daniell et al. 2016). A recurring problem in political analysis is to link public opinion to public policy. Public opinion may change public policy, but changes in policy may also shift public attitude toward the public policy, which could be regarded as a closed loop process (Coppock 1977; Burnstein 2003). Previous studies focusing on this relation involve different kinds of public policies, such as immigration policy (Luedtke 2005), energy policy (Bolsen and Cook 2009), redistributive policy (Harell et al. 2014), health policy (Jansen 2008; Bourguet et al. 2013; Barry et al. 2015; Seo et al. 2015; Giles and Adams 2015), criminal policy (Warr 1995), and environmental policy (Toteng 2001; Moyano et al. 2008; Calel 2009). In the late twentieth century, the public opinion from televisions had attracted much attention by scholars (Conway et al. 1981; Simon and Ostrom 1989). However, this is not the case at

© Springer Nature Singapore Pte Ltd. 2017
J. Chen et al. (Eds.): KSS 2017, CCIS 780, pp. 101–116, 2017.
https://doi.org/10.1007/978-981-10-6989-5_9

present. In recent years, with the rapid development of the Internet, effective platforms for an increasing number of the public to participate in arguing the policies have been established, such as the Chinese Weibo website. In general, according to Mcconnell (2010), the initial purpose of the establishment of public policy is to benefit most populations, but a complete success is seriously hard to achieve, which sometimes may cause failed public policies. In general, the failed public policies could be defined as the public policies that ended in failure. Some previous studies suggested the failure of public policies are always inevitable (Trouiller et al. 2002; Martone 2014), especially in those countries with centralized state power, such as China. When this happened, a great deal of people will give their opinions on the Internet, the majority of which are negative. These network public opinions may arouse detrimental impacts on the credibility of the government, to a large extent, which could be seen as a kind of risk to publish a public policy. Therefore, an ex-ante method that could predict the risk before the release of public policies is great of significance. However, issues researched by previous literature related to the network public opinion area are only restricted to ones within the afterwards phase, such as the dynamic transmitting mechanism (Liu et al. 2014; Wang et al. 2015; Kim et al. 2015), evaluation methodologies (Zhang 2015) and content analysis (Abbasi et al. 2008; Fu et al. 2012). It seems that policy makers prefer to be more interested in identifying the situation of network public opinions that will happen in the future. We claim that this is the case because no one previously developed such a way to research this problem. For this point of view, the duration could be cited as one of the most direct and simple indicators reflecting the degree of severity of Internet public opinions. From the public opinion perspective, the duration could be understood as the time from the beginning to the end of public opinions.

Therefore, this article tries to identify the factors that may affect the duration of network public opinion caused by the failure of public policies, and thus proposing a model that could effectively ex-ante predict the duration of network public opinions based on the multivariate regression model and the Cobb-Douglas production function. The contributions of the present paper are summarized as followings:

(1) This is the first paper focusing on the ex-ante forecasting practice relevant to the Internet public opinions;
(2) The present paper proposes a model that can be applied to accurately predict the duration of Internet public opinions caused by the failure of public policies before the release of policies.

## 2   Model

The output is selected as the duration of network public opinions, which has been cited as an important indicator by previous studies (Ma et al. 2014; Gu et al. 2014). When analyzing issues relevant to the public opinion, economic situation is always considered (Citrin et al. 1997; Watts and Dodds 2007). Several studies even confirmed a dominant role played by the state of economy in the reactions of the public (Edwards 2007; Stoutenborough et al. 2014), due to that it could significantly and directly influence the basic livings and behaviors of the people. Moreover, in terms of the

network public opinion, Internet resources are necessary conditions for its dissemination. In general, people from areas with richer network resources are more likely to
give their own view points on some specific events. Therefore, in this paper, the
duration of network public opinions is viewed as the output, and elements with respect
to the audience and the environment from both reality and the Internet are set as the
input variables, including the number of influenced audiences (both reality and Internet), the annual income per capita of influenced audiences (both reality and Internet),
GDP per capita, CPI index, the number of websites per capita of influenced area, and
the amount of international export broadband per capita of influenced area, in order to
represent entire circumstances. The organization of the entire index system is shown as
Fig. 1. In addition, we try to keep the models relatively simple, in order to confirm that
even such a sparse specification can make quite well performance with respect to
predictions of the duration of network public opinions.

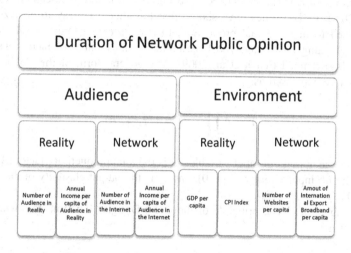

**Fig. 1.** Index system for the duration forecasting of network public opinions

Based on the Fig. 1, output taken as dependent variable, inputs taken as independent variables, take the initial form:

$$T = f(P_R, P_I, S_R, S_I, GDP, CPI, Web, BroadBand) \tag{1}$$

in which, $T$ is the duration of the network public opinion, $PR$ the number of audience
influenced by the public policy in reality, $SR$ the annual income per capita of this
population, $GDP$ the GDP of the target area of public policy, and $CPI$ the CPI index
during the contemporary period of public policy. In addition, $PI$ is the number of
audience in the Internet, $SI$ the annual income per capita of network population, while
$Web$ and $BroadBand$ measure the number of websites and the amount international
export broadband per capita of the target area, respectively.

Due to that the network population could be cited as a part separated from the realistic population, multiple identities and cross-cutting natures may exist in the relationship between these two groups. Therefore, these two variables are brought into consideration together as one factor. Similar process is applied in the annual income per capita of reality group and network group as well. In addition, in order to find a production model with effective validity to predict the duration of future network public opinions, two commonly used methods are applied in this paper. The first method is the multivariate linear regression method, taking the following form of the model:

$$T = \beta_0 + \beta_1 \frac{1}{2}(P_R + P_I) + \beta_2 \frac{1}{2}(S_R + S_I) + \beta_3 GDP + \beta_4 CPI \\ + \beta_5 Web + \beta_6 BroadBand + \varepsilon \tag{2}$$

Another method applied in this paper refers to the Cobb-Douglas function, which describes an empirical relation between two inputs: labor and fixed capital (Cobb and Douglas 1928). Early studies have shown that this model is easy to analyze, and it could be used to fit the actual production, to a large extent (Yuan et al. 2009). In addition, depending on the analytical requirement, the number of factors within the model can be extended (Yuan et al. 2009). The general form of the Cobb–Douglas function can be expressed as:

$$Y = A \prod_{i=1}^{n} X_i^{\alpha_i} (i = 1, 2, \ldots, n) \tag{3}$$

in which $n$ is the number of input variables, $A$ is the total productivity factor ($A > 0$), $X_i$ measures the $i$th input variable ($X_i > 0$), and $\alpha_i$ is the output elasticity of the $i$th input ($0 < \alpha_i < 1$).

According to the basic principle of the Cobb-Douglas function, following model could be set up:

$$T = A\sqrt{P_R P_I}^{\alpha_1} \sqrt{S_R S_I}^{\alpha_2} GDP^{\alpha_3} CPI^{\alpha_4} Web^{\alpha_5} BroadBand^{\alpha_6} \mu \tag{4}$$

Take the logarithm of each side of Eq. (4), the following equation is produced:

$$\ln T = \beta_0 + \beta_1 \ln(\frac{1}{2}(P_R + P_I)) + \beta_3 \ln(\frac{1}{2}(S_R + S_I)) + \beta_3 \ln GDP \\ + \beta_4 \ln CPI + \beta_5 \ln Web + \beta_6 \ln BroadBand + \varepsilon \tag{5}$$

## 3  Data Collection

Cases applied in this paper include the network public opinion caused by 23 public policies in China, which all ended in failure. Applying such a database mainly aims to ensure the representativeness of data and to achieve a relatively fair comparison. The source of data with respect to network public opinions is selected from the Weibo website, one of the most important communication platforms in China. In addition, we crawled the data by using the software named the Unotice Network Public Opinion Monitoring System located in the Institutes of Science and Development, Chinese Academy of Sciences, and the potential influencing factors are selected from various Chinese locality statistical yearbooks and the 26th–37th Statistical Report on Internet Development in China published by CNNIC (China Internet Network Information Center). Complete information of these policies and the duration of network public opinions caused by each policy are listed in Table 1, and the trends are shown as Fig. 2. In addition, due to that the data concerning the Internet aspect of the areas affected by each public policy could not be gained, following methods are applied.

**Table 1.** Information of 23 Chinese failed public policies and the duration of their aroused Internet public opinions

| Serial number | Policy location | Year | Policy content | Duration (Day) |
|---|---|---|---|---|
| 1 | Xianyang City, Shaanxi Province | 2010 | Reward citizens who picked up cigarette and exterminated rats | 5 |
| 2 | Guangzhou City, Guangdong Province | 2010 | Provide free public transportation for 30 days, during The Asian Games | 8 |
| 3 | Wuhan City, Hubei Province | 2010 | Reward citizens who report traffic violations | 4 |
| 4 | Shenzhen City, Guangdong Province | 2011 | Prohibit electric vehicles | 8 |
| 5 | Jiangmen City, Guangdong Province | 2011 | Ban dog order | 6 |
| 6 | Nanjing City, Jiangsu Province | 2011 | Levy name-adding taxes in property deals | 9 |
| 7 | Foshan City, Guangdong Province | 2011 | Ease of property purchases | 7 |
| 8 | Wuhu City, Anhui Province | 2012 | Exempt the deed tax for ordinary commodity housing deals, and provide additional monetary subsidies for housing purchases with the area less than 90 square meters | 4 |
| 9 | Shanghai City | 2012 | Restrict sales in property deals | 9 |

(*continued*)

Table 1. (continued)

| Serial number | Policy location | Year | Policy content | Duration (Day) |
|---|---|---|---|---|
| 10 | Zhoukou City, Henan Province | 2012 | The Flat Fen rehabilitation and funeral reform policy | 4 |
| 11 | Nationwide | 2013 | China's strictest-ever traffic regulations | 13 |
| 12 | Nationwide | 2013 | Levy highway cost | 12 |
| 13 | Beijing City | 2013 | Restrict electric vehicle purchases | 5 |
| 14 | Nationwide | 2013 | Cancel vacation time on Lunar New Year's Eve | 17 |
| 15 | Tianjin City | 2013 | Tail number limited policy | 8 |
| 16 | Guangzhou City, Guangdong Province | 2014 | Allow the Healthcare Insurance Assigned Hospitals to sell food and household articles | 7 |
| 17 | Guangzhou City, Guangdong Province | 2014 | The Babies' Safe Island policy | 7 |
| 18 | Hohhot City, Inner Mongolia Province | 2015 | Ease of property purchases | 8 |
| 19 | Nanjing City, Jiangsu Province | 2015 | "One person one seat" policy | 7 |
| 20 | Lianyungang City, Jiangsu Province | 2015 | The Adding Scores Lines to Deliver Student Documents policy | 4 |
| 21 | Chongqing City | 2015 | Medical price adjustment policy | 6 |
| 22 | Nationwide | 2016 | Circuit-breaker mechanism | 13 |
| 23 | Shenyang City, Liaoning Province | 2016 | Free down payment of building purchases for college students | 5 |

$$P_I = P_{IChina} * (P_R/P_{RChina}) \tag{6}$$

$$S_I = S_{IChina} * (S_R/S_{RChina}) \tag{7}$$

$$Web = Web_{China} * (GDP/GDP_{China}) \tag{8}$$

$$BroadBand = BroadBand_{China} * (GDP/GDP_{China}) \tag{9}$$

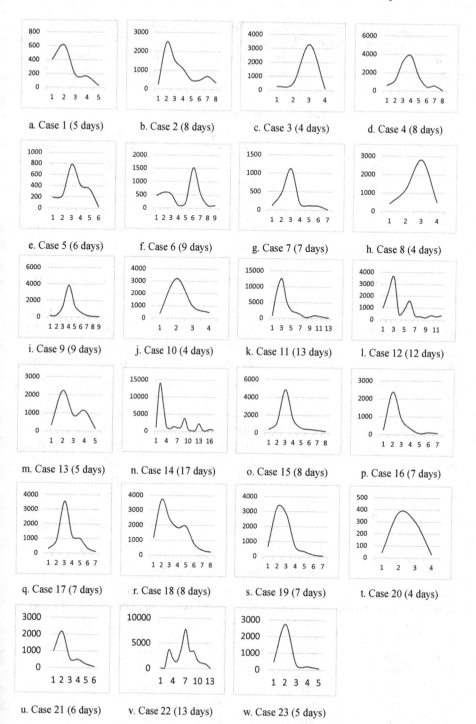

**Fig. 2.** Trend of network public opinion caused by 23 public policies (Note: horizontal axis is number of days; vertical axis is number of Internet public opinions)

where $P_{RChina}$, $P_{IChina}$, $S_{RChina}$, $S_{IChina}$, $GDP_{China}$, $Web_{China}$, $BroadBand_{China}$ measures the national population, national Internet population, national annual income per capita in reality, national annual income per capita with respect to the Internet, national GDP, number of national websites, and amount of nationwide international export broadband

**Table 2.** Values of realistic factors

| Serial number | Number of audiences in reality (10,000) | Annual income per capita (RMB) | GDP (10,000 RMB) | CPI Index |
|---|---|---|---|---|
| 1 | 520 | 18914 | 1099 | 1.055 |
| 2 | 806 | 30658 | 10748 | 1.042 |
| 3 | 837 | 20806 | 5565 | 1.046 |
| 4 | 299 | 36505 | 12950 | 1.064 |
| 5 | 394 | 23924 | 1831 | 1.062 |
| 6 | 811 | 31100 | 6146 | 1.062 |
| 7 | 375 | 30717 | 6210 | 1.055 |
| 8 | 383 | 23784 | 1874 | 1.032 |
| 9 | 1427 | 40188 | 20182 | 1.032 |
| 10 | 1126 | 16503 | 1574 | 1.020 |
| 11 | 136072 | 26955 | 588019 | 1.020 |
| 12 | 136072 | 26955 | 588019 | 1.021 |
| 13 | 151 | 40321 | 19801 | 1.027 |
| 14 | 136072 | 26955 | 588019 | 1.025 |
| 15 | 1472 | 32294 | 14370 | 1.025 |
| 16 | 832 | 42049 | 15420 | 1.025 |
| 17 | 842 | 39229 | 16707 | 1.024 |
| 18 | 300 | 35629 | 2894 | 1.023 |
| 19 | 819 | 39115 | 8821 | 1.020 |
| 20 | 507 | 21461 | 2141 | 1.012 |
| 21 | 2991 | 25147 | 14263 | 1.014 |
| 22 | 137462 | 26959 | 676708 | 1.017 |
| 23 | 811 | 21733 | 7280 | 1.021 |

**Table 3.** Values of network factors

| Serial number | Number of network audiences (10,000) | Annual income per capita (RMB) | Number of websites per capita | Amount of international export broadband per capita (MBPS) |
|---|---|---|---|---|
| 1 | 149 | 19186 | 58 | 16 |
| 2 | 275 | 33911 | 183 | 105 |
| 3 | 285 | 23128 | 91 | 52 |
| 4 | 108 | 35554 | 455 | 294 |
| 5 | 142 | 23301 | 49 | 32 |
| 6 | 292 | 30290 | 80 | 51 |

(continued)

**Table 3.** (*continued*)

| Serial number | Number of network audiences (10,000) | Annual income per capita (RMB) | Number of websites per capita | Amount of international export broadband per capita (MBPS) |
|---|---|---|---|---|
| 7 | 143 | 33960 | 207 | 125 |
| 8 | 145 | 23346 | 56 | 34 |
| 9 | 541 | 39448 | 161 | 97 |
| 10 | 469 | 18357 | 17 | 12 |
| 11 | 56400 | 27324 | 48 | 34 |
| 12 | 59100 | 27582 | 50 | 36 |
| 13 | 66 | 41259 | 1510 | 1077 |
| 14 | 61800 | 28326 | 52 | 55 |
| 15 | 669 | 33937 | 117 | 125 |
| 16 | 376 | 41294 | 206 | 220 |
| 17 | 381 | 38525 | 221 | 235 |
| 18 | 139 | 37248 | 90 | 124 |
| 19 | 378 | 40892 | 100 | 138 |
| 20 | 246 | 22100 | 46 | 61 |
| 21 | 1453 | 25896 | 25 | 33 |
| 22 | 68826 | 33192 | 31 | 39 |
| 23 | 406 | 26758 | 56 | 72 |

in China, respectively. Information of realistic factors and network factors calculated according to Eqs. (6)–(9) is shown in Tables 2 and 3.

## 4 Empirical Test

### 4.1 Regression Analysis

In order to test the performance of proposed models, cases 1–20 are set as the in-sample, and cases 21–23 are set as the out-sample. Based on the multivariate linear regression model and the Cobb-Douglas production function, six models with better performance are identified, in which Model 1 and Model 4 take the CPI index and the amount of international export broadband per capita as reality and the network environmental factors, Model 2 and Model 5 take the number of websites per capita as the network environmental factors, while Model 3 and Model 6 only consider realistic environmental elements. Estimation results through two different methodologies are shown in Tables 4 and 5.

Following the results shown in Tables 4 and 5, it can be noticed that all above models have well goodness of fit ability, and independent variables can explain the dependent variable very well (p-value < 0.01), in which the Model 5 and Model 6, represented by Cobb-Douglas function form, provide a better fit in comparison with other models according to their higher adjusted R square (0.878, 0.877) and F value

**Table 4.** Multivariate linear regression models estimation results

| Dependent variables: T | Model 1 | | Model 2 | | Model 3 | |
|---|---|---|---|---|---|---|
| Independent variables | Coefficient | VIF | Coefficient | VIF | Coefficient | VIF |
| Constant | −42.550 | / | −47.254 | / | −47.281** | / |
| Audience | 8.817E−005*** | 1.206 | 8.901E−005*** | 1.195 | 8.904E−005*** | 1.195 |
| Income | 0.000223*** | 1.455 | 0.000214*** | 1.355 | 0.000217*** | 1.387 |
| GDP | / | / | / | / | −2.853E−006** | 1.342 |
| CPI | 41.125 | 1.168 | 45.839 | 1.167 | 45.784** | 1.167 |
| Web | / | / | −.002*** | 1.310 | / | / |
| Broadband | −0.004*** | 1.441 | / | / | / | / |
| R square | 0.892 | | 0.892 | | 0.892 | |
| Adjusted R square | 0.864 | | 0.863 | | 0.863 | |
| F | 31.066 | | 30.869 | | 31.041 | |
| P-value | .000 | | .000 | | .000 | |

Note: ***Significant at 1%, **Significant at 5%, *Significant at 10%.

**Table 5.** Cobb-Douglas regression models estimation results

| Dependent variables: lnT | Model 4 | | Model 5 | | Model 6 | |
|---|---|---|---|---|---|---|
| Independent variables | Coefficient | VIF | Coefficient | VIF | Coefficient | VIF |
| Constant | −10.605*** | / | −10.661*** | / | −10.472*** | / |
| lnAudience | 0.167*** | 1.561 | 0.166*** | 1.405 | 0.165*** | 1.448 |
| lnIncome | 1.119*** | 4.397 | 1.132*** | 2.802 | 1.202*** | 3.757 |
| lnGDP | / | / | / | / | −0.115* | 4.321 |
| lnCPI | 0.249*** | 1.212 | 0.312*** | 1.268 | 0.296*** | 1.204 |
| lnWeb | / | / | −0.117** | 3.357 | / | / |
| lnBroadband | −0.092 | 5.025 | / | / | / | / |
| R square | 0.891 | | 0.904 | | 0.903 | |
| Adjusted R square | 0.863 | | 0.878 | | 0.877 | |
| F | 30.803 | | 35.202 | | 34.913 | |
| P-value | .000 | | .000 | | .000 | |

Note: ***Significant at 1%, **Significant at 5%, *Significant at 10%.

(35.202, 34.913). This outcome indicates that the factors selected in this paper is in relation with the duration of network public opinions. In addition, results of VIF (all smaller than 10) indicate that there is no multicollinearity in these six models. Furthermore, considering p-values of coefficients, significant and robust positive relationship between the duration and the number of audience, annual income per capita with respect to target area of public policies could be confirmed in all specifications. In terms of Model 1, Model 2, Model 4 and Model 5, considering both realistic and network environmental effects, the conclusion that significant and robust negative relation exists between the output variable and the number of websites per capita, the amount of international expand broadband can be proved as well. With respect to Model 3 and Model 6, in which only realistic environmental influences are considered, a significant and robust negative relation between the duration and the GDP per capita could be ensured. Additionally, these two models also detect a significant positive reaction between the CPI Index and duration as well.

In the case of network public opinions caused by public policies, the Cobb-Douglas functional form could be applied to predict the duration with more effectiveness compared to Multivariate linear regression model. The population, concerned by public policy is an important factor that influencing the duration of network public opinions, although the Internet has offered the opportunity for all peoples to communicate with each other around the world, which seemly indicates that only individuals who are impacted by policies will state their own views. In addition, it is interesting to observe that the annual income per capita is positively related to the duration of network public opinion. In other words, when there is an increase in the disposable income, persons will pay more attention to policies and more likely to express their viewpoints. This may primarily contribute to the enhancement of citizen-consciousness accompanied with the growth of the income, to some extent. However, in terms of the reality environmental factors, the macroeconomic environment reflects a relatively opposite phenomenon: an increase in the GDP and CPI Index respectively shorten and lengthen the duration of public opinions. This may due to that under the economic depression, individuals are more willing to blame to the government; while on the contrary, the confidence for the government from the public may boost up under the economic prosperity background. In the case of correlations between the network factors and the duration of public opinions, a negative relation could be observed. In general, the development of the Internet resources was supposed to enable the public to be more attentive and easily to participate in the discussion of policies. However, this is actually not the case. More than 1200 relevant laws and regulations of the Internet have been issued by the government in China by the end of the first quarter in 2016 (http://www.pkulaw.cn/), which largely standardized the behavior of Chinese netizens. Moreover, better refined Governance Mechanism systems focusing on Internet public opinions that could timely and effectively avoid the spread of Internet public opinion have been implemented by the Chinese government, and combined vigorous propagandas also promoted the online self-regulation awareness of the public, which leads to a more rational and objective attitude towards the failure of public policies.

## 4.2 Forecasting Analysis

Models gained through the regression analysis are shown as formula (10) to (15). In order to test the prediction performance of these models, case 21 to case 24 are set as the out-sample. Comparison results are displayed as Fig. 3 and Table 6. It can be noticed that, the average accuracy rate of the multivariate linear group and the Cobb-Douglas function group is 86.51% and 90.43%, and the average error of each group is 1.14 days and 0.58 days, respectively. It seems that the models with the

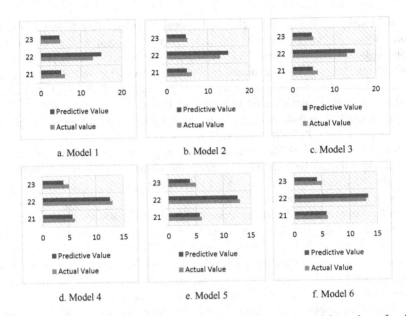

**Fig. 3.** Prediction results of six models (Note: horizontal axis is serial number of policies; vertical axis is number of days of network public opinion)

**Table 6.** Prediction results of six models

|  | Model | Case 21 | | Case 22 | | Case 23 | | Average | |
|---|---|---|---|---|---|---|---|---|---|
|  |  | Error | Error rate | Error | Error rate | Error | Error rate | Error | Accuracy rate |
| Multivariate linear regression models | Model 1 | 0.92 | 0.15 | 2.13 | 0.16 | 0.22 | 0.04 | 1.09 | 88.00% |
|  | Model 2 | 1.16 | 0.19 | 1.92 | 0.15 | 0.32 | 0.06 | 1.14 | 86.46% |
|  | Model 3 | 1.27 | 0.21 | 1.85 | 0.14 | 0.47 | 0.09 | 1.20 | 85.06% |
|  | Average | 1.12 | 0.19 | 1.97 | 0.15 | 0.34 | 0.07 | 1.14 | **86.51%** |
| Cobb-Douglas production function models | Model 4 | 0.43 | 0.07 | 0.47 | 0.04 | 1.04 | 0.21 | 0.65 | 89.47% |
|  | Model 5 | 0.35 | 0.06 | 0.43 | 0.03 | 1.08 | 0.22 | 0.62 | 89.79% |
|  | Model 6 | 0.24 | 0.04 | 0.32 | 0.02 | 0.87 | 0.17 | 0.48 | 92.02% |
|  | Average | 0.34 | 0.06 | 0.41 | 0.03 | 1.00 | 0.20 | 0.58 | **90.43%** |

Cobb-Douglas function form have better forecasting performance compared to multivariate linear regression based models, to a large extent. In other words, the Cobb-Douglas function could be viewed as a relatively more effective approach applied for the prediction of the duration of network public opinions caused by public policies before the release of policies.

**Model 1 :**

$$T = -42.55 + 0.00008817 \times Audience + 0.00023 \times Income + 41.125 \times CPI$$
$$- 0.004 \times Broadhand$$

$$(10)$$

**Model 2 :**

$$T = -47.254 + 0.00008901 \times Audience + 0.000214 \times Income + 45.839 \times CPI$$
$$- 0.002 \times Web$$

$$(11)$$

**Model 3 :**

$$T = -47.281 + 0.00008904 \times Audience + 0.000217 \times Income - 0.000002853 \times GDP$$
$$- 45.784 \times CPI$$

$$(12)$$

**Model 4 :**

$$\ln T = -10.605 + 0.167 \times \ln Audience + 1.119 \times \ln Income + 0.249 \times \ln CPI$$
$$- 0.092 \times \ln Broadhand$$

$$(13)$$

**Model 5 :**

$$\ln T = -10.661 + 0.166 \times \ln Audience + 1.132 \times \ln Income + 0.312 \times \ln CPI$$
$$- 0.117 \times \ln Web$$

$$(14)$$

**Model 6 :**

$$\ln T = -10.472 + 0.165 \times \ln Audience + 1.202 \times \ln Income - 0.115 \times \ln GDP$$
$$+ 0.296 \times \ln CPI$$

$$(15)$$

## 5  Conclusion

This paper puts forward an ex-ante forecasting theory related to the Internet public opinion area. The aim of this essay is to find factors that may affect the duration of Internet public opinions caused by the failure of public policies. A set of several factors with respect to four dimensions: audience and environment from reality and network, is taken as the elements of inputs, and the duration is set as the output. In addition, both the multivariate linear regression model and the Cobb-Douglas production function are applied to form the models. A group of Cobb-Douglas functional models that can sufficiently forecast the duration of network public opinions raised by the failure of public policies are proposed with about 91% accuracy rate in average, in order to provide some statistical evidence for the government to quantify the risk of publishing public policies before the release of policies, to a large extent. Results suggest that the Cobb-Douglas function is a promising technique for the research application in the areas of the network public opinion topic. Moreover, it is proved that the primary active population participating in the Internet public opinions caused by public policies ought to be the number of audience which directly affected by the policies. In addition, with the development of the Internet, the governance implemented by the Chinese government has achieved a significant success, to a large extent, not only standardizing the behavior of netizens, but also better refining the self-regulation awareness of the public. However, the failure of policies should not be only contributed to the opinions stated by the public, the Chinese government should also be responsible for it. Efficient rehearsal and in-depth exploration ought to be carried out before the implementation of public policies, aiming to ensure the success rate and utility of policies during drafting and implementing processes.

Additional studies should pay more attention to the empirical tests with more data and potential influencing factors, and other vital characteristics of network public opinions with respect to public policies should be also considered, in order to provide a more effective and accurate methodology or functional form to forecast future issues of Internet public opinions.

**Acknowledgement.** Thankfulness shall be expressed to the reviewers for their useful discussions and comments on this manuscript. The corresponding author of the present paper is supported by the National Natural Science Foundation of China under Grant Nos. 71573247, 91024010, 91324009 and also by the Youth Innovation Promotion Association Project under Grant No: 2014139.

# References

Abbasi, A., Chen, H., Salem, A.: Sentiment analysis in multiple languages: feature selection for opinion classification in web forums. ACM Trans. Inf. Syst. **26**(3), 12 (2008)

Barry, C.L., Mcginty, E.E., Vernick, J.S., Webster, D.W.: Two years after Newtown–public opinion on gun policy revisited. Prev. Med. **79**, 55–58 (2015)

Bolsen, T., Cook, F.L.: The polls-trends: public opinion on energy policy: 1974–2006. Public Opin. Q. **72**(2), 364–388 (2009)

Bourguet, J.R., Thomopoulos, R., Mugnier, M.L., Cassis, J.: An artificial intelligence-based approach to deal with argumentation applied to food quality in a public health policy. Expert Syst. Appl. **40**(11), 4539–4546 (2013)

Burnstein, P.: The impact of public opinion on public policy: a review and agenda. Polit. Res. Q. **56**(1), 29–40 (2003)

Burstein, P.: Public opinion, public policy, and democracy. In: Leicht, K.T., Jenkins, J.C. (eds.) Handbook of Politics: State and Society in Global Perspective. Springer, New York (2010). doi:10.1007/978-0-387-68930-2_4

Cobb, C.W., Douglas, P.H.: A theory of production. Am. Econ. Rev. **18**, 139–165 (1928)

Calel, R.: Environmental policy and public opinion: a note on instrument choice. Opticon 7, 1–5 (2009)

Citrin, J., Green, D.P., Muste, C., Wong, C.: Public opinion toward immigration reform: the role of economic motivations. J. Polit. 59(3), 858–881 (1997)

Conway, M.M., Wyckoff, M.L., Feldbaum, E., Ahern, D.: The news media in children's political socialization. Public Opin. Q. 45(2), 164–178 (1981)

Coppock, R.: Decision-making when public opinion matters. Policy Sci. 8(2), 135–146 (1977)

Daniell, K.A., Morton, A., Insua, D.R.: Policy analysis and policy analytics. Ann. Oper. Res. 236 (1), 1–13 (2016)

Edwards, M.: Public opinion regarding economic and cultural globalization: evidence from a cross-national survey. Rev. Int. Polit. Econ. 13, 587–608 (2007)

Feng, C.: Reexamine the traditional approaches of Chinese public policy participation-from the aspect of cyber democracy. J. Appl. Libr. Inf. Sci. 2, 29–32 (2012)

Fu, T., Abbasi, A., Zeng, D., Chen, H.: Sentimental spidering: leveraging opinion information in focused crawlers. ACM Trans. Inf. Syst. 30(4), 24 (2012)

Giles, E.L., Adams, J.M.: Capturing public opinion on public health topics: a comparison of experiences from a systematic review, focus group study, and analysis of online, user-generated content. Front. Public Health 3, 200 (2015)

Gu, Q., He, X., Wang, X.: Study on evolution trends of network public opinion based on hyperlink analysis. J. Digit. Inf. Manag. 12(6), 421–428 (2014)

Harell, A., Soroka, S., Ladner, K.: Public opinion, prejudice and the racialization of welfare in Canada. Ethn. Racial Stud. 37(14), 2580–2597 (2014)

Liu, Y., Li, Q., Tang, X., Ma, N., Tian, R.: Superedge prediction: what opinions will be mined based on an opinion supernetwork model? Decis. Support Syst. 64(3), 118–129 (2014)

Luedtke, A.: European integration, public opinion and immigration policy. Eur. Union Polit. 6 (1), 83–112 (2005)

Jansen, M.W.J.: Integration between practice, policy and research in public health: results of a multiple case study. Expert Syst. Appl. 37(5), 3986–3999 (2008)

Kim, K., Baek, Y.M., Kim, N.: Online news diffusion dynamics and public opinion formation: a case study of the controversy over judges' personal opinion expression on SNS in Korea. Soc. Sci. J. 52(2), 205–216 (2015)

Ma, Y.P., Shu, X.M., Shen, S.F., et al.: Study on network public opinion dissemination and coping strategies in large fire disasters. Procedia Eng. 71, 616–621 (2014)

Martone, K.: The impact of failed housing policy on the public behavioral health system. Psychiatr. Serv. 65(3), 313–314 (2014)

Marchi, G.D., Lucertini, G., Tsoukiàs, A.: From evidence-based policy making to policy analytics. Ann. Oper. Res. 236(1), 15–38 (2016)

Mcconnell, A.: Policy success, policy failure and grey areas in-between. J. Public Policy 30(3), 345–362 (2010)

Moyano, E., Paniagua, A., Lafuente, R.: Environmental policy, public opinion and global climate change in southern Europe: the case of Andalusia. Open Environ. J. 2, 62–70 (2008)

Seo, S., Chun, S., Newell, M., et al.: Korean public opinion on alcohol control policy: a cross-sectional International Alcohol Control study. Health Policy 119, 33–43 (2015)

Simon, D.M., Ostrom, C.W.: The impact of televised speeches and foreign travel on presidential approval. Public Opin. Q. 53(1), 58–82 (1989)

Stoutenborough, J.W., Liu, X., Vedlitz, A.: Trends in public attitudes toward climate change: the influence of the economy and climategate on risk, information, and public policy. Risk Hazards Crisis Public Policy 5, 22–37 (2014)

Toteng, E.N.: Urban environmental management in Botswana: toward a theoretical explanation of public policy failure. Environ. Manag. 28(1), 19–30 (2001)

Trouiller, P., Olliaro, P., Torreele, E., Orbinski, J., Laing, R., Ford, N.: Drug development for neglected diseases: a deficient market and a public-health policy failure. Lancet **359**(9324), 2188–2194 (2002)

Wang, G., Liu, Y., Li, J., Tang, X., Wang, H.: Superedge coupling algorithm and its application in coupling mechanism analysis of online public opinion supernetwork. Expert Syst. Appl. **42** (5), 2808–2823 (2015)

Warr, M.: Public opinion on crime and punishment. Public Opin. Q. **59**(2), 296–310 (1995)

Watts, D.J., Dodds, P.S.: Influentials, networks, and public opinion formation. J. Consum. Res. **34**(4), 441–458 (2007)

Yuan, C., Liu, S., Wu, J.: Research on energy-saving effect of technological progress based on Cobb-Douglas production function. Energy Policy **37**(8), 2842–2846 (2009)

Zhang, C.: Research on recognition algorithm of network public opinion in view of evaluation. In: Proceedings of the 2015 2nd International Conference on Electrical, Computer Engineering and Electronics. Atlantis Press (2015)

# Modeling of Interdependent Critical Infrastructures Network in Consideration of the Hierarchy

ChengHao Jin[1(✉)], LiLi Rong[1], and Kang Sun[2]

[1] Institute of Systems Engineering, Dalian University of Technology,
Dalian 116023, China
lffxl@163.com, llrong@dlut.edu.cn
[2] Center for Studies of Marine Economy and Sustainable Development,
Liaoning Normal University, Dalian 116029, China
sunkangdl@163.com

**Abstract.** With the development of socio-economic, interdependencies between critical infrastructures become much closer, that resulting in the fragility of the system. This paper proposes a model of interdependent critical infrastructures network in consideration of the hierarchy. Physical interdependence among critical infrastructures is taken into consideration. Even more specifically, the interdependent network is constructed by analyzing the energy, water supply, telecommunication and transportation, together with the consideration of the hierarchy structure. Moreover, within the interdependent network, the determining method of edges weight is developed on the basis of the supply capacity and quantity of critical infrastructures system elements, which helps to describe the regional characteristics. Finally, the interdependent network with the hierarchy is constructed by taking energy, water supply, telecommunication and transportation of a city in China as the objects. And the analysis on the structural characteristics of the network shows that energy nodes have a greater influence, which suggests that the energy related critical infrastructures need more attention.

**Keywords:** Critical infrastructures · Interdependent · Physical interdependence · Hierarchy · Complex network

## 1 Introduction

According to the report of the U.S. President's Commission on Critical Infrastructure Protection (PCCIP), the critical infrastructure system is defined as "a network of independent, mostly privately owned, manmade systems and processes that function collaboratively and synergistically to produce and distribute a continuous flow of essential goods and services" [1]. Although there are different definitions and classifications of Critical Infrastructure (CI), the main body of CI consists of electricity, telecommunications, transportation, water supply and water treatment, oil and gas, banking and financial system, government service and emergency service system [2]. With the development of socio-economic, the interactions among critical

J. Chen et al. (Eds.): KSS 2017, CCIS 780, pp. 117–128, 2017.
https://doi.org/10.1007/978-981-10-6989-5_10

infrastructures become much closer, that resulting in the fragility of the system [3], such as the major blackout in India. Caused simply by a relay failure, the blackout leads to the region paralyzed for a long time, with the direct results of the blackout included water supply interruption, oil supply interruption, communications interruption, and serious traffic congestion, these events, adding together, leading to the region paralyzed for a long time. Another example of the fragility was the 2008 Chinese winter storms, which leaded to large-scale transmission facilities damages, railway transport disruption and road transport disruption. Furthermore, those disruptions blocked the energy supply's distribution channels, and caused a dilemma where power supply restoring, traffic resuming and energy infrastructure repairing depended on each other. Therefore, scholars have carried out many researches on modeling of interdependent critical infrastructure. On this basis, the vulnerability and robustness of interdependent critical infrastructure, and other related research were also studied, so as to provide some measures for protections.

In terms of the classification of the interdependencies among critical infrastructures, scholars have not formed a unified view on it yet. Rinaldi et al. divides the interdependencies between critical infrastructures into physical interdependence, information interdependence, geographic interdependence, and logical interdependence [4]. Other scholars have also proposed different classification methods, such as spatial dependency and functional dependency [5]; physical interdependency, information interdependence, geographic interdependence, policy interdependence and social interdependence [6]; input interdependence, bidirectional interdependence, shared interdependence, exclusive interdependence and geographic interdependence [7]. Through comparative analysis, Ouyang points out that the classification proposed by Rinaldi et al. has a better applicability [8].

There are many related researches focusing on modeling of interdependent critical infrastructures at system level. Gao et al. investigated the interdependencies between energy, electricity, transportation, water supply and telecommunication systems [9]. Lauge et al. used expert questionnaires to study the interdependencies between energy, information, telecommunication technologies, water supply and others critical infrastructures [10]. Min et al. analyzed the vulnerability of interdependent critical infrastructures by constructing a system dynamics model, which included agricultural production, food supply, water supply and other critical infrastructures [11]. Pinnaka et al. used the complex network approach to construct an interdependent network, which consists of telecommunications, energy, transportation and other critical infrastructures, for the sake of studying the robustness of the network [12].

As for related studies on modeling of interdependent critical infrastructures at system elements level, their research objects are usually composed of two different categories. For example, power and telecommunication system [13], electricity and natural gas system [14], electricity power and water supply [15]. Part of these studies focus on two or more interdependent subsystems in a critical infrastructure system to construct the interdependent critical infrastructure network, such as power grids and telecommunication networks of the electricity power system [16]; electricity power, telecommunication and railway of the electrified railway system [17]. Nevertheless, on account of these researches conducting at system elements level, there is very few modeling researches in the situation of relying on three or more categories interdependent critical infrastructures. For example, electricity power, heating and natural gas systems [18].

In addition, critical infrastructures system usually has a hierarchy structure [19]. The more the categories are, and the lower the level under consideration is. Likewise, the more quantity of system elements are, then it will be required more relevant expertise to clarify the interdependencies among system elements.

In summary, modeling of interdependent critical infrastructures at system elements level focus on interdependencies between system elements. However, this type of modeling method cannot reflect regional differences of interdependent critical infrastructures. Modeling of interdependent critical infrastructures at system elements level generally considers two-category systems. Therefore, from the perspective of physical interdependence, this paper sets the research objectives which respectively are the energy, telecommunication, transportation and water supply. It also considers the hierarchy structure of critical infrastructures, for constructing the interdependent network at system elements level. The determining method of edges weight is developed on the basis of the supply capacity and quantity of critical infrastructures system elements. In this way, it will be a great help to describe the regional characteristics to a certain extent. The rest of the paper is organized as follows. Section 2 develops the analysis of the interdependencies between critical infrastructures; Sect. 3 presents the interdependent network model applying the hierarchy; Sects. 4 and 5 presents the case analysis and the conclusion respectively.

## 2   The Interdependence Between Critical Infrastructures

Rinaldi et al. points out that systems like power, energy, telecommunications, transportation and water supply are typically interdependent [4]. Therefore, this paper analyzes the hierarchy of these systems from the point view of physical interdependence. To be more specifically, this paper analyzes the material and energy dependence between system elements.

### 2.1   The Hierarchy Structure of Critical Infrastructures

The losses statistics of critical infrastructures in "Statistical System of Large-scale Natural Disasters" is one of the most important parts as for direct economic losses, and it also displays the main compositions of critical infrastructures system elements. The losses statistics of critical infrastructures are published by National Disaster Reduction Committee Office of Ministry of Civil Affairs of the People's Republic of China. Therefore, based on the "Statistical System of Large-scale Natural Disasters", combing with national standards and industry standards which are published by National Ministry of China, the hierarchy structure of energy, transportation, water supply and telecommunication, as well as system elements within these systems, are determined and shown in Fig. 1.

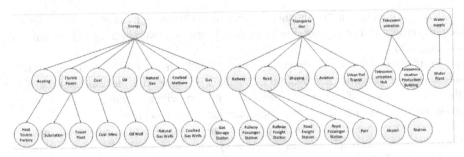

**Fig. 1.** The hierarchy structure of critical infrastructures

## 2.2 The Interdependence Between Critical Infrastructures

Firstly, this paper selects the bottom level of system elements as the object under analysis; Secondly, we analyze the material and energy interdependencies among these selected system elements, upon which physical interdependence are determined, as shown in Fig. 2. These analyzes are developed in accordance with the relevant national standards and industry standards of critical infrastructures which are published by National Ministries of China. We actually analyze the material and energy interdependence between system elements that belong to different levels, and the hierarchy structure of critical infrastructures. But limited by the length of the paper, the conclusion of this analysis is shown in Fig. 3.

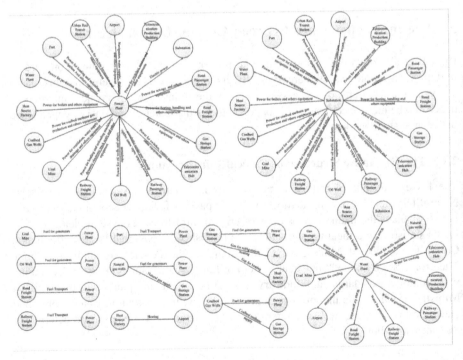

**Fig. 2.** The material and energy dependence between system elements in the bottom level

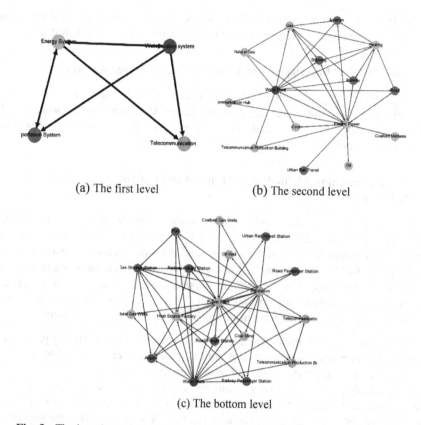

(a) The first level                               (b) The second level

(c) The bottom level

**Fig. 3.** The interdependent network of critical infrastructures with different levels

# 3  Modeling of Interdependent Network by Considering the Hierarchy

The PageRank algorithm proposed by Google's founder Larry Page is a method used to identify the importance of web pages [20]. This algorithm constructs a directed graph $G = (V, E)$, where vertex $V$ represents the set of all pages, edge $E$ denotes the links between the pages. It uses traditional citation analysis for reference to some extent: When page $Va$ is linked to page $Vb$, $Vb$'s PageRank score will partly rely on $Va$'s contribution to it. Moreover, $Va$'s contribution to $Vb$'s PageRank actually positively depends on the importance of $Va$ itself, i.e., the greater the importance of page $Va$ is, the higher the contribution of page $Vb$ can get from page $Va$. Because the web pages in the network possess links that point to each other, the PageRank score can calculated in an iterative way. Finally, the collection of web pages will be sorted according to their score. Inspired by the PageRank algorithm, this paper constructs critical infrastructures as a directed graph, and abstracts the system elements into nodes and interdependencies between system elements as edges. Furthermore, the supply capacity and the quantity of nodes in the network reflect the importance of critical infrastructures to some extent, and also reveal regional differences of critical infrastructures to a certain degree.

## 3.1    The Model of Interdependent Critical Infrastructure Network

By taking the hierarchy structure of critical infrastructures into consideration, using the system elements from the energy, telecommunication, transportation and water supply as nodes, and using the interdependence between these elements as edges, this paper establishes an interdependent network, as shown in Fig. 3.

## 3.2    The Determining Method of the Edge Weight of Interdependent Network

The above analysis shows that, node $Va$'s importance grows with the quantity of other nodes which it is associated with, or another way of saying, linked with. Thus, the amount of nodes that are associated with $Va$ and the supply capacity level of $Va$ are both treated as factors contributing to $Va$'s node importance, and the value of it is also influenced by the edge weights. Taking supply capacity and quantity of critical infrastructures in a specific region into account, this paper proposes an edges weight calculation method for interdependent network. The algorithm is as follows.

(i) Classifying supply ability levels of system elements. The classifying process not only takes the hierarchy structure of the critical infrastructures into consideration, but also uses relevant standards, books and other literatures as reference.

(ii) Calculating comprehensive quantity by using weights assigned to each supply capacity level and quantity of system elements under each capacity level as inputs. Firstly, the weights for each supply capacity level are determined. Then, the quantity of system elements under each supply capacity level is obtained. The quantity information together with previously calculated weights codetermine the comprehensive quantity of system elements.

The weight formula for supply capacity level $i$ is given below:

$$W(i) = (2i - 1)/2i \tag{1}$$

where $i$ is the supply capacity level of the system element $CIpCq$.

The comprehensive quantity of system element $C(CIpCq)$ is determined as follows:

$$C(CIpCq) = \sum_{\substack{i = 1 \\ k = 1}} k \times W(i) \tag{2}$$

where $k$ is the quantity of $CIpCq$ at supply capacity level $i$.

(iii) Determination of edges weight. First, comprehensive quantity of system elements in interdependent network is calculated using the following formula:

$$T(CIpCq) = \sum_{CIsCt \in In(CIpCq)} T(CIsCt) + \sum_{CImCn \in Out(CIpCq)} T(CImCn) \tag{3}$$

where $T(CIpCq)$ is the comprehensive quantity of the system element $CIpCq$, and it indicates this node's importance; $\sum_{CIsCt \in In(CIpCq)} T(CIsCt)$ stands for the sum of comprehensive quantity of all system elements that are linked with $CIpCq$ through input edges; $\sum_{CIsCt \in Out(CIpCq)} T(CImCn)$ is the sum of comprehensive quantity of all system elements that are linked with $CIpCq$ through output edges. Second, the comprehensive quantity of the system element $CIpCq$ is regarded as this node's importance. Finally, the comprehensive quantity value of each system element is equally distributed to its output edges, and the value assigned to each edge can be regarded as edge weight.

In addition, as the level of the hierarchy becomes lower, the system elements will go through continuous subdivision. On the basis of the hierarchy structure of critical infrastructures, it will merge the comprehensive quantity of its subclasses step by step, i.e. system elements that are at lower level of the hierarchy. Thus, the comprehensive quantity of a system element (node) at higher level of the hierarchy can be calculated. The system elements of each hierarchy level can then be obtained, and the corresponding edges weight in interdependent network also can be calculated.

(iv) Modeling the interdependent network by considering the weight. Utilizing results from (2) and (3), interdependent network model with the weight is constructed.

### 3.3    The Influence Node Analysis on Interdependent Networks

Critical infrastructures may fail, either due to external disturbances or internal disturbances. The failure within one system can spread to other systems through the interdependencies among critical infrastructures. When the efforts of preventing cascade failure do not work out, the entire system may breakdown [21]. Obviously, the failure of a node with greater influence will impact plenty of other nodes, potentially along with causing serious consequences. Thus, in order to provide some measures for protection, the nodes with greater influences in interdependent network should be analyzed.

From the point view of network structure, this paper analyzes the node's betweenness by Dijkstra algorithm, and identifies the nodes with greater influences. The network topology parameters are defined as follows.

In this paper, node's betweenness is calculated through comparing the number of shortest paths between two nodes that passes through a certain node $i$ with the number of all the shortest paths. Or in other words, node betweeness of node $i$ characterizes the influence of $i$ on other nodes [22], and the calculation method is shown below:

$$Bv(i) = \sum_{s \neq i \neq t} Gst(i)/Gst \qquad (4)$$

where $Gst$ represents the shortest path quantity between node $Vs$ and node $Vt$; $Gst(i)$ represents the shortest path between $Vs$ and $Vt$ through $Vi$, and the shortest path can be obtained by Dijkstra algorithm.

# 4   The Case Analysis

Using the energy, telecommunication, water supply, and transportation of a city in China as the case analysis, an interdependent network model with the hierarchy structure is constructed, and nodes with much greater influences are analyzed.

## 4.1   The Hierarchical of Interdependent Network Model

Based on the "Statistical System of Large-scale Natural Disasters", electricity, energy, telecommunication, water supply and transportation of a city in China are used as research objects, and the hierarchy structure is analyzed. Firstly, the comprehensive quantity of system elements in the bottom level of critical infrastructures is calculated; Secondly, according to the hierarchy structure of critical infrastructures, merges are conducted step by step, and then the comprehensive quantity of system elements is obtained. Due to space limitations, classifying of different supply capacity levels, and the quantity of the system elements in the bottom level are all shown in the appendix.

Judged from coal mine production safety supervision data and coal mine production licenses data, it can be seen that the city does not permit any production license of coal mines. Thus, it can be adjudged there are no coal mine, and no coalbed gas field as well. Besides, the city dose not possess any oil fields or gas fields either. Since it is located in southern China, we do not consider the heating system (Fig. 4).

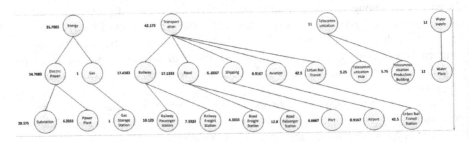

**Fig. 4.** The hierarchy structure and the comprehensive quantity of critical infrastructures in a city of China

Firstly, as discussed before, the material dependence and energy dependence are analyzed. Combining with the hierarchy structure of critical infrastructures, they are used to determine the physical interdependencies between system elements. Secondly, the system elements are treated as nodes, while the interdependencies between system elements are treated as edges, based on which the interdependent network is then constructed.

# 5   Analysis of Node Influence of Interdependent Networks

For the interdependent network with the hierarchy, we use formulas (3) and (4) to compute influences of nodes, as follows (Table 1):

**Table 1.** The betweenness of nodes in interdependent critical infrastructures networks

| node $Bv(i)$ | node $Bv(i)$ | node $Bv(i)$ | node $Bv(i)$ |
|---|---|---|---|
| *The first hierarchical* | | | |
| Energy 0.7778 | Water supply 0.5556 | Telecom 0.3333 | Transportation 0.5556 |
| *The second hierarchical* | | | |
| Electric power 0.7857 | Gas 0.25 | Water plant 0.25 | Telecom Hub 0.1429 |
| Railway 0.25 | Urban rail transit 0.0814 | Road 0.25 | Aviation 0.1071 |
| Shipping 0.25 | Telecom production building 0.1429 | | |
| *The bottom level* | | | |
| Power plant 0.686 | Substation 0.2326 | Gas storage station 0.2326 | Water plant 0.2093 |
| Telecom Hub 0.1047 | Urban rail transit station 0.0814 | Road passenger station 0.0814 | Road freight station 0.2093 |
| Airport 0.0814 | Port 0.2093 | Telecom production building 0.1047 | Railway passenger station 0.0814 |
| Railway freight station 0.2093 | | | |

As we can see, for the first hierarchy, the betweenness value of the energy (0.7778) is the highest in the interdependent network, and then water supply and transportation rank the second, which indicates that the energy node has the greatest influence, while the water supply and transportation node have slightly less influences on other nodes. In the second tier of the hierarchy structure, the betweenness value of the electric power (0.7857) is much higher than other nodes, and the gas supply, water plant, railway, road and shipping also have a relatively high betweenness value. It indicates that the electric power has the greatest influence, while the gas supply, water plant, railway, road, and shipping subsystems also have relatively significant influences. In the third tier of the hierarchy structure, the betweenness value of the power plant (0.686) is the highest, and the substation, gas storage station, water plant, railway freight station, the road freight station and the port also exhibit a high betweeness value. The results show that the node with highest influence is the power plant, and substations, gas stations, water plants, rail freight station, road freight station, port nodes also have relatively prominent influences. In summary, the energy nodes in interdependent networks have the most influencing, which need to be given more attention.

# 6 Conclusion

Aiming at modeling of interdependent critical infrastructures network, this paper proposes a model in consideration of the hierarchy structure from the perspective of physical interdependence. Using the energy, telecommunication, water supply and transportation as research objects, combining with the hierarchy structure of critical infrastructures, the interdependent critical infrastructures network is established. This paper also delivers the method of determining edges weight within the interdependent critical infrastructures network, which takes the supply capacity and quantity of the system elements into account. Among all the nodes in interdependent critical infrastructures network, the nodes that have relatively prominent influences should be paid more attention to. The model proposed in this paper discusses the interdependencies between multiple critical infrastructure categories at the system elements level. It also can reflect regional differences of interdependent critical infrastructure to a certain extent. The model actually can help provide pertinent measures for the protection of critical infrastructures.

**Acknowledgments.** Particular thanks to JingWen Zhang and KeSheng Yan. This work is partially supported by the National Natural Science Foundation of China under Grant No. 71371039.

# Appendix

See Table 2.

**Table 2.** The classification of supply capacity and quantity of the critical infrastructure system elements of a city in China

| The bottom level of the system elements | Supply capacity level | Each supply capacity level weight | | Quantity | The comprehensive weights |
|---|---|---|---|---|---|
| | | Weight range | Weight value | | |
| Power plant | Large | (2/3, 1) | 5/6 | 2 | 6.3333 |
| | Medium | (1/3, 2/3] | 1/2 | 6 | |
| | Small | (0, 1/3] | 1/6 | 10 | |
| Substation | Ultra high voltage transmission substation | [3/4, 1) | 7/8 | 0 | 28.375 |
| | Extra high voltage transmission substation | [2/4, 3/4) | 5/8 | 9 | |
| | High voltage transmission substation | [1/4, 2/4) | 3/8 | 34 | |
| | High voltage distribution substations | (1/4, 0) | 1/8 | 180 | |

(*continued*)

**Table 2.** (*continued*)

| The bottom level of the system elements | Supply capacity level | Each supply capacity level weight | | Quantity | The comprehensive weights |
|---|---|---|---|---|---|
| | | Weight range | Weight value | | |
| Gas storage station | | (0, 1) | 1/2 | 91 | 45.5 |
| Water plant | Under level I | (0, 1/5) | 1/10 | 42 | 12 |
| | Level I | [1/5, 2/5) | 3/10 | 4 | |
| | Level II | [2/5, 3/5) | 1/2 | 5 | |
| | Level III | [3/5, 4/5) | 7/10 | 2 | |
| | Above level III | [4/5, 1) | 9/10 | 3 | |
| Telecommunication Hub | | (1/2, 1) | 3/4 | 7 | 5.25 |
| Telecommunication production building | | (0, 1/2] | 1/4 | 23 | 5.75 |
| Railway passenger station | Extra large station | (3/4,1) | 7/8 | 3 | 10.125 |
| | Large station | (2/4, 3/4] | 5/8 | 0 | |
| | Medium station | (1/4, 2/4] | 1/2 | 15 | |
| | Small station | [1/4, 0) | | | |
| Railway freight station | Principal station | [5/6, 1) | 11/12 | 1 | 7.3333 |
| | First level station | [4/6, 5/6) | 9/12 | 3 | |
| | Second level station | [3/6, 4/6) | 7/12 | 1 | |
| | Third level station | [2/6, 3/6) | 5/12 | 3 | |
| | Fourth level station | [1/6, 2/6) | 3/12 | 9 | |
| | Fifth level station | (0, 1/6) | 1/12 | 1 | |
| Urban rail transit station | First level station | (2/3, 1) | 1/2 | 85 | 42.5 |
| | Second level station | (1/3, 2/3] | | | |
| | Third level station | (0, 1/3] | | | |
| Road passenger station | First level station | [4/5, 1) | 9/10 | 6 | 12.8 |
| | Second level station | [3/5, 4/5) | 7/10 | 8 | |
| | Third level station | [2/5, 3/5) | 1/2 | 0 | |
| | Fourth level station | [1/5, 2/5) | 3/10 | 1 | |
| | Fifth level station | (0, 1/5) | 1/10 | 15 | |
| Road freight station | First level station | (2/3, 1) | 5/6 | 0 | 4.333 |
| | Second level station | (1/3, 2/3] | 1/2 | 1 | |
| | Third level station | (0, 1/3] | 1/6 | 23 | |
| Airport | First level station | [5/6, 1) | 11/12 | 1 | 0.9167 |
| | Second level station | [4/6, 5/6) | 9/12 | 0 | |
| | Third level station | [3/6, 4/6) | 7/12 | 0 | |
| | Fourth level station | [2/6, 3/6) | 5/12 | 0 | |
| | Fifth level station | [1/6, 2/6) | 3/12 | 0 | |
| | Sixth level station | (0, 1/6) | 1/12 | 0 | |
| Port | Main port | (2/3, 1) | 5/6 | 8 | 6.6667 |
| | Area important port | (1/3, 2/3] | 1/2 | 0 | |
| | Ordinary port | (0, 1/3] | 1/6 | 0 | |

# References

1. President's Commission on Critical Infrastructure Protection (PCCIP): Critical foundations: protecting America's infrastructures: the report of the President's commission on critical infrastructure protection. U.S. Government Printing Office, Washington, D.C. (1997)

2. Jones, A.: Critical infrastructure protection. Comput. Fraud Secur. **12**(4), 11–15 (2007)
3. Liu, X., Peng, H., Gao, J.: Vulnerability and controllability of networks of networks. Chaos, Solitons Fractals **80**, 125–138 (2015)
4. Rinaldi, S.M., Peerenboom, J.P., Kelly, T.K.: Identifying, understanding, and analyzing critical infrastructure interdependencies. IEEE Control Syst. **21**(6), 11–25 (2001)
5. Zimmerman, R.: Social implications of infrastructure network interactions. J. Urban Technol. **8**(3), 97–119 (2001)
6. Lee, E.E., Mitchell, J.E., Wallace, W.A.: Restoration of services in interdependent infrastructure systems: a network flows approach. IEEE Trans. Syst. Man Cybern. Part C Appl. Rev. **37**(6), 1303–1317 (2007)
7. Zhang, P., Peeta, S.: A generalized modeling framework to analyze interdependencies among infrastructure systems. Transp. Res. Part B: Methodol. **45**(3), 553–579 (2011)
8. Ouyang, M.: Review on modeling and simulation of interdependent critical infrastructure systems. Reliab. Eng. Syst. Saf. **121**, 43–60 (2014)
9. Gao, J., Buldyrev, S.V., Havlin, S., et al.: From a single network to a network of networks. Nat. Sci. Rev. **1**(3), 346–356 (2014)
10. Lauge, A., Hernantes, J., Sarriegi, J.M.: Critical infrastructure dependencies: a holistic, dynamic and quantitative approach. Int. J. Crit. Infrastruct. Prot. **8**, 16–23 (2015)
11. Min, H.S.J., Beyeler, W., Brown, T., et al.: Toward modeling and simulation of critical national infrastructure interdependencies. IEEE Trans. **39**(1), 57–71 (2007)
12. Pinnaka, S., Yarlagadda, R., Çetinkaya, E.K.: Modelling robustness of critical infrastructure networks. In: Design of Reliable Communication Networks (DRCN), pp. 95–98 (2015)
13. Rueda, D.F., Calle, E.: Using interdependency matrices to mitigate targeted attacks on interdependent networks: a case study involving a power grid and backbone telecommunications networks. Int. J. Crit. Infrastruct. Prot. **16**, 3–12 (2016)
14. Ouyang, M.: Critical location identification and vulnerability analysis of interdependent infrastructure systems under spatially localized attacks. Reliab. Eng. Syst. Saf. **154**, 106–116 (2016)
15. Zhang, Y., Yang, N.: Upmanu lall.: modeling and simulation of the vulnerability of interdependent power-water infrastructure networks to cascading fail users. J. Syst. Sci. Syst. Eng. **25**(1), 102–118 (2016)
16. Hu, J., Yu, J., Cao, J., et al.: Topological interactive analysis of power system and its communication module: a complex network approach. Phys. A Stat. Mech. Its Appl. **416**, 99–111 (2014)
17. Zhang, J., Song, B., Zhang, Z., et al.: An approach for modeling vulnerability of the network of networks. Phys. A: Stat. Mech. Its Appl. **412**, 127–136 (2014)
18. Augutis, J., Jokšas, B., Krikštolaitis, R., et al.: The assessment technology of energy critical infrastructure. Appl. Energy **162**, 1494–1504 (2016)
19. Agarwal, J.: Improving resilience through vulnerability assessment and management. Civ. Eng. Environ. Syst. **32**(1–2), 5–17 (2015)
20. Kamvar, S.D., Haveliwala, T.H., Manning, C.D., et al.: Extrapolation methods for accelerating pagerank computations. In: International Conference on World Wide Web, pp. 261–270. ACM (2003)
21. Adachi, T., Ellingwood, B.R.: Serviceability of earthquake-damaged water systems: effects of electrical power availability and power backup systems on system vulnerability. Reliab. Eng. Syst. Saf. **93**(1), 78–88 (2008)
22. Albert, R., Barabási, A.L.: statistical mechanics of complex networks. Rev. Mod. Phys. **74**(1), 18–35 (2002)

# Emergency Attribute Significance Ranking Method Based on Information Gain

Ning Wang[✉], Haiyuan Liu, Huaiming Li, Yanzhang Wang,
Qiuyan Zhong, and Xuehua Wang

Faculty of Management and Economics, Dalian University of Technology,
Dalian 116024, China
wn@dlut.edu.cn

**Abstract.** In emergency management, it is important to measure the significance of emergency attributes. This paper proposes a method for ranking the significance of attributes. Firstly, it builds the emergency decision table based on rough set theory. Secondly, it uses information gain to measure the objective significance of emergency attributes. Further, the information gain is combined with the prior knowledge of experts to rank the total significance of emergency attributes. This method could help decision-makers identify the significance of attributes, providing a support for the decision-makers to take emergency response measures.

**Keywords:** Emergency management · Rough set · Significance ranking · Information gain

## 1 Introduction

Emergency decision-making is typically characterized by time limitations, partial or incomplete information, and decision pressure resulting from potentially serious outcomes [1]. Quickly collect the information of emergencies and take rapid response measures could greatly reduce the hazards caused by emergencies. Confronted with the complex and redundant information of emergency scenarios, find the key attributes is important for the decision-makers to optimize the allocation of resources and decrease the response time of emergencies. Therefore, to identify the key attributes, it's necessary to rank the significance of attributes.

Rough set, as a theory of dealing with imprecise and incomplete mathematical data [2–4], is based on the classification mechanism. It regards knowledge as the data division constituted by the equivalence relations in a specific space. Some scholars have introduced rough set into measuring the significance of attributes [5, 6] to express the influence of each attribute on decision-making in the emergency environment. However, rough set can't express the prior knowledge of experts in this field [7], even may ignore the practical meaning of the attributes [8]. But in the field of emergency management, experts usually have the prior knowledge of many attributes, which is also quite important for identifying the key attributes of emergencies.

Information gain [9], as the content of information theory, can be used to measure the amount of information that an attribute contains. The larger the information gain

© Springer Nature Singapore Pte Ltd. 2017
J. Chen et al. (Eds.): KSS 2017, CCIS 780, pp. 129–135, 2017.
https://doi.org/10.1007/978-981-10-6989-5_11

value of an attribute, the larger the difference between it and others, and the larger its influence on the classification.

Therefore, this paper combines information gain and experts' prior knowledge to determine the significance of attributes. To begin with, based on the knowledge acquisition of rough set theory, it builds the emergency decision table. Secondly, calculate the information gain values of each attribute, forming the objective significance of the attributes. Then experts score the significance of each attribute, forming the subjective significance of attributes. Further, the final ranking result of attributes' significance is determined by taking the weighted average of the objective significance and the subjective significance. Finally, the experimental analysis demonstrates the reasonableness and effectiveness of the method.

## 2  Theories

Decision Table. In rough set theory, decision table is seen as a decision-making information systems, noted as a four-tuple $T = (U, C \cup D, V, f)$, where $U$ is a non-empty finite set of objects; $C$ is condition attribute set and $D$ is decision attribute set; $f = \{f_a | f_a : U \rightarrow V\}$ is an assignment function; $V$ is value domain [3].

Information Gain. Suppose $U$ is a set of $u$ samples, and there are $m$ categories in the category attribute, $u_i (u_i \subseteq U)$ is the number of samples separately belonging to $m$ categories, $u_i / u$ is the probability estimates that a sample belongs to a certain classification. The information entropy for classifying a given sample is:

$$I(u_1, u_2, \ldots, u_m) = -\sum_{i=1}^{m} \frac{u_i}{u} \log_2 \frac{u_i}{u} \tag{1}$$

The value domain of attribute $A$ is $(a_1, a_2, \ldots, a_v)$, which divides the $U$ into $\{U_1, U_2, \ldots, U_v\}$, and the value of $A$ in $U_j$ is $a_j$. Suppose $U_j$ contains $U_{ij}$ sample for the classification $i$. Therefore, the entropy of $A$ is:

$$E(A) = -\sum_{j=1}^{v} \frac{u_{1j} + \ldots + u_{mj}}{u} I(u_{1j}, \ldots, u_{mj}) \tag{2}$$

Base on the division of $A$, the information gain value is obtained:

$$Gain(A) = I(u_1, \ldots, u_m) - E(A) \tag{3}$$

$Gain(A)$ contains the amount of information that attribute $A$ can provide for the classification of the decision table. Accordingly, we rank the objective significance of condition attribute based on its information gain value, namely the larger the information gain value, objectively, the more important the attribute is.

# 3 Construction of Emergency Decision Table

Decision table describes the relationship between the condition attributes and decision attributes, which usually contains a large amount of sample information, and each of sample represents a basic decision rule. In the original data of emergencies, there may be incomplete, incorrect or noisy or inconsistent data. Data cleaning process can be used to fill the missing attribute values, to remove noise and to identify outliers, and to correct inconsistencies in the data. In addition, there are huge differences between the attributes of emergencies, therefore, we should make data transform, including standardization, discretization and so on. To abstract the samples of decision table from many historical data, the original data of emergencies is required to be processed by data cleaning and data transformation, and the specific techniques are detailed in the literature [10].

In addition, in the process of extracting emergency attributes, through distinguishing the causal relationship among attributes, the attributes of each emergency sample can be divided into condition attributes and decision attributes, namely the condition attributes set $C$ and decision attribute set $D$ of the emergency decision table. The construction process of the emergency decision table is shown in Fig. 1.

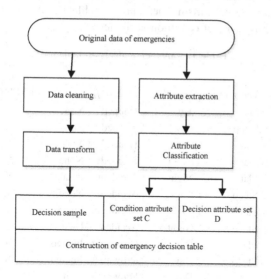

**Fig. 1.** Construction process of emergency decision table

# 4 Method of Ranking the Significance of Emergency Attributes

The target of ranking attribute significance is to provide the relative significance of each attribute and help decision-makers to identify key attributes. The total significance of the condition attribute is composed of the objective significance and the subjective significance, where the objective significance is determined by information gain values, and the subjective significance is determined based on the prior knowledge of experts. The method of ranking the significance of emergency attributes is shown in follows.

Input: emergency decision table.

Output: the ranking result of the significance of condition attributes.

Steps:

- Calculate the information gain value of condition attribute $A_i$, noted $Gain(A_i)$;
- Mapping $Gain(A_i)$ into the interval [0,1]: $Gain(A_i)' = \frac{Gain(A_i) - \min_{Gain(A)}}{\max_{Gain(A)} - \min_{Gain(A)}}$;
- $Gain(A_i)'$ is the objective significance of $A_i$, noted as $Osig_{A_i}$;
- Experts score the significance of $A_i$ in [0,1] and the normalization of processing to get the subjective significance of $A_i$, noted as $Ssig_{A_i}$;
- $Sig_{A_i} = \lambda Osig_{A_i} + (1 - \lambda)Ssig_{A_i}$ is the total significance of $A_i$, where $\lambda \in [0, 1]$ is the given weight of the objective significance;
- Based on the value of $Sig_{A_i}$, the ranking result is obtained.

## 5   Experiment

In order to construct the emergency decision table, we cite the data, existing in [5], to construct a decision table of tourism emergencies (Table 1). On this basis, we discuss how to rank the significance of condition attributes.

**Table 1.** Emergency decision table.

| U | C | | | | | D |
|---|---|---|---|---|---|---|
| | Weather ($A_1$) | Terrain ($A_2$) | Time ($A_3$) | Type ($A_4$) | Casualty ($A_5$) | 110 center ($d_1$) |
| 1 | Sunny | Flat | day | Trapped | None | N |
| 2 | Sunny | Steep | Day | Trapped | None | N |
| 3 | Rainy | Flat | Day | Trapped | None | N |
| 4 | Rainy | Flat | Night | Trapped | None | N |
| 5 | Rainy | Flat | Night | Accident | Injured | Y |
| 6 | Sunny | Steep | Day | Accident | None | Y |
| 7 | Sunny | Flat | Night | Accident | Injured | Y |
| 8 | Rainy | Steep | Night | Accident | Died | Y |
| 9 | Rainy | Flat | Day | Drowning | Injured | N |
| 10 | Sunny | Steep | Day | Drowning | Injured | N |
| 11 | Sunny | Flat | Day | Drowning | None | Y |
| 12 | Rainy | Steep | Day | Drowning | Injured | N |
| 13 | Rainy | Flat | Night | Attacked | None | Y |
| 14 | Sunny | Steep | Day | Attacked | Injured | Y |
| 15 | Rainy | Flat | Night | Attacked | Injured | Y |
| 16 | Sunny | Steep | Day | Attacked | Died | Y |
| 17 | Rainy | Flat | Night | Falling | Injured | N |
| 18 | Rainy | Steep | Day | Falling | Died | Y |
| 19 | Rainy | Steep | Night | Falling | Injured | N |
| 20 | Sunny | Steep | Day | Falling | Injured | N |

1. Calculate the objective significance:

$$I(U) = -\frac{10}{20}\log_2\frac{10}{20} - \frac{10}{20}\log_2\frac{10}{20} = 1$$

$$E(A_1) = \frac{9}{20}\left(-\frac{5}{9}\log_2\frac{5}{9} - \frac{4}{9}\log_2\frac{4}{9}\right) + \frac{11}{20}\left(-\frac{5}{11}\log_2\frac{5}{11} - \frac{6}{11}\log_2\frac{6}{11}\right) = 0.99$$

$$Gain(A_1) = I(U) - E(A_1) = 0.01$$

$$E(A_2) = \frac{10}{20}\left(-\frac{5}{10}\log_2\frac{5}{10} - \frac{5}{10}\log_2\frac{5}{10}\right) + \frac{10}{20}\left(-\frac{5}{10}\log_2\frac{5}{10} - \frac{5}{10}\log_2\frac{5}{10}\right) = 1$$

$$Gain(A_2) = I(U) - E(A_2) = 0$$

$$E(A_3) = \frac{12}{20}\left(-\frac{5}{12}\log_2\frac{5}{12} - \frac{7}{12}\log_2\frac{7}{12}\right) + \frac{8}{20}\left(-\frac{5}{8}\log_2\frac{5}{8} - \frac{3}{8}\log_2\frac{3}{8}\right) = 0.97$$

$$Gain(A_3) = I(U) - E(A_3) = 0.03$$

$$E(A_4) = \frac{4}{20}\left(-\frac{4}{4}\log_2\frac{4}{4} - 0\right) + \frac{4}{20}\left(-\frac{4}{4}\log_2\frac{4}{4} - 0\right) + \frac{4}{20}\left(-\frac{1}{4}\log_2\frac{1}{4} - \frac{3}{4}\log_2\frac{3}{4}\right)$$
$$+ \frac{4}{20}\left(-\frac{4}{4}\log_2\frac{4}{4} - 0\right) + \frac{4}{20}\left(-\frac{1}{4}\log_2\frac{1}{4} - \frac{3}{4}\log_2\frac{3}{4}\right) = 0.32$$

$$Gain(A_4) = I(U) - E(A_4) = 0.68$$

$$E(A_5) = \frac{6}{20}\left(-\frac{2}{6}\log_2\frac{2}{6} - \frac{4}{6}\log_2\frac{4}{6}\right) + \frac{10}{20}\left(-\frac{4}{10}\log_2\frac{4}{10} - \frac{6}{10}\log_2\frac{6}{10}\right)$$
$$+ \frac{4}{20}\left(-\frac{4}{4}\log_2\frac{4}{4} - 0\right) = 0.76$$

$$Gain(A_5) = I(U) - E(A_5) = 0.26$$

After normalizing $Gain(A_i), i = 1, \ldots, 5$, we got the objective significance of each condition attribute, shown in Table 2.

**Table 2.** Objective significance

| $A_4$ | $A_5$ | $A_3$ | $A_1$ | $A_2$ |
|---|---|---|---|---|
| 1 | 0.34 | 0.04 | 0.015 | 0 |

2. Calculate the subjective significance:

Experts in this field scored the significance of each condition attribute, forming their subjective significance, suppose the scores of condition attributes are shown in Table 3.

**Table 3.** Subjective significance

| $A_4$ | $A_5$ | $A_3$ | $A_1$ | $A_2$ |
|---|---|---|---|---|
| 1 | 0.4 | 0.1 | 0.05 | 0.01 |

3. Calculate the total significance:

Suppose the weight of the objective significance is 0.5, then the total significance of each condition attribute can be obtained, shown in Table 4.

**Table 4.** Total significance

| $A_4$ | $A_5$ | $A_3$ | $A_1$ | $A_2$ |
|---|---|---|---|---|
| 1 | 0.37 | 0.07 | 0.03 | 0.005 |

Finally, based on the total significance of each condition attribute, we get the ranking result of the significance of condition attributes:

$$A_4 \succ A_5 \succ A_3 \succ A_1 \succ A_2$$

## 6  Conclusions

Based on the knowledge acquisition of rough set theory, we built the emergency decision table. Information gain, as a measure of the significance of attributes, was introduced in this paper. We combined the information gain with the prior knowledge of experts to determine the significance of emergency attributes. This method takes the objectively historical data into consideration and reflects the prior knowledge of experts in this field. It achieves the unification of objective information gain values and subjective prior knowledge, which avoids ignoring the practical meaning of emergency attributes. It is helpful to provide decision support for decision-makers in the field of emergency management.

**Acknowledgments.** This work was supported by the National Natural Science Foundation of China under Grant 71373034.

## References

1. Xu, X., Du, Z., Chen, X.: Consensus model for multi-criteria large-group emergency decision making considering non-cooperative behaviors and minority opinions. Decis. Support Syst. **79**, 150–160 (2015)
2. Pawlak, Z., Skowron, A.: Rough sets: some extensions. Inf. Sci. **177**, 28–40 (2007)
3. Pawlak, Z., Wong, S.K.M., Ziarko, W.: Rough sets: probabilistic versus deterministic approach. Int. J. Man Mach. Stud. **29**, 81–95 (1988)

4. Wang, J., Miao, S., Zhou, Y.: Review on rough set and its application. Pattern Recog. Artif. Intell. **9**, 341–356 (1996)
5. Gao, T., Du, J., Wang, S.: Tourism emergency attribute reduction based on rough set. J. SE. Univ. (Natural Science Edition), 163–167 (2009)
6. Kim, K., Yamashita, E.Y., Pant, P.: Hit-and-run crashes: use of rough set analysis with logistic regression to capture critical attributes and determinants. Transp. Res. Rec. J. Transp. Res. Board **2083**, 114–121 (2008)
7. Cao, X., Liang, J.: The method of ascertaining attribute weight based on rough sets theory. Chin. J. Manage. Sci. **10**, 98–100 (2002)
8. Bao, X., Zhang, J., Liu, C.: A new method of ascertaining attribute weight based on rough sets conditional information entropy. Chin. J. Manage. Sci. **17**, 131–135 (2009)
9. Quinlan, J.R.: Induction of decision tree. Mach. Learn. **1**, 81–106 (1986)
10. Han, J., Kamber, M.: Data mining: concepts and techniques. Data Min. Concepts Models Methods Algorithms Second Ed. **5**, 1–18 (2000)

# Predicting Hashtag Popularity of Social Emergency by a Robust Feature Extraction Method

Qianqian Li[1] and Ying Li[2(⊠)]

[1] Institutes of Science and Development, Chinese Academy of Sciences,
Beijing, China
[2] College of Computer Science and Technology, Jilin University,
Changchun, China
liying@jlu.edu.cn

**Abstract.** Social emergency information is usually disseminated and driven by a hot topic described succinctly with a hashtag in social media. In China, hashtag prediction for social emergencies is more and more practical for E-governance. How to predict the hashtag popularity for social emergency has become a considerably important task. However, previous research mainly focused on commercial hashtag prediction, such as marketing and promotion. For the hashtag popularity prediction, the core issue is to identify the key features for improving prediction accuracy. To the best of our knowledge, there is few research focus on the feature extraction of hashtag for social emergency. In addition, we extract features for hashtag popularity prediction from "seed information" by avoiding excessive crawling. The "seed information" are the microblogs under a hashtag for a 24-h period since the hashtag was published. Based on the "seed information", the user-based and content-based features are derived, which facilitate the spread of social emergency information. Furthermore, recursive feature elimination (RFE) analysis and nine machine learning classification models are integrated to determine the optimal features among all possible feature combinations. The effectiveness and robustness of our proposed features are verified.

**Keywords:** Social emergency · Hashtag · Information diffusion · Prediction

## 1 Introduction

Recent research in the social media field has demonstrated its utility for spreading information and this is especially true with regard to social emergencies. Hashtag popularity prediction has attracted increasing attention in recent years. Early research on predicting the hashtag was mainly based on historical data [8,11,12]. Due to the lack of historical data for a new hashtag, researchers turned to hashtag popularity prediction for newly emerging hashtags [13]. In order to

J. Chen et al. (Eds.): KSS 2017, CCIS 780, pp. 136–149, 2017.
https://doi.org/10.1007/978-981-10-6989-5_12

improve the prediction performance, new features, such as the hashtag spreading network, were introduced, which requires excessive external crawling. In addition, hashtag popularity prediction is more applied to commercial field, such as the prediction of video popularity [4, 14] and photo sharing [3].

With Sina Weibo becoming increasingly popular in China, social emergency information can be disseminated and driven by a hot topic described succinctly with a hashtag. The dissemination of public opinion in emergencies mainly takes the form of micro topics in the social media, where the micro topic appears in the form of "# topic keyword #". Commonly, after an event occurs, a registered user initiates a post on the subject of the event and other users discuss the topic and publish related information. The topics have a "chat room" function in the social media for the proliferation of public opinion regarding a sudden event to achieve information aggregation and user interaction. Hashtag popularity prediction is an important part of predicting the dissemination scale of public opinion, which is based on understanding the initial state of the public opinion in a sudden event to predict the potential development trends of future public opinion. For the relevant decision-makers, prejudging public opinion is a prerequisite for the early warning and judgment of public opinion, which enhances the scientific level of social governance. Therefore, it is of great practical significance to predict the dissemination scale of public opinion on the Internet for social emergency.

In addition, recent research on social networks has showed that "seed users" tend to be more centrally located in the network, which means that "seed users" play an important role for spreading information [5]. In this paper, we incorporate the features of the "seed users", who are early participants in the hashtag diffusion process. Due to the high complexity of crawling data and constructing social networks, for simplicity, the key features based on "seed information" are devised, which achieves a satisfactory performance. Distinct from a previous research focus on commercial campaigns, such as marketing, advertising on Twitter, YouTube, Flicker, etc., we focus on the hashtag popularity prediction for social emergencies in the Chinese microblogging platform Sina Weibo. The novel hashtag features for social emergencies from user-based and contentbased perspectives are constructed. RFE and multiple machine learning models are integrated to obtain the critical features with minimum number and optimal performance. All possible feature combinations for various prediction models are verified.

The most valuable contributions of this paper can be summarized as follows:

(1) The most influential factors for hashtag popularity prediction were considered, which reduce the computation costs of crawling data and processing data. The informationoriginating from the earliest published user, seed diffusion users, and early microblogs posted by the initial participants are used to construct features.
(2) Most research on social hashtag popularity has focused on commercial hashtags, video, photo, and music sharing. Our research aims at improving the prediction effectiveness and accuracy of the hashtag popularity for social emergencies in China. The prediction model combines user-based (including

the first user and the seed users) and content-based (semantic and sentiment) information for influencing the hashtag dissemination for social emergencies.

(3) The proposed features achieve relatively high accuracy and are robust for different machine learning methods. We verify all possible feature combinations for various prediction models. The results indicate a strong robustness of the features.

This paper first presents features for social emergency hashtag popularity prediction. Second, data collection and preprocessing processes are described. Third, the verification of effectiveness and the robustness of features are presented. The research framework is detailed in Fig. 1.

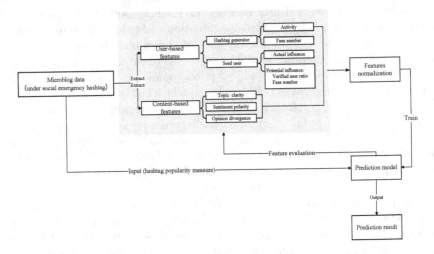

**Fig. 1.** Framework for social emergency hashtag prediction

## 2   Related Work

Since the emergence of online social media, a great deal of research has investigated hashtag diffusion, which can be used in promotions, advertisement, policy-making, etc. Most of the related research has investigated two aspects of hashtags: the temporal patterns of hashtag popularity and the prediction of hashtag popularity.

### 2.1   Temporal Patterns of Hashtag Popularity

Several research studies have focused on the temporal patterns of hashtag popularity. The discovery of these patterns is usually focused on pattern research for the spread of content-based information. There are two types of hashtag temporal prediction, one is the determination of temporal shape patterns [17] and the other is based on certain features of hashtag evolution and does not extract explicit shape patterns. [2] distinguished internal and external influences by classifying the information bursts into exogenous subcritical, exogenous critical, endogenous critical, and endogenous subcritical. [10] focused on peaks in the popularity of hashtags and defined three classes for the evolution of hashtagpopularity, which included continuous activity, periodic activity, and activity concentrated around an isolated peak.

## 2.2   Prediction of Hashtag Popularity

The prediction of hashtag popularity can be divided into two types. One focuses on the detection of bursts. For example, researchers have determined whether the hashtag popularity will burst or not, and when the popularity will burst [9,16]. The other type of prediction focuses on the popularity volume. [15] predicted the number of tweets annotated by hashtags on a weekly basis. The features were extracted mainly from the hashtag itself (e.g., characterlength, numberofwords, orthography in a hashtag). [13] proposed content features from hashtag strings and contextual features from a social graph formed by users who have adopted the hashtag in Twitter and predicted the number of users who adopted the hashtag on a daily basis. [7] combined temporal factors and user comments to construct a time-aware bipartite graph and proposed a regularizationbased ranking algorithm to predict the popularity of videos, photos, and music.

The existing research has not fully considered the characteristics of the initially posting microblogs under a hashtag, e.g., dissemination influence, content guidance, etc. This paper categorizesthe topic participants into three categories: source user, seed users, and general users. In the early stage of the topic formation, the source user and the seed users mainly influence the dissemination of the topic. This paper focuses on the characteristics of these two groups as well as topic distribution, emotional tendency, and viewpoint difference for forecasting the scale of dissemination of the topic. The topic's popularity is measured by the number of microblogs for each hashtag of the topic for social emergencies and the prediction of the topic's popularity prediction is transformed into a classification problem. The details will be described in the following section.

# 3   Feature Extraction for Social Emergency Hashtags

## 3.1   User-Based Features

The user characteristics are divided into source-user and seed-user characteristics. The source user refers to the topic's initiator (i.e., topic host), who is the first one publish the hashtag. The seed users are users initially involved in the topic; the information they publish is in the form of "# topic keyword #" and subsequently, the followers of the seed users will participate in the discussion of the topic for the first time; this information cascades to form a public opinion.

Considering the differences between source user and seed users with regard to topic dissemination, the following characteristics are extracted:

– Source user

  **Source user activity:** opinion leaders influence others; therefore, they must have a higher degree of participation in the social media; an important measure is the average number of daily microblogs, referring to a certain period of time $(t, t + \Delta t)$ and the number of microblogs the user publishes per unit time (day).

$$x_1 = \frac{N_{source_u}(t)}{\Delta t} \tag{1}$$

  where $N_{source_u}(t)$ is the total amount of information published by the host in a given time.

  **Source user influence:** the user influence represents the influence a user has on other users; The number of followers of the user $(x_2)$ is an intuitive and commonly-used index.

– Seed users

  Seed users are the first users to reach the hashtag. They are important participants to promote the generation of the public opinion and they are the first-level participants that cause the dissemination of the information cascade. The measurement of the seed users is divided into two parts: **potential influence** and **actual influence**.

  The **potential influence** includes the percentage of the $V$ (certified user) users $(x_3)$ in seed users and the total number of followers of the seed users $(x_4)$, expressed as:

$$x_3 = \frac{N_{V\_seed}}{N_{seed}} \tag{2}$$

$$x_4 = \sum_{i=1}^{N_{seed}} u_{i\_followers} \tag{3}$$

  $N_{seed}$ is the total amount of seed users, $N_{V\_seed}$ is the number of users who are $V$ users in the seed users, including individually certificated orange $V$ users and officially certificated blue $V$ users. $u_{i\_followers}$ is the number of followers of the ithseed user.

  The **actual influence** of the seed user is the popularity measure of the information that is published on that topic. Assuming that the seed user issues $M$ messages under a topic, the popularity of each message is:

$$I_a(m_i) = w_1 * \frac{v_r}{\overline{v_r}} + w_2 * \frac{v_c}{\overline{v_c}} + w_3 * \frac{v_l}{\overline{v_l}} \tag{4}$$

  $v_r$, $v_c$, and $v_l$ are the amount of the corresponding forwarded microblogs, the amount of comments, and the amount of likes respectively. $\overline{v_r}$, $\overline{v_c}$, and $\overline{v_l}$ are the average amount of the forwarded tweets of all messages by the seed user, the average amount of comments, and the average amount of likes; $w_1$, $w_2$, and $w_3$ are the weights of the three dimensions, where $w_1 + w_2 + w_3 = 1$. Thus, the actual influence of $M$ messages tweeted by the seed user is:

$$x_5 = \sum_{i=1}^{M} I_a(m_i) \tag{5}$$

## 3.2   Content-Based Features

Based on the information published by the seed users, this paper extracts the topic clarity, sentiment tendency, and opinion divergence features.

– Topic clarity

The topic's distribution represents whether emerging topics of public opinion information are concentrated or scattered. Usually, there is a clear discussion objective in social emergency topics. Once a social emergency occurs, the public will offer opinions on the emergency. Focused comments on the topic indicate a clear theme, which will facilitate the hashtag dissemination. By using statistical topic models, the text of the topic number is quantified into a low-dimensional representation by parameter estimation. Currently, Latent Dirichlet Allocation (LDA) is a commonly used topic model [1] and is an unsupervised machine learning method. Assuming that every word in each microblog is to choose a topic at a certain probability, LDA uses a bag method (bag of words) and takes all the microblogs published by the seed users as a word frequency vector, which is used to generate a probability distribution of the information on certain topics. The document-topic probability matrix is expressed by $p\left(\theta\right)$:

$$
p\left(\theta\right) = \begin{matrix} & \begin{matrix} T_1 & \cdots & T_k \end{matrix} \\ \begin{matrix} m_1 \\ \cdots \\ m_N \end{matrix} & \begin{pmatrix} x_{11} & \cdots & x_{1k} \\ \cdots & \cdots & \cdots \\ x_{11} & \cdots & x_{Nk} \end{pmatrix} \end{matrix} \tag{6}
$$

$T_1, \ldots, T_k$ are the topics; $m_i$ is the $i^{th}$ microblog under the topic; $x_{i,j}$ indicates the probability that the $i^{th}$ microblog will belong to topic $j$, where $\sum_{j=1}^{k} x_{i,j} = 1$. Given the threshold $\xi = 0.5$, when $\exists x_{i,j} \in \{x_{i,j}\}_{j=1,\ldots,k}$ : $x_{i,j} \geq \xi$, then the topic of the microblog $i$ is more concentrated. Otherwise, the topic of the microblog is considered to be scattered.

The percentage of the concentrated topics in all the seed information is considered as topic clarity measurement:

$$
f\left(X_i\right) = \begin{cases} 1 & \exists x_{ij} \geq \xi, j = 1, \ldots, k \\ 0 & else \end{cases}
$$

$$
x_6 = \frac{\sum_{i=1}^{N} f(X_i)}{N} \tag{7}
$$

– Sentiment polarity

This article constructs the user sentiment words dictionary for social emergency, and provide the weight of the sentiment words. The weight is range from 1 to 9, the value 1 indicates the most negative sentiment; the value 9 indicates the most positive sentiment. By using ICTCLAS [18], all the microblogs can be appropriately segmented, neglecting any numbers, auxiliary words, stop words, or symbols. For all information published by the seed user, if $N$ sentimental words are included, the average sentimental score is calculated by:

$$x_7 = \frac{\sum_{i=1}^{N} h_{\omega_i} * f_i}{\sum_{i=1}^{N} f_i} \qquad (8)$$

$f_i$ is the frequency of the $i^{th}$ sentiment word that appears in the seed information; $\omega_i$ is the $i^{th}$ sentiment word, and $h_{\omega_i}$ is the word's weight. The value of the sentiment score is set to $[1, 9]$; when $x_7 \in [1, 4)$, the sentiment for the topic is negative; when $x_7 \in [4, 6]$, the sentiment for the topic is neutral; when $x_7 \in (6, 9]$, the sentiment for the topic is positive (Fig. 2).

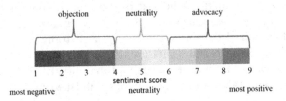

**Fig. 2.** Sentiment score category

- Opinion divergence
  The above sentiment analysis calculates the average of all the emotional sentiments for the seed information. Sometimes online users might not read all the seed information before publishing opinion. Therefore, this paper calculates the difference between the numbers of microblogs of positive and negative sentiment and the difference is expressed by the following equation:

$$x_8 = \frac{N_+ - N_-}{N} \qquad (9)$$

where $x_8$ is in a range of $[-1, 1]$. $N_+$ is the number of microblogs with positive sentiment; $N_-$ is the number of microblogs with negative sentiment. When the number of microblogs that have positive sentiment are higher than the numbers with negative sentiment $(N_+ > N_-)$, $0 < x_8 < 1$; when the number of microblogs with negative sentiment of the seed users are higher than those with positive sentiment $((N_- > N_+)$, $-1 < x_8 < 0$.

## 4    Data Collection

The dataset is collected from Sina Weibo, the biggest social media platform of China. Based on the "Hotspot public opinion event list" (2016.5.1–2016.9.31) published daily by Sina Micro Public Opinions, public events related to social emergencies are selected. Using the "interest home page" function of the Sina Weibo search page, the related hashtag for social emergencies can be found (Table 1). Microblog data and user data were crawled under each social emergency hashtag. We collected data for the first 24-h period since the hashtag was

**Table 1.** Selected social emergency hashtags from Sina Weibo

| No. | Hashtag in Chinese | Hashtag Translation |
|---|---|---|
| 1 | # 北京房山地震 # | Beijing Fangshan Earthquake |
| 2 | # 常州中学污染 # | Changzhou Middle School Pollution |
| 3 | # 陈仲伟医生一路走好 # | Death of the dentist Chen Zhongwei |
| 4 | # 防抗台风莫兰蒂 # | Super Typhoon Meranti |
| 5 | # 福建暴雨 # | Fujian Rainstorm |
| 6 | # 福建泰宁山体滑坡 # | Landslide in Fujian's Taining Country |
| 7 | # 关注强降雨 # | Heavy Rainfall Events |
| 8 | # 广西梧州 5.4 级地震 # | AMagnitude-5.4 Earthquake in Guangxi's Cangwu County |
| 9 | # 广州暴雨 # | Rainstorm in Guangzhou |
| 10 | # 贵州大方山体滑坡 # | Landslide in Guizhou's Dafang County |
| 11 | # 哈尔滨冰雹 # | Heavy Rain and Hail Blasted Harbin |
| 12 | # 河南雎县校车货车相撞 # | School Bus-Truck Collision in Henan's Suixian Country |
| 13 | # 湖南暴雨 # | Heavy Rainfall in Hunan |
| 14 | # 湖南高速大巴起火 # | Bus Caught Fire in Hunan's Yizhang County |
| 15 | # 济聊高速公路车祸 # | Bus-Truck Crash in Jinan-Liaocheng Expressway |
| 16 | # 江苏交警失联后死亡 # | Mysterious Death of Police Officer Zhang |
| 17 | # 金波去世 # | Sudden Death of Jin Bo, the Deputy Editor-in-Chief ofChina's Leading Online Forum Tianya |
| 18 | # 金山水上飞机撞桥 # | Fatal Seaplane Crash |
| 19 | # 老虎袭人致 1 死 1 伤 # | A Woman Attacked by a Tiger at Badaling Wildlife World in Beijing |
| 20 | # 丽水山体滑坡 # | Landslide in Zhejiang's Lishui Country |
| 21 | # 南昌暴雨 # | Rainstorm in Nanchang |
| 22 | # 南京暴雨 # | Rainstorm in Nanning |
| 23 | # 女大学生深夜面试失联 # | A Female College Student Becomes Lost following an Interview |
| 24 | # 女孩被骗光学费离世 # | A Girl Died after Being Cheated on Tuition |
| 25 | # 女孩两次跳河自杀 # | A Girl Jumped into a River Twice to commit Suicide |
| 26 | # 屁股被警棍打开花 # | Abuse of Police Power |
| 27 | # 青海门源地震 # | Earthquake in Qinghai |
| 28 | # 山东非法疫苗 # | Illegal Vaccine |
| 29 | # 上海浦东机场爆炸 # | Shanghai Airport Bombing |
| 30 | # 上海中环线被压断 # | An Overloaded Truck Overturns on the Middle Ring Road in Shanghai |
| 31 | # 邵东医生被大锤砸身亡 # | A Doctor is Killed by a Hammer |
| 32 | # 审讯室内戴头盔死亡 # | A Man Died in an Interrogation Room |
| 33 | # 四川绵阳地震 # | Earthquake in Sichuan's Mianyang Country |
| 34 | # 四川强降雨 # | Heavy Rainfall in Sichuan |
| 35 | # 台风妮妲来了 # | Typhoon Nida |
| 36 | # 魏则西百度推广事件 # | Death of Wei Zexi |
| 37 | # 武汉开启看海模式 # | Heavy Rainstorm in Wuhan |
| 38 | # 西安南郊电厂爆炸 # | Power Plant Explosion in Xi'an |
| 39 | # 新乡特大暴雨 # | Extraordinary Rainstorm in Xinxiang |
| 40 | # 邢台暴雨 # | Rainstorm in Hebei's Xingtai Country |
| 41 | # 盐城遭龙卷风袭击 # | Tornado in Jiangsu's Yancheng Country |
| 42 | # 央视曝毒跑道黑手 # | Toxic Track |
| 43 | # 幼儿园霉大米事件 # | Mildew Rice in Kindergarten |
| 44 | # 云南个旧连续地震 # | Earthquake in Yunnan's Gejiu Country |
| 45 | # 浙江温州楼房倒塌 # | Building Collapses in Zhejiang's Wenzhou Country |
| 46 | # 追砸运钞车被打死 # | A Man was shot after Hita Cash-in-Transit |

published. As we described in the introduction, this "seed information" plays an important role in accelerating the spread of information.

The final number of all microblogs under each hashtag is taken as the output. In other words, given a new hashtag, our task is to predict the number of microblogs containing the hashtag. It is worth noting that predicting the exact value of total microblogs containing the hashtag is difficult and unnecessary. As a result, we convert the problem into a classification problem. We define three ranges with different sizes: $[0, \phi]$, $[\phi, 10\phi]$ and $[10\phi, +\infty]$. The $\phi$ is a controllable parameter, which controls the size of the defined ranges. With the three defined ranges, the problem can be formulated as a classification problem. We refer to these as being marginally popular, popular, and very popular. Note that we focus on social emergencies. In addition, according to a judicial interpretation given by the Supreme People's Court and the Supreme People's Procuratorate of China, a crime occurs if online falsehoods are retweeted more than 500 times. Although the 500 times definition is for retweeting behavior, we consider it is a meaningful threshold. As a result, in this paper, we define $\phi = 500$.

# 5  Experiments

## 5.1  Classification Evaluation

The prediction methods were evaluated by determining the accuracies based on a confusion matrix (Table 2). In this paper, the confusion matrix consists of nine cases. The accuracy can be calculated as follows:

$$\text{Accuracy} = \frac{f_{11} + f_{22} + f_{33}}{f_{11} + f_{12} + f_{13} + f_{21} + f_{22} + f_{23} + f_{31} + f_{32} + f_{33}} \tag{10}$$

The accuracies of the prediction methods were calculated by performing a 10-fold cross validation.

**Table 2.** Confusion matrix

|  |  | Predicted Value | | | Accuracy |
|---|---|---|---|---|---|
|  |  | Class 1 | Class 2 | Class 3 |  |
| Actual Value | Class 1 | $f_{11}$ | $f_{12}$ | $f_{13}$ | $\dfrac{f_{11}}{f_{11} + f_{12} + f_{13}}$ |
|  | Class 2 | $f_{21}$ | $f_{22}$ | $f_{23}$ | $\dfrac{f_{22}}{f_{21} + f_{22} + f_{23}}$ |
|  | Class 3 | $f_{31}$ | $f_{32}$ | $f_{33}$ | $\dfrac{f_{33}}{f_{31} + f_{32} + f_{33}}$ |

The major advantage of $k$-fold cross-validation is its suitability for small datasets. The dataset is split into $k$ sets. For each of the $k$ "folds", the following procedure was followed:

(1) The prediction model was trained by using k−1 of the folds as the training dataset;
(2) The prediction accuracy was calculated on the test dataset, which was the remainder of the dataset. The prediction model accuracy reported by kfold cross-validation is the average of the accuracy values calculated for each iteration.

## 5.2  Feature Evaluation

– Method 1: RFE.
 In order to verify the effectiveness of our proposed features, we apply a feature evaluation, which is a necessary process in data mining for avoiding overfitting. The goal of the feature evaluation is to find a subset of features that can

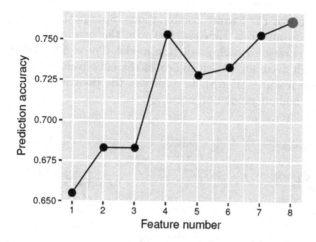

**Fig. 3.** RFE feature selection result

result in an accurate model. Recursive feature elimination (RFE), a powerful feature selection algorithm, is applied in this paper. RFE implements a backward selection of the features based on feature importance ranking. According to the ranking order, the less important features are sequentially eliminated prior to modeling. RFE is a greedy optimization algorithm for selecting the bestperforming subset of a number of features [6].

In the RFE calculation, a specific model must be specified in the rfe control function, which controls the feature selection algorithm. In this paper, a random forest was used in the rfe control function. The optimum number of features is determined by the highest classification accuracy. Based on the RFE result, eight features were selected (Fig. 3).

– Method 2: prediction on all combinations of features.

The RFE method is based on the idea of repeatedly constructing a prediction model. As a result, the RFE heavily depends on the model used for the feature selection. In order to eliminate the bias of the model selection, we selected nine machine learning prediction methods: Bagging, Boosting, Classification and Regression Tree (CART), Flexible Discriminant Analysis (FDA), Logistic Regression (LR), Multivariate Adaptive Regression Spline (MARS), Multinomial Regression (MR), Neural Network (NN), and Support Vector Machine (SVM). In this study, the caret package of the R program was used. The above machine learning methods is implemented in caret package. In addition, we tested all possible combinations of the eight features. That is to say, for the feature number equal to 1, we ran 9 prediction methods on $C_8^1 = 8$ kinds of features; for the feature number equal to 2, we ran 9 prediction methods on $C_8^2 = 28$ kinds of feature combinations, etc. In total, the nine prediction methods were applied $C_8^1 + C_8^2 + C_8^3 + C_8^4 + C_8^5 + C_8^6 + C_8^7 + C_8^8 = 8 + 28 + 56 + 70 + 56 + 28 + 8 + 1 = 255$ kinds of feature combinations. In general, Fig. 4 shows that the prediction accuracy improves steadily with an increasing number of features. The number labels in the graph are the average values for the feature combinations. Specifically, when we used the nine prediction methods for eight features, the prediction results achieved maximum accuracy for each prediction model. The eight features achieved a relatively high accuracy, which indicates the effectiveness and robustness of our proposed features (Fig. 5). In addition, the boosting prediction method achieved the highest prediction accuracy.

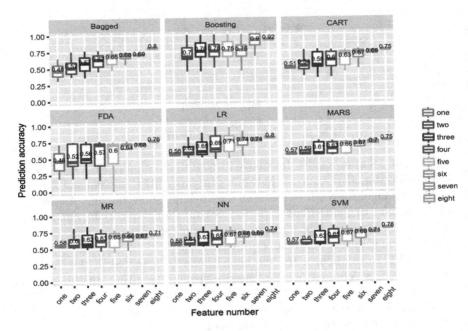

**Fig. 4.** Boxplotof prediction results sorted by feature number

Note: Lines inside the boxes indicate the median prediction accuracy, whereas the boxes themselves show the interquartile ranges. The numbers in the boxes indicate the average values of the prediction accuracy.

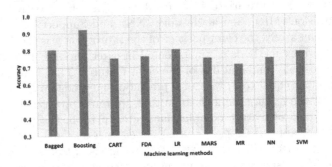

**Fig. 5.** Predictionresult for the selected eight features

# 6   Discussion

In this paper, we proposed effective and robust features to predict the popularity of social emergency hashtags in a Chinese microblogging platform. The features were extracted from two perspectives: user-based and content-based. Different from previously commercial hashtag feature extractions, we focus on the factors that drive the spread of information for a social emergency. Therefore, although the perspectives were not new, the specific innovation in the calculation can be manifested in three ways: (1) "seed users" have a key role in spreading the initial information and we identify their actual and potential influences. (2) We use a LDA model to identify the representation of the topic in each microblog and we present a new way to measure the topic's clarity by extracting the features from a document-topic vector. Topic clarity was first calculated by using the Kullback-Leibler divergence [13]. However, this method relies mostly on historical information and, in addition, the authors were focused on the prediction of the number of users who adopted the hashtag. (3) Based on the sentiment polarity, the opinion divergence is calculated. The information that is exposed to a new user can affect his/her attitude. The opinion divergence of the seed information can impact the extent to which the new user accepts the positive or negative impact.

Our study can benefit social management by providing the ability to predict the opularity of a social emergency hashtag in a timely manner. When a social emergency occurs, a new hashtag will be published and the spread of information will increase. Decision-makers can predict whether the discussion of the social emergency will evolve into a big public opinion event. In addition, our method does not require crawling complex data and can be applied easily.

# 7   Conclusions

This study investigated the hashtag popularity prediction for social emergenciesin Sina Weibo, which is the biggest microblogging platform in China. Information regarding social emergencies is mostly spread in the form of hashtags in China. We propose features related to accelerating the dissemination of information regarding social emergencies and we formulate the problem as a classification task. The main focus of our research was to identify the effective features and verify the robustness of the features while minimizing the costs for crawling data. We crawled the informationregardinga social emergency topic for a 24-h period after the hashtag was published. We assumed that the information published by the seed users played an important role in spreading information on social emergencies. We proposed user-based features and content-based features extracted from the data. We conducted the experiments using nine classification methods, evaluated the features using an RFE method, and ran the nine machine learning methods based on the number of features. Our experiments demonstrated that the eight proposed features achieved the best prediction accuracy. We also demonstrated the robustness of our proposed features and a good prediction accuracy for the nine classification methods.

148    Q. Li and Y. Li

**Acknowledgements.** This work is supported by the National Natural Science Foundation of China (Grant No. 71403262, 71774154, 71573247, 71503246).

# References

1. Blei, D.M., Ng, A.Y., Jordan, M.I.: Latent Dirichlet allocation. J. Mach. Learn. Res. **3**, 993–1022 (2003)
2. Crane, R., Sornette, D.: Robust dynamic classes revealed by measuring the response function of a social system. Proc. Natl. Acad. Sci. **105**(41), 15649–15653 (2008)
3. Ferrara, E., Interdonato, R., Tagarelli, A.: Online popularity and topical interests through the lens of instagram. In: ACM Conference on Hypertext and Social Media. pp. 24–34 (2014)
4. Figueiredo, F., Benevenuto, F., Almeida, J.M., Fabr, F.F., Almeida, B.J.M.: The tube over time: characterizing popularity growth of youtube videos. In: Proceedings of the 4th ACM International Conference on Web Search and Data Mining. pp. 745–754 (2011)
5. González-Bailón, S., Borge-Holthoefer, J., Rivero, A., Moreno, Y.: The dynamics of protest recruitment through an online network. Sci. R. **1**, 1–7 (2011)
6. Guyon, I., Weston, J., Barnhill, S., Vapnik, V.: Gene selection for cancer classification using support vector machines. Mach. Learn. **46**(1–3), 389–422 (2002)
7. He, X., Gao, M., Kan, M.Y., Liu, Y., Sugiyama, K.: Predicting the popularity of web 2.0 items based on user comments. In: The International ACM SIGIR Conference on Research and Development in Information Retrieval. pp. 233–242 (2014)
8. Jeon, J., Croft, W.B., Lee, J.H., Park, S.: A framework to predict the quality of answers with non-textual features. In: International ACM SIGIR Conference on Research and Development in Information Retrieval. pp. 228–235 (2006)
9. Kong, S., Mei, Q., Feng, L., Ye, F., Zhao, Z.: Predicting bursts and popularity of hashtags in real-time. In: Proceedings of the 37th International ACM SIGIR Conference on Research and Development in Information Retrieval. pp. 927–930 (2014)
10. Lehmann, J., Gonçalves, B., Cattuto, C., Ramasco, J.J., Cattuto, C.: Dynamical classes of collective attention in Twitter. In: Proceedings of the 21st International Conference on World Wide Web. pp. 251–260. WWW 2012, NY, USA. ACM, New York (2012)
11. Liu, Q., Agichtein, E., Dror, G., Gabrilovich, E., Maarek, Y., Dan, P., Szpektor, I.: Predicting web searcher satisfaction with existing community-based answers. In: Proceeding of the International ACM SIGIR Conference on Research and Development in Information Retrieval, SIGIR 2011, Beijing, China. pp. 415–424 July 2011
12. Liu, Y., Huang, X., An, A., Yu, X.: ARSA: a sentiment-aware model for predicting sales performance using blogs. In: International ACM SIGIR Conference on Research and Development in Information Retrieval. pp. 607–614 (2007)
13. Ma, Z., Sun, A., Cong, G.: On predicting the popularity of newly emerging hashtags in Twitter. J. Am. Soc. Inf. Sci. Technol. **64**(7), 1399–1410 (2013)
14. Pinto, H., Almeida, J.M., Gonçalves, M.A.: Using early view patterns to predict the popularity of Youtube videos. In: Proceedings of the 6th ACM International Conference on Web Search and Data Mining, New York, USA. pp. 365–374. ACM, New York February 2013

15. Tsur, O., Rappoport, A.: What's in a hashtag?: content based prediction of the spread of ideas in microblogging communities. In: Proceedings of the 5th ACM International Conference on Web Search and Data Mining. pp. 643–652. ACM (2012)
16. Wang, S., Yan, Z., Hu, X., Yu, P.S., Li, Z.: Burst time prediction in cascades. In: 29th AAAI Conference on Artificial Intelligence. pp. 325–331 (2015)
17. Yang, J.S.U., Leskovec, J.S.U.: Patterns of temporal variation in online media categories and subject descriptors. In: ACM International Conference on Web Search and Data Minig. pp. 1–13 (2011)
18. Zhang, H.P., Yu, H.K., Xiong, D.Y., Liu, Q.: HHMM-based Chinese lexical analyzer ICTCLAS. In: Proceedings of the 2nd SIGHAN Workshop on Chinese Language Processing. pp. 184–187. Association for Computational Linguistics (2003)

# Mining Online Customer Reviews for Products Aspect-Based Ranking

Chonghui Guo[1,2(✉)], Zhonglian Du[1], and Xinyue Kou[1]

[1] Institute of Systems Engineering, Dalian University of Technology,
Dalian, China
dlutguo@dlut.edu.cn, duzhonglian@mail.dlut.edu.cn
[2] State Key Laboratory of Software Architecture (Neusoft Corporation),
Shenyang, China

**Abstract.** Massive online reviews contain a lot of useful information that can not only provide purchasing decision support for consumers, but also allow producers and suppliers to understand the competitive market. This paper proposes a new aspect-based online reviews mining method, which combines both textual data and numerical data. Firstly, the probability distribution of topics and words is constructed by LDA topic model. With word cloud images, the keywords are visualized and corresponding relationship between LDA topics and product reviews is analyzed. The weight of each aspect is calculated based on the probability distribution of documents and topics. Then, the dictionary-based approach is used to calculate the objective sentiment values of the product. The subjective sentiment tendency from different consumers because of their different individual needs are also taken into consideration. Finally, the directed graph model is constructed and the importance of each node is calculated by improved PageRank algorithm. The experimental results illustrate the feasibility of proposed mining method, which not only makes full use of massive online reviews, but also considers individual needs of consumers. It provides a new research idea for online customer review mining and personalized recommendation.

**Keywords:** Online review mining · LDA topic model · Improved PageRank algorithm · Personalized recommendation

## 1 Introduction

With the rapid development of web 2.0, the role of Internet users has changed greatly from previous information recipients to information producers. Specifically, the development of online shopping platforms has attracted more and more consumers to share their shopping experience and product feedback on the Internet. This information is presented in forms of semi-structured or unstructured data, such as text, numbers, images, etc., providing more objective references for other potential consumers. A survey found that 63% of online shoppers consistently search and read reviews before making a purchasing decision. Among them, 64% spend 10 min reading reviews, whereas 33% spend 30 min or more. Although many scholars have adopted content-based recommendation, collaborative filtering and knowledge-based recommendation methods for

© Springer Nature Singapore Pte Ltd. 2017
J. Chen et al. (Eds.): KSS 2017, CCIS 780, pp. 150–161, 2017.
https://doi.org/10.1007/978-981-10-6989-5_13

user recommendation research, there are relatively few user recommendations based on online review information [1, 2]. At the very beginning, scholars used numerical scoring information from online reviews to sort products. Mary et al. selected 8 million product reviews and 1.5 million merchant reviews to evaluate quality and rank different products. She found that the average value has a better performance [3]. However, due to fuzziness of the numerical scores, it can't well reflect users' sentiment expression of products' unique characteristics. That is, products with same scores can't be distinguished from each other [4]. As valuable user-generated contents, review texts can better reflect users' emotional expression. Khan et al. proposed a new text sentiment analysis method (eSAP) for decision support systems and carried out experiments with data sets from seven different fields to verify the effectiveness of the method [5]. In order to solve the problem of unbalanced sentiment classification in Chinese product review, Tian et al. proposed a new method based on topic sentence to improve the accuracy of classification [6].

However, text reviews can't cover comprehensive information, most of which are written only based on a certain aspect or some aspects of products. Recently, models which integrate numerical scores and review texts have attracted a lot of attention [7, 8]. From the aspect of product evaluation method, the fusion of different data is pretty simple and the weight is mostly determined by subjective decision or weighted average method. Yang et al. proposed a method based on multi-source heterogeneous information. However, only online reviews are taken into consideration and subjective opinions of consumers are neglected while calculating weights. Zhang et al. proposed a method to calculate weights based on the validity of comments and importance of time, of which the former one is decided by the votes from other consumers and the latter one is decided by the time this review was published. However, while calculating the overall score of the product, the same weights are given to different aspects which cannot reflect their different importance for consumers [10]. Some other methods, such as intuitionistic fuzzy set theory, TOPSIS are also used to do products ranking [11, 12].

There are different types of online reviews, such as numeric ratings and text descriptions. Such heterogeneous information brings customers more complexity to make purchase decisions. Zhang et al. used the directed graph model to evaluate relative advantages of products, but there was no analysis of node weights in the network [13]. The node represents comprehensive sentiment value of the product. The direction of the edge represents relative advantage between two products. The weight of the side represents the value of relative advantage.

In this paper, a directed graph model is proposed to integrate such rich and heterogeneous information. The modified PageRank method is used to calculate the value of each node and to rank different products. The rest of this paper is organized as follows. In Sect. 2, a method to mine and integrate text and numerical information is proposed. A directed graph model is constructed and the importance of each node is calculated by improved PageRank algorithm. In Sect. 3, an experimental study based on automobile online reviews is taken to illustrate the feasibility of proposed method. Finally, in Sect. 4, contribution of this paper and future work is summarized.

## 2   Research Methods

In order to make full use of user-generated contents for more effective ranking of different products, this paper selects aspect-based online text reviews and data information to do analysis. The directed graph model is used for information fusion, and improved PageRank method is deployed to find the node value, which refers to the final scores of different products. The research framework is shown in Fig. 1.

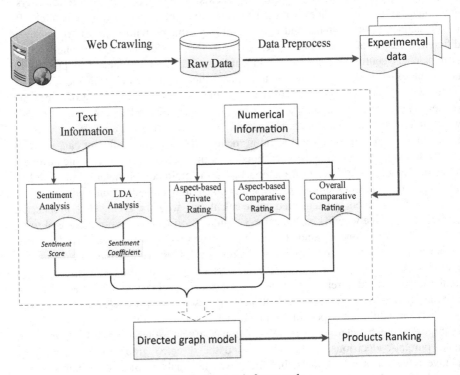

**Fig. 1.**  Research framework

### 2.1   Text Information Processing Method

In this section, we first formulate the problem of aspect-based products ranking. Then we describe the method of sentiment intensity calculation and aspect-based weight determination.

#### 2.1.1   Sentiment Intensity Calculation Method

A large number of online aspect-based reviews are crawled from the third-party website. The set $P = \{P_1, P_2, \ldots, P_i, \ldots, P_I\}$ denotes the set of I alternative products, where $P_i$ is the $i$th alternative product. The original corpus is denoted as $C_0 = \{r_{i1}, r_{i2}, \ldots, r_{ij}, \ldots, r_{IN}\}$, where $r_{ij}$ is the $j$th online review of alternative product $P_i$. Firstly, the original corpus needs to be preprocessed, including word segmentation and stop words removing. The result is a word set denoted as $W_{ij} = \{w_{ij1}, w_{ij2}, \ldots, w_{ijm}, \ldots, w_{IJM}\}$, where $w_{ijm}$ means the $m$th word segmentation result in the $j$th online review of alternative product $P_i$.

In this paper, the sentiment word lexicon established by the information retrieval team of Dalian University of Technology is used as the basis of text analysis [14]. The sentiment word lexicon is denoted by $SW_0 = \{sw_1, sw_2, \ldots, sw_q, \ldots, sw_Q\}$, where $sw_q$ denotes the $q$th sentiment word. The sentiment intensity of sentiment word is denoted by $SV_0 = \{sv_1, sv_2, \ldots, sv_q, \ldots, sv_Q\}$, where $sv_q$ denotes the sentiment intensity of the $q$th sentiment word in the lexicon. Let $W_{ij}^k = \{W_{ij}^1, W_{ij}^2, \ldots, W_{ij}^k, \ldots, W_{ij}^n\}$, where $W_{ij}^k$ denotes the word segmentation result of the $j$th online review concerning the $k$th aspect of product $P_i$. Let $S_{ij}^k = \{S_{ij}^1, S_{ij}^2, \ldots, S_{ij}^k, \ldots, S_{ij}^n\}$, where $S_{ij}^k$ denotes the sentiment intensity of the $k$th aspect in the $j$th online review of alternative product $P_i$. The sentiment intensity calculation process is given below.

Step 1: If the $i$th word of $W_{ij}^k$ appears in $SW_0$ and its corresponding segment word is $sw_i$, then the sentiment intensity and sentiment polarity are denoted as $sv_i$ and $sp_i$ respectively.

Step 2: Take the location of $W_{ij}^{kn}$ as center and two characters as window size to determine whether there is any negative words within this range. If there is not, the modified sentiment intensity is calculated by $sv_i' - sv_i = 0$; otherwise, it is calculated by $sv_i' + sv_i = 0$.

Step 3: Determine the sentiment polarity $sp_i$. If $sp_i = 1$, which means the review is positive, then $S_{ij}^k = S_{ij}^k + sv_i'$; otherwise, $S_{ij}^k = S_{ij}^k - sv_i'$.

In this paper, we assume that weights of reviews from different customers for the $k$th aspect of the $i$th product is equal, that is, $S_{ij}^k$ and $S_{ij+1}^k$ are of no individual differences. Therefore, the sentiment intensity of $k$th aspect about alternative product $P_i$ can be calculated as:

$$S_i^k = \frac{1}{n} \sum_{j=1}^{n} S_{ij}^k \tag{1}$$

### 2.1.2   Aspect-Based Weight Determination Method

In order to obtain the overall sentiment intensity of product $P_i$, we need to calculate the weight of each aspect. LDA (Latent Dirichlet Allocation) is a typical topic model, and it is a generative directed graph model. It is mainly used to deal with discrete data, and has a wide range of applications in information retrieval, natural language processing and other fields.

The relationship of LDA variables is described in Fig. 2, where hollow circles represent hidden variables and solid circles represent observable variables. Only word frequency $w_{t,n}$ of word $n$ in document $t$ is observable, it depends on topic of word $n$ in document $t$ $Z_{t,n}$ and word frequency corresponding to topic $k$ $\beta_k$. Meanwhile, $Z_{t,n}$ depends on the topic distribution $\theta_t$ and $\theta_t$ depends on the parameter $\alpha$ of Dirichlet Allocation, while $\beta_k$ depends on parameter $\eta$. Accordingly, the probability distribution of LDA is given as follows:

$$p(W, z, \beta, \theta | \alpha, \eta) = \prod_{t=1}^{T} p(\theta_t | \alpha) \prod_{i=1}^{K} p(\beta_k | \eta) (\prod_{n=1}^{N} P(w_{t,n} | z_{t,n}, \beta_k) P(z_{t,n} | \theta_t)) \qquad (2)$$

where $p(\theta_t | \alpha) = \dfrac{\Gamma(\sum_k \alpha_k)}{\prod_k \Gamma(\alpha_k)} \prod_k \theta_{t,k}^{\alpha_k - 1}$ and $p(\beta_k | \eta)$ usually obey $K$-dimensional and $N$-dimensional Dirichlet Allocation with parameters $\alpha$ and $\eta$. The aspect-based weight calculation process is given below.

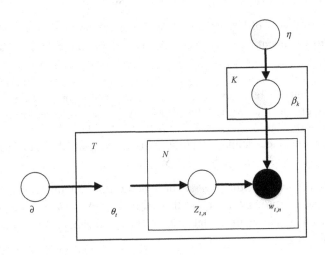

**Fig. 2.** LDA model

Step 1: With prior knowledge, parameters $K$, number of topics, can be determined. The probability matrix of subject and word can be obtained from $\beta_k$, based on which the word cloud for each subject can be generated. Using text data visualization, we can display each subject more intuitively and summarize realistic meaning of each subject with consideration of realistic background;

Step 2: From $\theta_t$, we can know the probability matrix of document and topic, and then determine the proportion $\alpha\theta_t$ of each topic in a document. Let $\alpha_k$ be the weight of each topic:

$$\alpha_k = \frac{1}{T} \sum_{t=1}^{T} \alpha\theta_t \qquad (3)$$

Step 3: Compute comprehensive sentiment value of a product $TS_i$. It is consisted of two parts: one is the objective sentiment value based on consumer's product reviews, i.e.,

$$S_i = \sum_{k=1}^{k} \alpha_k S_i^k \tag{4}$$

The other one is subjective sentiment value based on the intention of consumers who have the intension to purchase, i.e.,

$$S_i' = \sum_{k=1}^{k} \beta_k S_i^k \tag{5}$$

Where $\beta_k$ represents personal preference for all aspects of the product. Therefore, the comprehensive sentiment value of a product $P_i$ can be computed as:

$$TS_i = \delta S_i + (1-\delta)S_i' \tag{6}$$

## 2.2   Numerical Information Processing Method

Numerical information is adopted by many scholars because of its easy access and intuitive comment form. Numerical value represents the satisfaction degree of the consumer to the product. This paper will analyze it from two aspects: numerical rating based on itself and the numerical rating based on the comparison.

a. The numerical rating based on itself

In this paper, we assume the weight from different customers for the kth aspect of the product is equal. $R_{ij}^k$ denotes the numerical rating concerning jth aspect in the kth online review of alternative product $P_i$. So, the comprehensive numerical rating of product $P_i$ concerning the jth aspect can be computed as:

$$R_{ij} = \frac{1}{k}\sum_{k=1}^{K} R_{ij}^k \tag{7}$$

In summary, the comprehensive numerical rating of product $P_i$ can be computed based on the topic weight $\alpha_k$.

$$TR_i = \sum_{k=1}^{K} \alpha_k R_{ij} \tag{8}$$

b. The numerical rating based on the comparison

Nowadays, many third-party platforms have launched comparison plates for homogeneous products. Let $CR_{ij}$ be comparative rating concerning the jth aspect of alternative product $P_i$. Let $PR_i$ be the superiority of product $P_i$ over other same level products. If $P_i$ is superior to other products in the $j$th aspect, then $PR_{ij} > 0$. If it is inferior to other products, $PR_{ij} < 0$. If they are same in the $j$th aspect, $PR_{ij} = 0$. So, the relative comparative superiority of product $P_i$ over $P_m$ can be computed as:

$$Q_{im} = \frac{1}{J} \frac{\sum_{j=1}^{J} CR_{ij}}{\sum_{j=1}^{J} CR_{mj}} \tag{9}$$

## 2.3    Information Fusion Network Construction Method

Construct a directed graph model $G(V, E, Q^v, Q^E)$. Node $V$ denotes the product. Edge $E$ denotes the directed connection between products, and $Q^v$ denotes weight of node $V$, which refers to the comprehensive sentiment value $TS_i$. $Q^E$ denotes weight of edge $E$, that is, the relative comparative superiority $q_{im}$, which can be computed as:

$$q_{im} = \frac{|Q^i - Q^m|}{\min(Q^i, Q^m)} \tag{10}$$

If $q_{im} > 1$, the edge is directed from $P_m$ to $P_i$. If $q_{im} < 1$, the direction is reversed. If $q_{im} = 1$, there is no connection between these two nodes.

There are many ways to calculate the importance of network nodes. The classical search engine page ranking algorithm PageRank is based on the idea that pages from high-quality web pages must be quality web pages [15]. Similarly, one product with comparative advantage from another high evaluation product will surely be of high

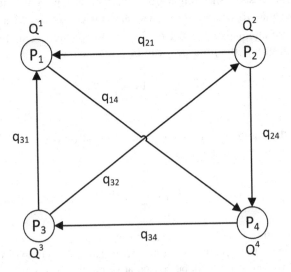

Fig. 3.  Directed graph example

evaluation. We use improved PageRank algorithm to calculate importance of each node. Assume that there are $n$ nodes in a directed graph, n = 1, 2, 3, ..., N, the number of output edges of node $n$ is denoted as $L(n)$. Therefore, the importance of node A can be computed as:

$$Q^A = Q^A + \frac{Q^i q_{iA}}{L(i)} \qquad (11)$$

Where the node $i$ means all nodes that have a directed chain to node A. Figure 3 shows the importance of the nodes. Therefore, the importance of the node 1 can be computed as:

$$Q^1 = Q^1 + \frac{Q^2 q_{21}}{2} + \frac{Q^3 q_{31}}{2} \qquad (12)$$

# 3  Experimental Study

With the rapid development of China's economy, people's living standards have been improved significantly. Automobile has become a necessity in people's life. It is shown that the third-party consumer reviews platforms have a better reputation and recognition [16].

## 3.1  Data Collection and Preprocessing

In order to verify the practicality and effectiveness of proposed method, two third-party consumer review platforms, autohome (http://www.autohome.com.cn/) and Pcauto Network (http://www.pcauto.com.cn/) are selected to do online reviews analysis. According to products' level and price, we choose compact SUV automobiles with prices between 100,000 and 300,000 yuan. We choose four brands automobiles, Mazda, Honda, Trumpchi and Roewe, and then use web crawler to obtain the four kinds of products' data. The detailed data is shown in Table 1.

**Table 1.** Data set

| | Total number of aspect-based reviews | Total number of comprehensive reviews | $f_1$ | $f_2$ | $f_3$ | $f_4$ | $f_5$ | $f_6$ | $f_7$ | $f_8$ |
|---|---|---|---|---|---|---|---|---|---|---|
| $P_1$ | 2338 | 307 | 297 | 280 | 287 | 298 | 301 | 294 | 291 | 290 |
| $P_2$ | 2140 | 282 | 264 | 263 | 260 | 273 | 274 | 269 | 276 | 261 |
| $P_3$ | 1384 | 179 | 175 | 171 | 174 | 172 | 172 | 174 | 176 | 170 |
| $P_4$ | 2381 | 316 | 297 | 296 | 302 | 294 | 301 | 293 | 300 | 298 |

$P_1$: Mazda CX-5 2015, 2.5L (Deluxe Edition automatic four-wheel drive); $P_2$: Trumpchi GS4 2016, (Deluxe Edition 235T G-DCT); $P_3$: Roewe RX5 2016 (30T automatic four-wheel drive version of the Internet); $P_4$: Honda CR-V 2015, 2.4L (Deluxe Edition two-wheel drive). $f_1$: Appearance; $f_2$: Car interior; $f_3$: Space; $f_4$: Configuration; $f_5$: Power; $f_6$: Cross-country; $f_7$: Oil consumption; $f_8$: comfort.

To preprocess the text data, we apply Stanford Parser to segment Chinese text. An external dictionary is added to the word segmentation to avoid ambiguities because of separation of field words. The external dictionary includes 2401 words selected from automobile field words in Sogou cell lexicon.

## 3.2  Data Experiment

The comprehensive reviews are decomposed into aspect-based review corpus, based on which sentiment analysis is carried out. The sentiment intensity values of product $P_i$ based on different aspects $S_{ij}^k$ are obtained. And then the sentiment value of product $P_i$ towards the $k$th aspect $S_i^k$ can be calculated by $S_i^k = \frac{1}{n}\sum_{j=1}^{n} S_{ij}^k$. The sentiment value based on aspect of the four kinds of products are calculated and the result is shown in Table 2.

**Table 2.** Aspect-based sentiment values of different products

|       | $f_1$ | $f_2$ | $f_3$ | $f_4$ | $f_5$ | $f_6$ | $f_7$ | $f_8$ |
|-------|-------|-------|-------|-------|-------|-------|-------|-------|
| $P_1$ | 4.77  | 3.37  | 2.29  | 2.84  | 3.78  | 7.00  | 3.16  | 2.51  |
| $P_2$ | 4.95  | 2.91  | 2.97  | 2.22  | 3.13  | 5.18  | 2.57  | 1.94  |
| $P_3$ | 6.80  | 5.88  | 4.33  | 3.97  | 3.70  | 4.99  | 2.92  | 2.68  |
| $P_4$ | 4.57  | 2.48  | 3.00  | 2.63  | 3.90  | 4.02  | 3.55  | 2.68  |

According to the method proposed in this paper, the preprocessed text review data is used to determine the weight based on LDA topic model. Take Mazda review data as an example. Given the K = 8, then the document and topic probability matrix $\theta_t$ can be obtained. Meanwhile, with LDA topic model, the distribution of keywords under each topic can be determined and then value of word frequency f can be set. Let f = 40, 163 keywords are kept, from which a 8 × 163 topic-keywords probability matrixes $\beta_k$ can be generated.

$$
\beta_k = \begin{matrix} & Word1 & Word2 & \cdots & Word163 \\ Topic1 & 0.0034 & 0.0146 & \cdots & 0.0023 \\ Topic2 & 0.0043 & 0.0092 & \cdots & 0.0022 \\ \vdots & \vdots & \vdots & & \vdots \\ Topic8 & 0.0047 & 0.0104 & \cdots & 0.0026 \end{matrix}
$$

With $\beta_k$, we can make word cloud to visualize display the subject words, from which we can get the realistic meaning of LDA topic in automobile reviews. By Eq. (4), the objective sentiment values of different products based on online reviews can be obtained. The results are shown in Table 3.

**Table 3.** Topic-product correspondence and sentiment value analysis

|       |         | Topic1 | Topic2 | Topic3 | Topic4 | Topic5 | Topic6 | Topic7 | Topic8 |
|-------|---------|--------|--------|--------|--------|--------|--------|--------|--------|
| $P_1$ |         | $f_1$  | $f_2$  | $f_7$  | $f_6$  | $f_3$  | $f_5$  | $f_8$  | $f_4$  |
|       | $\alpha_k$ | 0.012 | 0.305 | 0.119 | 0.317 | 0.021 | 0.017 | 0.074 | 0.135 |
|       | $S_i^k$ | 4.772 | 2.071 | 3.160 | 4.997 | 2.292 | 3.788 | 2.508 | 2.844 |
|       | $S_i$   | 0.0573 | 0.6318 | 0.3760 | 1.5840 | 0.0481 | 0.0644 | 0.1856 | 0.3839 |
| $P_2$ |         | $f_3$  | $f_6$  | $f_4$  | $f_8$  | $f_1$  | $f_7$  | $f_2$  | $f_5$  |
|       | $\alpha_k$ | 0.088 | 0.041 | 0.031 | 0.18 | 0.048 | 0.246 | 0.338 | 0.028 |
|       | $S_i^k$ | 2.968 | 5.184 | 2.794 | 3.943 | 4.950 | 3.574 | 4.915 | 3.128 |
|       | $S_i$   | 0.2612 | 0.2126 | 0.0866 | 0.7098 | 0.2376 | 0.8793 | 1.6612 | 0.0876 |
| $P_3$ |         | $f_4$  | $f_7$  | $f_5$  | $f_8$  | $f_6$  | $f_2$  | $f_3$  | $f_1$  |
|       | $\alpha_k$ | 0.023 | 0.057 | 0.009 | 0.065 | 0.047 | 0.461 | 0.281 | 0.046 |
|       | $S_i^k$ | 3.972 | 2.916 | 3.704 | 2.682 | 4.994 | 5.883 | 4.330 | 6.804 |
|       | $S_i$   | 0.0914 | 0.1662 | 0.0333 | 0.1743 | 0.2347 | 2.7119 | 1.2166 | 0.3130 |
| $P_4$ |         | $f_1$  | $f_6$  | $f_2$  | $f_4$  | $f_3$  | $f_7$  | $f_5$  | $f_8$  |
|       | $\alpha_k$ | 0.025 | 0.102 | 0.326 | 0.241 | 0.017 | 0.164 | 0.044 | 0.081 |
|       | $S_i^k$ | 4.557 | 4.016 | 4.481 | 3.627 | 2.997 | 3.551 | 3.899 | 2.684 |
|       | $S_i$   | 0.1139 | 0.4096 | 1.4608 | 0.8740 | 0.0509 | 0.5823 | 0.1715 | 0.2174 |

Based on analysis in Sect. 2.3, we can construct the directed graph model after calculating the weights of each network node, the direction and weight of each edge. The result is shown in Fig. 4.

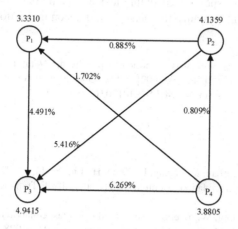

**Fig. 4.** Directed graph model of four products

Based on the improved PageRank algorithm, the final scores for different products can be obtained. By Eq. (10), the final scores of the four products are 3.3713, 4.1464, 5.2842 and 3.8805. So, the final ranking result is $P_3 > P_2 > P_4 > P_1$.

Competition car ranking information provided by Pacific automotive network shows that the actual ranking result is $P_3 > P_1 > P_2 > P_4$. It means the result calculated by proposed method in this paper is relatively consistent with the performance of each product in the competition ranking. However, the proposed method is not totally consistent with the actual result. It is because the text review information is taken into consideration in the method proposed in this paper, which also verifies the effectiveness of the proposed method.

## 4 Conclusion

In this paper, a new method of mining aspect-based online reviews is proposed. Firstly, different from other studies with single data, this paper uses both numerical data and review text data. Secondly, LDA topic model in text mining field is used to solve probability distribution of document-topic and topic-keyword, based on which the word cloud about the topic is obtained. Through the visualization of text data, each topic's content is shown. The virtual topic of LDA topic model is endowed with practical significance and the weights of aspects are calculated. With full use of massive online review knowledge, the customers' objective sentiment values are finally digged out.

At the same time, considering that different consumers have different preferences towards different aspects of products, their subjective sentiment orientation weights are taken into account. The probability of text and of personal preference are integrated to calculate the comprehensive sentiment value of the product, which makes the result more persuasive and reliable. The proposed algorithm of this paper provides a new theoretical method for aspect-based online text review mining. What's more, developing decision support systems for personalized recommendation is also of great practical significance.

**Acknowledgements.** This work was supported in part by the Natural Science Foundation of China [Grant Numbers 71771034, 71031002] and the Open Program of State Key Laboratory of Software Architecture [Item number SKLSAOP1703].

## References

1. Yang, X., Liang, C., Zhao, M., Wang, H., Ding, H., Liu, Y., Li, Y., Zhang, J.: Collaborative filtering-based recommendation of online social voting. IEEE Trans. Comput. Soc. Syst. 1–13 (2017)
2. Park, Y., Park, S., Jung, W., Lee, S.G.: Reversed CF: a fast collaborative filtering algorithm using a k-nearest neighbor graph. Expert Syst. Appl. **42**, 4022–4028 (2015)
3. Mary, M., Natalie, G., Zach, R.: Star quality: aggregating reviews to rank products and merchants. In: International Conference on Weblogs and Social Media, ICWSM 2010, Washington, D.C., USA, May 2010
4. Ghose, A., Ipeirotis, P.G.: Estimating the helpfulness and economic impact of product reviews: mining text and reviewer characteristics. IEEE Trans. Knowl. Data Eng. **23**, 1498–1512 (2011)

5. Khan, F.H., Qamar, U., Bashir, S.: eSAP: a decision support framework for enhanced sentiment analysis and polarity classification. Inf. Sci. **367–368**, 862–873 (2016)
6. Tian, F., Wu, F., Chao, K.M., Zheng, Q., Shah, N., Lan, T., Yue, J.: A topic sentence-based instance transfer method for imbalanced sentiment classification of Chinese product reviews. Electron. Commer. Res. Appl. **16**, 66–76 (2016)
7. Hu, N., Koh, N.S., Reddy, S.K.: Ratings lead you to the product, reviews help you clinch it? The mediating role of online review sentiments on product sales. Decis. Support Syst. **57**, 42–53 (2014)
8. Najmi, E., Hashmi, K., Malik, Z., Rezgui, A., Khan, H.U.: CAPRA: a comprehensive approach to product ranking using customer reviews. Computing **97**, 843–867 (2015)
9. Yang, X., Yang, G., Wu, J.: Integrating rich and heterogeneous information to design a ranking system for multiple products. Decis. Support Syst. **84**, 117–133 (2016)
10. Zhang, K., Cheng, Y., Liao, W.K., Choudhary, A.: Mining millions of reviews: a technique to rank products based on importance of reviews. In: International Conference on Electronic Commerce, pp. 1–8 (2011)
11. Liu, Y., Bi, J.W., Fan, Z.P.: Ranking products through online reviews: a method based on sentiment analysis technique and intuitionistic fuzzy set theory. Inf. Fusion **36**, 149–161 (2017)
12. Chen, K., Kou, G., Shang, J., Chen, Y.: Visualizing market structure through online product reviews: integrate topic modeling, TOPSIS, and multi-dimensional scaling approaches. Electron. Commer. Res. Appl. **14**, 58–74 (2015)
13. Zhang, Z., Guo, C., Goes, P.: Product comparison networks for competitive analysis of online word-of-mouth. ACM Trans. Manag. Inf. Syst. **3**, 1–22 (2013)
14. Xu, L.H., Lin, H.F., Pan, Y., et al.: The construction of emotional lexical ontology. J. Inf. Sci. **27**, 180–185 (2008)
15. Page, L.: The PageRank citation ranking: bringing order to the web. In: Stanford Digital Libraries Working Paper, vol. 9, pp. 1–14 (1998)
16. Gu, B., Park, J., Konana, P.: Research note: the impact of external word-of-mouth sources on retailer sales of high-involvement products. Inf. Syst. Res. **23**, 182–196 (2012)

# An Empirical Analysis of the Chronergy of the Impact of Web Search Volume on the Premiere Box Office

Ling Qu, Guangfei Yang$^{(\boxtimes)}$, and Donghua Pan

School of Management Science and Engineering,
Dalian University of Technology, Dalian 116024, China
gfyang@dlut.edu.cn

**Abstract.** This paper studies the chronergy of the effects of pre-release search volume for movies and stars and the heterogeneity in the effects of our focal variables across the various movies. We sample the panel data of 593 movies released in China and our estimation is based on multiple sets of multiple linear regression contrast models. The main findings are: (1) only search volume for movies within 3 weeks before the release of movies has positive and significant effect on premiere box office; (2) the search volume for leading actors generated until 3 weeks before the movies' release affects the premiere box office significantly; (3) these effects vary with genre and showing time. On the basis of these empirical results, we put forward some suggestions on the use of search volume data in box office studies and film marketing.

**Keywords:** Premiere box office · Web search volume · Chronergy · Heterogeneity

## 1  Introduction

With people's pursuit of spiritual culture, the film industry which has both economic attributes and social and cultural property is booming nowadays. As a product with short life cycle and high participation, the risk of the film industry has been paid close attention by researchers. One of the most important parts is the study of the box office which almost means the success or failure of a movie.

At present, there have been a lot of researches on box office, which can be divided into two aspects. On one hand, some of the researchers focused on factors affecting box office or in other words, factors which can improve the box office forecast, such as the basic attributes of a movie (e.g., genre, showing time, competition, MPAA rating, sequel, number of screens), other influencing factors: advertising inputs [1] star power [2], word of mouth [3, 4], pre-release movie piracy [5], and the combination among them [6]. On the other hand, some researchers concentrated on modeling methods for forecasting box office. Sharda and Delen [7] indicated that neural network works well in predicting box-office success of motion pictures. Lee and Chang [8] computed the probability of a movie's box-office success using the Bayesian belief network (BBN) model, which improved forecasting accuracy compared to artificial neural

© Springer Nature Singapore Pte Ltd. 2017
J. Chen et al. (Eds.): KSS 2017, CCIS 780, pp. 162–174, 2017.
https://doi.org/10.1007/978-981-10-6989-5_14

network and decision tree. Zhang [9] built an effective prediction model by a multi-layer BP neural network (MLBP) with multi-input and multi-output. Parimi and Caragea [10] and Kim et al. [11] applied machine learning algorithms to box office forecasting. Wang and Jia [12] built a dynamic simultaneous-equations model linking online search and box office revenues. Ghiassi et al. [13] developed a model whose forecasting accuracy was as high as 94.1% based on a dynamic artificial neural network, etc.

In recent years, as the Internet grows, researchers focus more on Web search volume which has been shown to play an important role in forecasting [14–16] and also use it as a new explanatory variable to improve the predictive power of the box-office forecasting models [17, 18]. However, although the results all show that movie's Web search volume does increase the accuracy of the prediction of its box office, different researchers use the search volume across different time period in their models. Researches similarly use the search volume in period t − 1 to calculate the box office receipts in period t when forecasting the revenue of the later running period. Nevertheless, when predicting premiere box office, some studies use the search volume in the period of one week before the premiere while some studies prefer the search volume in a longer period before release, and some studies even didn't specifically define the time period of the search volume data they used. There is no study showing which one is better. What's more, there are hardly any studies considering the impact of the director's and main actors' Web search volume. Therefore, we want to explore the Web search volume within which period of time can maximize its role in the box office forecast and whether the directors' and main actors' Web search volume also play an important role while predicting box office.

This paper collected Web search volume and basic data of movies released on the mainland of China between January 2013 and June 2016 and did an empirical analysis using multiple linear regression models which were commonly used to study the impact of certain factors on the box office [1, 5, 19, 20]. Our results show that the premiere box office is significantly related with the last four weeks' search volume for the movie before the movie is released and there is a varied relationship between the search volume for directors and actors and premiere box office with the genre and showing time.

Our study contributes to the literature on understanding the varied impact of Web search volume before a movie is released on its premiere box office. At the same time, we specify a valid scope for the chronergy of the search volume, which can be used as a reference for the research on the box-office forecasting modeling using search volume. What's more, we first consider the relationship between Web search volume for directors and actors and premiere box office, which explains how the behavior of going to the movie happens and provides a new way to measure the impact of stars on box office. We believe these results can provide useful guidance to both related researchers and industry managers.

## 2 Research Hypotheses

### 2.1 Hypotheses Regarding the Web Search Volume for Movies

Recent research shows that search engine use is a major activity of netizens. To a certain extent, Web search volume recorded by search engine reflects people's interest and needs. Carneiro and Mylonakis [21] proved that Google Flu Trends can detect regional outbreaks of influenza 7–10 days before conventional Centers for Disease Control and Prevention surveillance systems. Baker and Fradkin [22] demonstrated the utility of Google job search as an instant feedback tool for policy makers. What's more, Goel et al. [16] observed that search volume is generally predictive of consumer behavior. Kulkarn et al. [14] developed a model of pre-launch search activity to forecast product sales using online search data. Choi and Varian [15] showed that search engine data from Google Trends can be used to predict near-term market outcomes. Handa and Judge [17] found the potential of Google trends data for improving the accuracy of cinema admission forecasting models. Okazaki et al. [18] found the difference of prediction accuracy and essential information to predict box office between the search-based model and baseline model.

Although there have been many researchers focusing on the impact of Web search volume on box-office forecasting, the selection of search volume data varies from person to person. Kulkarni et al. [14] focused on 9 weeks of search data—8 weeks of pre-launch and the week of launch (1 week of post-launch). Handa and Judge [17] took the first value for each month from the Google Trends data for their regressions. Miao and Ma [20] and Wang and Jia [23] used the average of the Web search volume for a movie over the week before it is released. For the difference in the selection of search volume data, there is no specific explanation. According to inter-temporal decision theory, people value a near-future reward more than a distant-future reward, even the latter is larger [24]. This theory suggests that people are more willing to search a movie which will be released in the near rather than the distant future under other conditions being equal. Another research showed that most people pay attention to a movie about four weeks in advance because of the release of movie trailers [25]. Therefore, we make the following assumptions about "the near future":

**H1a.** The search volume for movies during the period of pre-launch are positively correlated with the premiere box office.
**H1b.** Only the effect of the search volume for movies during about four weeks of pre-launch on the premiere box office is significant.
**H1c.** The effect of the pre-release search volume for movies on premiere box office varies across the various movies.

## 2.2 Hypotheses Regarding the Web Search Volume for Stars

Many studies indicated that directors power is not important compared with star power. Albert [2] confirmed that the previous movie impression of the actors can still influence the box office of the present movies that they participate. Prior studies also have identified the positive effect of star power on the box office [3]. The star power or star value has been generally used as an explanatory variable in the study of movie box office modeling, and there has been a variety of ways to assess the impact of stars on box office revenues except for the search volume. Ravid [26] used stars' historical box office success, Liu [27] and Ainslie et al. [28] chose the visibility and acclaim such as standings in industry-produced power lists, and Basuroy et al. [29] selected Academy Award wins and nominations to measure star power. However, to a certain extent, these measures are one-sided. Since previous studies have proved the role of Web search volume for movies in box-office forecasting research, why do not we consider the effect of the search volume for the stars? Our study concentrates on stars' ability to attract consumer represented by Web search volume for stars rather than their previous success or awards. And we wonder if it's a good way to measure star power by the search volume for stars. In the paper, we divide stars into directors and the top five leading actors of movies.

**H2a.** The pre-release Web search volume for directors is positively correlated with the premiere box office.
**H2b.** The pre-release Web search volume for actors is positively correlated with the premiere box office.
**H2c.** The effect of the pre-release search volume for stars on premiere box office varies across the various movies.

# 3 Data and Variables

We collect our data from two sources: EBOT a database of Ent Group and Baidu Index official website (http://index.baidu.com/). Our data are made up of 593 movies (after removing samples with incomplete data) released on the mainland of China between January 2013 and June 2016. We obtain basic characteristics of these movies from EBOT which is a database of Ent Group including information of over 300,000 films and television productions, data of Chinese cinemas and films (www.cbooo.cn), including directors, top five starring, release date, genre, screens, country, technical effects, and box office revenue. In addition, we collect Web search volume from Baidu Index official website.

Tables 1 and 2 gives the definitions, descriptions and key summary statistics of all of the variables in our study.

**Table 1.** Description of variables

| Variable | Description | Source |
|---|---|---|
| Premiere box office | The Chinese box office revenue of a movie on an opening day | Ent group |
| Genre | The genre of the movie | Ent group |
| Competition | The number of movies released in the same week | Ent Group |
| Number of screens | The proportion of the movie's screens in all movies shown in theaters nationwide on its opening day | Ent group |
| Web search volume (10 weeks before release) | Weekly average search volume before a movie is released | Baidu index |
| Technical effects | Whether a movie is screened in the form of 3D and IMAX or 2D | Ent group |
| Showing time | Whether a movie is released on festivals or not | Ent group |

**Table 2.** Descriptive statistics of the variables (N = 593)

|  | Mean | Median | Std. dev. | Min | Max |
|---|---|---|---|---|---|
| Premiere box office (¥ 0000) | 1330.087 | 211.4 | 2900.314 | 0.600 | 27224.800 |
| Competition | 6.764 | 7 | 2.326 | 1.000 | 12.000 |
| Number of screens | 0.1052 | 0.0595 | 0.1085 | 0.0002 | 0.7416 |
| Major genre (%) | Action: 14.3, Love: 32.5, Suspense thriller: 15.7, Comedy: 14.8, Cartoon: 11.5, Drama: 6.4 | | | | |
| Technical effects (%) | 2D: 17.2, 3D&IMAX: 82.8 | | | | |
| Showing time (%) | Non-festivals: 17.9, Festivals: 82.1 | | | | |

## 3.1 Dependent Variable

Our objective is to analyze the chronergy of the impact of Web search volume on the premiere box office with the release time approaching and explore the influence of Web search volume for starring and directors on the opening day box office. Box-office gross revenue and weekly box office were two of the most frequently used dependent variables in the previous literature [1, 5, 6, 8]. Studies have pointed out that the success of movie's box office is strongly related with its opening week box office. However, once the film is released, word of mouth and rating will affect its box office deeply, due to the film is a kind of high-involvement products. Therefore, we want to explore the relationship between Web search volume and the premiere box office rather than the opening week box office in the absence of word of mouth and rating.

## 3.2 Independent Variables

In our study, we considered the following variables which have been mentioned most in the previous researches as attributes affecting the premiere box office. We abandoned some variables based on combining the reality of Chinese market and the data source.

**Genre.** In our study, we included genre as an independent variable as most studies have done. We classified all movies into 8 content categories. For movies with multiple type attributes, we choose the most significant category as their genre. The categories included Science Fiction, Action, Love, Suspense thriller, Comedy, Drama, Cartoon, Documentary, represented by 1 to 8, respectively.

**Competition.** Sharda and Delen [7] judged the intensity of competition according to which month the movie is released, but this way is not eloquent enough. We think the competition for a movie is produced by movies released during the same period which divides the box office. Consequently, we used the number of movies released in the same week to show the simple competitiveness of the movie.

**Number of Screens.** Previous studies have shown that the box office performance is closely related to the number of screens [30]. The bigger the number of screens is, the more likely the moviegoers are to watch the movie. So, we got the proportion of the movie's screens in all movies shown in theaters nationwide on its opening day, due to the limit of the data source. And we defined it as a continuous explanatory variable.

**Web Search Volume.** We obtained daily Web search volume of the 593 movies from Baidu Index (https://index.baidu.com/), a website where one can examine the daily search index of keywords entered into the Baidu search engine. Baidu Index is the search frequency weighted sum of each keyword search in Baidu web search, scientifically analyzed and calculated based on netizens' search volume data of each keyword on Baidu search engine. Depending on the source of the Baidu search, the search index is divided into the PC search index and the mobile search index. In our study, we choose the sum of the PC search index and the mobile search index as the Web search volume. In previous studies, only the search volume for movies was used as an explanatory variable in their empirical models. Our study wants to explore the Web search volume for directors and actors too. Due to the use of search engine in online activities has been very prevalent, the search volume data suffer less from selection bias and represent the ordinary netizens better than blogging. Additionally, the search volume data can be collected and obtained with no need of much data cleaning or coding as analyses of online reviews. Therefore, we only carried out basic statistical processing on Baidu Index data collected.

**Technical Effects.** We also defined a variable to capture the technical effects of a movie. According to actual condition, we used a discrete variable to describe the technical merit. Three different technical effects categories: 2D, 3D, and IMAX are divided into two types which are represented by numbers 1 and 2, respectively.

**Showing Time.** The release time is also an important effect factor for a movie to its premiere box office success. Accordingly, we defined a binary variable to show the effect of showing time. It is whether the release time is during festivals. There are 18 festivals mentioned: New Year's Day, Spring Festival, Tomb Sweeping Day, Labor Day, Dragon Boat Festival, Mid-Autumn Festival, National Day, Valentine's Day, the Lantern Festival, Women's Day, April Fool's Day, Mother's Day, Children's Day, Star Festival, Teachers' Day, Double Ninth Festival, Thanksgiving Day, Christmas. If a movie is released on weekends (include Friday) or festivals, the value of 1 will be given to the variable.

# 4  Results

Tables 3 and 4 report the values, standard errors and significance of each parameter in the multiple linear regression models we obtained in our experiments. We examine the relationship between pre-release Web search volume during the different period and the premiere box office (Tables 3 and 4). Then, we investigate the differences in these relationships across movies, focusing on movies with different genre and showing time (Fig. 1 and Table 5).

## 4.1    Average Effects

**Effects of Pre-release Web Search Volume for Movies on the Premiere Box Office.** In our study, we first build multiple linear regression models (1) using all the 593 movies to explore the correlation between the pre-release Web search volume for movies and the premiere box office revenues.

$$\log(\text{premiereboxoffice}_i) = \alpha_0 + \alpha_1 \log(\text{avgWSVmovie}_{it})$$
$$+ \alpha_3 \text{technicaleffects}_i + \alpha_4 \text{showingtime}_i \qquad (1)$$
$$+ \alpha_5 \text{screens}_i + \alpha_6 \text{compection}_i + u_i$$

The meaning of each variable is shown in Table 1. It should be noted that the independent variable $\text{avgWSVmovie}_{it}$ is the weekly average Web search volume for movie i during the week t before it is released (e.g., when t = −1, $\text{avgWSVmovie}_{it}$ means the weekly average Web search volume for movie i over the week before it's released). $\text{technicaleffects}_i$ and $\text{showingtime}_i$ are dummy variables and control variables. Moreover, the variable $\text{premiereboxoffice}_i$ and $\text{avgWSVmovie}_{it}$ are log-transformed, for accounting for the highly skewed distributions of popularity.

The results of the regression of the models are in Table 3.

**Table 3.** Impact of pre-release Web search volume for movies on premiere box office

| | t = 0 | t = −1 | t = −2 | t = −3 | t = −4 | t = −5 | t = −6 | t = −7 | t = −8 |
|---|---|---|---|---|---|---|---|---|---|
| *Coefficient (t-value)* | | | | | | | | | |
| constant | 1.412*** | 0.233 | 1.233*** | 1.307*** | 1.354*** | 1.372*** | 1.380*** | 1.387*** | 1.393*** |
| | (20.86) | (1.498) | (14.41) | (16.06) | (17.58) | (18.46) | (18.89) | (19.22) | (19.67) |
| *Technical effects* | | | | | | | | | |
| 2D | . | . | . | . | . | . | . | . | . |
| 3DandIMAX | 0.178** | 0.196*** | 0.189*** | 0.189*** | 0.186*** | 0.183** | 0.181** | 0.180** | 0.180** |
| | (3.212) | (3.736) | (3.435) | (3.405) | (3.346) | (3.304) | (3.265) | (3.252) | (3.251) |
| *Showing time* | | | | | | | | | |
| Non-festivals | . | . | . | . | . | . | . | . | . |
| Festivals | 0.249*** | 0.195*** | 0.239*** | 0.245*** | 0.247*** | 0.246*** | 0.248*** | 0.248*** | 0.248*** |
| | (4.810) | (3.946) | (4.652) | (4.761) | (4.777) | (4.763) | (4.800) | (4.803) | (4.800) |
| Number of screens | 6.586*** | 5.206*** | 6.257*** | 6.371*** | 6.437*** | 6.462*** | 6.477*** | 6.490*** | 6.493*** |
| | (33.56) | (20.89) | (28.76) | (29.38) | (29.63) | (29.69) | (29.80) | (29.70) | (29.52) |
| Competition | −0.030*** | −0.024** | −0.030*** | −0.030*** | −0.030*** | −0.030*** | 0.030*** | 0.030*** | 0.030*** |
| | (3.525) | (2.880) | (3.494) | (3.508) | (3.448) | (3.452) | (3.474) | (3.489) | (3.449) |
| LOG (avgWSVmovie$_{it}$) | | 0.307*** | 0.056*** | 0.035* | 0.023 | 0.018 | 0.015 | 0.013 | 0.012 |
| | | (8.315) | (3.386) | (2.300) | (1.592) | (1.317) | (1.165) | (1.006) | (0.939) |
| Adjusted R² | 0.706 | 0.736 | 0.711 | 0.708 | 0.707 | 0.706 | 0.706 | 0.706 | 0.706 |
| F value | 356.302*** | 331.896*** | 292.406*** | 288.180*** | 286.293*** | 285.744*** | 285.485*** | 285.249*** | 285.161*** |

*p<0.05, **p<0.01, ***p<0.001

It can be seen that the coefficients of $\text{technicaleffects}_{\text{3D\&IMAX}i}$, $\text{showingtime}_{\text{Holidays}i}$, $\text{screens}_i$, and $\text{compection}_i$ are all positive and significant, which suggests that they all have positive and significant effects on the premiere box office. This is consistent with previous findings. Importantly, the coefficient of $\log(\text{avgWSVmovie}_{it})$ is also positive and significant just when t is −1, −2 and −3 and the adjusted R² is improved from

0.706 to even 0.736 with the addition of $\log(avgWSVmovie_{it})$. This result shows that the pre-release search volume for movies is positively related to the premiere box office (H1a), but just the effect of the weekly average search volume within 3 weeks before the movie is released is significant and only models with the search volume for movies during 4 weeks of pre-launch perform better than model without the search volume data (H1b). It means that the addition of search volume data can improve the accuracy of the box office model which is consistent with existing research findings [12, 17, 20], but the search period should be set within three or four weeks before the release which is consistent with Panaligan and Chen [25].

**Effects of Pre-release Web Search Volume for Stars on the Premiere Box Office.** In order to verify Hypothesis 2a and 2b, we build new multiple linear regression models (2) with all sample data.

$$
\begin{aligned}
\log(premiereboxoffice_i) = {} & \alpha_0 + \alpha_1 \log(avgWSVmovie_{it}) + \alpha_3 technicaleffects_i \\
& + \alpha_4 showingtime_i + \alpha_5 screens_i + \alpha_6 compection_i \\
& + \alpha_7 \log(avgWSVdirector_{it}) + \alpha_8 \log(avgWSVactor1_{it}) \\
& + \alpha_9 \log(avgWSVactor2_{it}) + \alpha_{10} \log(avgWSVactor3_{it}) \\
& + \alpha_{11} \log(avgWSVactor4_{it}) + \alpha_{12} \log(avgWSVactor5_{it}) + u_i
\end{aligned}
$$

$$(2)$$

where similarly with $avgWSVmovie_{it}$, $avgWSVdirector_{it}$ and $avgWSVactor1_{it}$ are the weekly average Web search volume for director and the first leading actor of movie i during the week t before it is released, and the definitions of other variables are identical to that of the model (1).

Table 4 shows us the results of the estimates of the model (2).

In addition to the nearly same conclusion, we can see the effect of variables (e.g., $screens_i$ and $avgWSVmovie_{it}$) which is the same as those in the model (1) from the coefficients and significance of these variables. The comparison of the adjusted $R^2$ of models in Tables 3 and 4 shows us that the addition of search volume for stars can also make the box office model perform better, which indicates the effect of pre-release search volume for stars on the opening box office. However, we find that for all movies, only the effect of $avgWSVactor1_{it}$ is positive and significant just when t is $-8$ to $-4$, at the same time, the effect of $avgWSVmovie_{it}$ is not significant. This reveals support for H2b and nonsupport for H2a. The results suggest us that moviegoers' choices of the movie begin with their interest of the main actors of the movie, and this interest then develops into the interest of the movie. Our conclusion is similar to that of Ainslie et al. [28] which pointed out that actors have a direct effect on consumer choice by leading viewers to focus on a movie earlier in its release. And our results show that people don't pay much attention to directors of a movie. We think it is caused by the public's strong desires for seeing stars or idols they like.

**Table 4.** Impact of pre-release Web search volume for stars on premiere box office

| | $t=0$ | $t=-1$ | $t=-2$ | $t=-3$ | $t=-4$ | $t=-5$ | $t=-6$ | $t=-7$ | $t=-8$ |
|---|---|---|---|---|---|---|---|---|---|
| Coefficient (standard error) | | | | | | | | | |
| Constant | 1.412*** (20.86) | 0.437** (0.132) | 1.172*** (0.089) | 1.236*** (0.086) | 1.274*** (0.083) | 1.295*** (0.081) | 1.306*** (0.080) | 1.307*** (0.080) | 1.311*** (0.079) |
| Technical effects | | | | | | | | | |
| 2D | . | . | . | . | . | . | . | . | . |
| 3Dand IMAX | 0.178** (3.212) | 0.207*** (0.053) | 0.207*** (0.056) | 0.206*** (0.056) | 0.205*** (0.056) | 0.203*** (0.056) | 0.198*** (0.056) | 0.198*** (0.056) | 0.198*** (0.056) |
| Showing time | | | | | | | | | |
| Non-festivals | . | . | . | . | . | . | . | . | . |
| Festivals | 0.249*** (4.810) | 0.187*** (0.050) | 0.225*** (0.052) | 0.232*** (0.052) | 0.233*** (0.052) | 0.232*** (0.052) | 0.235*** (0.052) | 0.236*** (0.052) | 0.235*** (0.052) |
| Number of screens | 6.586*** (33.56) | 5.078*** (0.268) | 5.994*** (0.247) | 6.118*** (0.247) | 6.188*** (0.246) | 6.220*** (0.246) | 6.253*** (0.244) | 6.262*** (0.245) | 6.273*** (0.246) |
| Competition | −0.030*** (3.525) | −0.024** (0.008) | −0.029*** (0.009) | −0.029*** (0.009) | −0.029** (0.009) | −0.028** (0.009) | 0.029** (0.009) | 0.029** (0.009) | 0.028** (0.009) |
| LOG (avgWSV movie$_{it}$) | | 0.310*** (0.039) | 0.068*** (0.021) | 0.044* (0.019) | 0.026 (0.018) | 0.019 (0.018) | 0.017 (0.017) | 0.013 (0.017) | 0.012 (0.017) |
| LOG (avgWSV director$_{it}$) | | 0.011 (0.014) | 0.015 (0.015) | 0.016 (0.015) | 0.017 (0.015) | 0.017 (0.015) | 0.015 (0.015) | 0.015 (0.015) | 0.016 (0.015) |
| LOG (avgWSV actor1$_{it}$) | | 0.017 (0.020) | 0.033 (0.020) | 0.036 (0.020) | 0.044* (0.020) | 0.042* (0.021) | 0.041* (0.020) | 0.048* (0.020) | 0.050* (0.020) |
| LOG (avgWSV actor2$_{it}$) | | 0.010 (0.019) | −0.002 (0.020) | −0.003 (0.020) | −0.010 (0.019) | −0.010 (0.020) | −0.010 (0.020) | −0.014 (0.020) | −0.015 (0.019) |
| LOG (avgWSV actor3$_{it}$) | | −0.029 (0.018) | −0.025 (0.018) | −0.026 (0.019) | −0.024 (0.019) | −0.024 (0.019) | −0.025 (0.018) | −0.023 (0.018) | −0.027 (0.018) |
| LOG (avgWSV actor4$_{it}$) | | 0.005 (0.017) | 0.013 (0.018) | 0.013 (0.017) | 0.010 (0.017) | 0.013 (0.018) | 0.012 (0.017) | 0.010 (0.018) | 0.014 (0.016) |
| LOG (avgWSV actor5$_{it}$) | | 0.008 (0.015) | 0.009 (0.015) | 0.008 (0.016) | 0.011 (0.016) | 0.012 (0.016) | 0.013 (0.016) | 0.012 (0.016) | 0.011 |
| Adjusted $R^2$ | 0.706 | 0.737 | 0.713 | 0.710 | 0.709 | 0.708 | 0.708 | 0.708 | 0.708 |
| F value | 356.302*** | 151.749*** | 134.693*** | 132.634*** | 132.103*** | 131.652*** | 131.400*** | 131.549*** | 131.721*** |

*p<0.05, **p<0.01, ***p<0.001

## 4.2 Heterogeneity of Effects

The average effects reported in Subsect. 4.1 vary substantially across movies. Figure 1 and Table 5 help us understand the heterogeneity in the effects of our focal variables across the various movies. And the results reveal strong support for H1c and H2c.

**Effects of Genre on Heterogeneous Pre-release Web Search Volume Effectiveness.** In our study, there are eight categories of movies that are divided, and four kinds of genre (Action, Love, Comedy and Suspense thriller) are analyzed. The other kinds of the genre are not concerned because their sample size is not large enough after dividing the category. We use the new sample data to build the model (2) again and Fig. 1. shows the different significance of the main variables we focus on in our models across various movies.

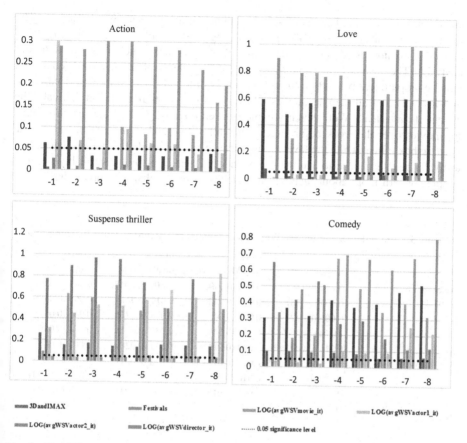

**Fig. 1.** The different pre-release Web search volume effectiveness across genre (Action, Love, Suspense thriller and Comedy)

From the figure, we can see that the technical effects have a significant impact on Action movies and Suspense thrillers and showing time affects Action movies and Love movies. What's more, the effect of search volume for director on Suspense thrillers is obviously significant while the effect of search volume for leading actors is very insignificant. Nevertheless, the other three categories of movies are just the opposite. The results suggest that for different movies viewers' attention is different: (1) for Action movies, Love movies, and Comedies, people may want to see more about star actors, while for Suspense thrillers, they consider more about movie content and production. (2) the form of 3D and IMAX has more positive influence on Action movies and Suspense thrillers. (3) on festivals, people prefer to see Action movies which can bring people to a strong tension of thrilling action and audio-visual enjoyment and Love movies which are more loved by couples and young people.

**Effects of Showing Time on Heterogeneous Pre-release Web Search Volume Effectiveness.** We divide movies into two categories according to whether they were released on festivals. Then we build the model (2) using two sets of data and the coefficients and standard errors of search volume variables are shown in Table 5 (The results of variables $\log(avgWSVactor3_{it})$, $\log(avgWSVactor4_{it})$ and $\log(avgWSVactor5_{it})$ are not listed, due to they are all insignificant which is helpless to our analysis). The results (see Table 5) show that the effects of pre-release Web search volume on premiere box office are not obviously different between movies released on non-festivals and all sample movies. However, for movies released on festivals, the effects of pre-release Web search volume weaken obviously. Only the search volume for movies over the week before release has a positive and significant impact on the opening box office, while the level of significance has declined. This means that during festivals, people choose to see the premiere with weaker planning, probably because people concern more about relaxing by watching a movie than the movie itself.

**Table 5.** The different pre-release Web search volume effectiveness across showing time

| | Weeks | LOG ($avgWSVmovie_{it}$) | LOG ($avgWSVdirector_{it}$) | LOG ($avgWSVactor1_{it}$) | LOG ($avgWSVactor2_{it}$) |
|---|---|---|---|---|---|
| Released on festivals | −1 | 0.244 (0.115) * | 0.008 (0.040) | −0.002 (0.068) | 0.063 (0.056) |
| | −2 | 0.004 (0.058) | 0.005 (0.042) | 0.016 (0.071) | 0.065 (0.056) |
| | −3 | 0.006 (0.052) | 0.008 (0.043) | 0.024 (0.071) | 0.066 (0.056) |
| | −4 | −0.011 (0.052) | 0.014 (0.044) | 0.031 (0.071) | 0.064 (0.055) |
| Released on non-festivals | −1 | 0.310 (0.041) *** | 0.010 (0.014) | 0.019 (0.020) | 0.002 (0.021) |
| | −2 | 0.074 (0.022) *** | 0.015 (0.015) | 0.032 (0.021) | −0.012 (0.021) |
| | −3 | 0.045 (0.021) * | 0.016 (0.016) | 0.035 (0.021) | −0.013 (0.021) |
| | −4 | 0.028 (0.020) | 0.016 (0.016) | 0.043 (0.021) * | −0.021 (0.021) |

*p<0.05, **p<0.01, ***p<0.001

## 5   Conclusion and Discussion

Our study reveals the chronergy of the effects of pre-release Web search volume on the premiere box office and the heterogeneity in the effects of pre-release Web search volume across the movies of different types and different release times. For all movie samples, the pre-release search volume for movies has a positive and significant effect on premiere box office but just for the search volume within 1 to 4 weeks especially one week before the movie is released. Simultaneously, although only the effect of search volume for the first leading actor is released is significant and the effect is continued until three weeks before the release of the movie, the pre-release search volume for stars (directors and leading actors) can indeed improve the accuracy and fit degree of box office models. Then we divide movies into 4 categories and 2 categories by genre and release time, respectively. The results show that for Action movies, Love movies, and Comedies, people care more about actors, while for Suspense thrillers, people are concerned more about directors.

The objective of this paper is not to explore how the pre-release search volume affects the premiere box office but to make recommendations helping people make better use of Web search volume to predict box office and develop better publicity strategies. From our findings, we suggest that the pre-release search volume for movies should be incorporated into premiere box office models which have been proposed by previous studies. But what we want to say is that the effect of the search volume is weakened with the advance of time so that it's better to choose search volume data within three weeks before the movie is released among which the search volume just over the week before movie's release is best. However, the search volume for stars is different. What's more, the effects of genre and technical effects on heterogeneous pre-release Web search volume effectiveness indicate that publishers should implement different publishing strategies and publicity strategies for different types of movies. For example, suspense thrillers should be shown more in form of 3D and the director of them should be more publicized; love movies which are released on festivals and actively promoted by leading actors before released will perform better.

One data limitation of the current study is that the variables in our box office model are not comprehensive because of the limit of data sources. We just consider five traditional influencing factors lacking some other factors which have been confirmed (e.g., budget, advertisement expenditure). If we can also incorporate these variables into the model, the results may be more persuasive. Another limitation is that we didn't explain how to design the search volume can improve the premiere box office model effect as much as possible. This is what we need to do in the future.

**Acknowledgement.** This work is supported by the National Natural Science Foundation of China (71671024, 71601028, 71421001), Fundamental Research Funds for the Central Universities (DUT17JC37, DUT17JC12), Humanity and Social Science Foundation of the Ministry of Education of China (15YJCZH198), Economic and Social Development Foundation of Liaoning (2016lslktzizzx–01), and Economic and Social Development Foundation of Dalian (2016dlskzd007).

# References

1. Gopinath, S., Chintagunta, P.K., Venkataraman, S.: Blogs, advertising, and local-market movie box office performance. Manag. Sci. **59**(12), 2635–2654 (2013)
2. Albert, S.: Movie stars and the distribution of financially successful films in the motion picture industry. J. Cult. Econ. **23**(4), 325–329 (1999)
3. Basuroy, S., Chatterjee, S., Ravid, S.A.: How critical are critical reviews? The box office effects of film critics, star power, and budgets. J. Mark. **67**(4), 103–117 (2003)
4. Chintagunta, P.K., Gopinath, S., Venkataraman, S.: The effects of online user reviews on movie box-office performance: accounting for sequential rollout and aggregation across local markets. Mark. Sci. **29**(5), 944–957 (2010)
5. Ma, L., Montgomery, A., Singh, P.V., Smith, M.D.: An empirical analysis of the impact of pre-release movie piracy on box-office revenue. Soc. Sci. Electron. Publ. **25**(3), 590–603 (2014)
6. Elberse, A., Eliashberg, J.: Demand and supply dynamics for sequentially released products in international markets: the case of motion pictures. Mark. Sci. **22**(3), 329–354 (2003)
7. Sharda, R., Delen, D.: Predicting box-office success of motion pictures with neural networks. Expert Syst. Appl. **30**(2), 243–254 (2006)

8. Lee, K.J., Chang, W.: Bayesian belief network for box-office performance: a case study on Korean movies. Expert Syst. Appl. Int. J. **36**(1), 280–291 (2009)
9. Zhang, L., Luo, J., Yang, S.: Forecasting box office revenue of movies with BP neural network. Expert Syst. Appl. **36**(3), 6580–6587 (2009)
10. Parimi, R., Caragea, D.: Pre-release box-office success prediction for motion pictures. In: Perner, P. (ed.) MLDM 2013. LNCS, vol. 7988, pp. 571–585. Springer, Heidelberg (2013). doi:10.1007/978-3-642-39712-7_44
11. Kim, T., Hong, J., Kang, P.: Box office forecasting using machine learning algorithms based on sns data. Int. J. Forecasting **31**(2), 364–390 (2015)
12. Wang, L., Jia, J.M.: Forecasting box office performance based on online search: evidence from Chinese movie industry. Syst. Eng.-Theory Pract. **34**(12), 3079–3090 (2014)
13. Ghiassi, M., Lio, D., Moon, B.: Pre-production forecasting of movie revenues with a dynamic artificial neural network. Expert Syst. Appl. Int. J. **42**(6), 3176–3193 (2015)
14. Kulkarni, G., Kannan, P.K., Moe, W.: Using online search data to forecast new product sales. Decis. Support Syst. **52**(3), 604–611 (2012)
15. Choi, H., Varian, H.: Predicting the present with Google trends. Econ. Rec. **88**(s1), 2–9 (2012)
16. Goel, S., Hofman, J.M., Lahaie, S., Pennock, D.M., Watts, D.J.: Predicting consumer behavior with web search. Proc. Natl. Acad. Sci. U.S.A. **107**(41), 17486–17490 (2010)
17. Judge, G., Hand, C.: Searching for the picture: forecasting UK cinema admissions making use of Google trends data. Appl. Econ. Lett. **19**(11), 1051–1055 (2012)
18. Okazaki, M., Nagasaka, R., Miyata, R.: Prediction of the box-office with publicly available information and web search volume. In: International Conference on Intelligent Informatics and Biomedical Sciences, pp. 418–419. IEEE Computer Society (2015)
19. Chintagunta, P.K., Gopinath, S., Venkataraman, S.: The effects of online user reviews on movie box-office performance: accounting for sequential rollout and aggregation across local markets. Mark. Sci. **29**(5), 944–957 (2010)
20. Miao, R., Ma, Y.: The dynamic impact of web search volume on product sales—an empirical study based on box office revenues (2015)
21. Carneiro, H.A., Mylonakis, E.: Google trends: a web-based tool for real-time surveillance of disease outbreaks. Clin. Infect. Dis. **49**(10), 1557 (2009)
22. Baker, S., Fradkin, A.: What drives job search? Evidence from Google search data. Discussion Papers 10-020 (2011)
23. Wang, L., Jia, J.M.: Forecasting box office performance based on online search: evidence from Chinese movie industry. Syst. Eng.-Theory Pract. **34**(12), 3079–3090 (2014)
24. Trope, Y., Liberman, N.: Temporal construal. Psychol. Rev. **110**(3), 403 (2003)
25. Panaligan, R., Chen, A.: Quantifying movie magic with Google search. Google Whitepaper— Industry Perspectives+User Insights (2013)
26. Ravid, S.A.: Information, blockbusters and stars? A study of the film industry. J. Bus. **72**(4), 463–492 (2000)
27. Liu, Y.: Word-of-Mouth for movies: its dynamics and impact on box office revenue. J. Mark. **70**(3), 74–89 (2006)
28. Ainslie, A., Drèze, X., Zufryden, F.: Modeling movie life cycles and market share. Mark. Sci. **24**(3), 508–517 (2005)
29. Basuroy, S., Desai, K.K., Talukdar, D.: An empirical investigation of signaling in the motion picture industry. J. Mark. Res. **43**(2), 287–295 (2006)
30. Neelamegham, R., Chintagunta, P.: A bayesian model to forecast new product performance in domestic and international markets. Mark. Sci. **18**(2), 115–136 (1999)

# Societal Risk and Stock Market Volatility in China: A Causality Analysis

Nuo Xu[1,2] and Xijin Tang[1,2(✉)]

[1] Academy of Mathematics and Systems Science, Chinese Academy of Sciences,
Beijing 100190, China
xunuo1991@amss.ac.cn, xjtang@iss.ac.cn
[2] University of Chinese Academy of Sciences, Beijing 100049, China

**Abstract.** A variety of societal contradictions and conflicts are exposed in China along the process of economic and social transformation. Online societal risk perception is acquired by public searching behavior which has been mapped into respective societal risks based on indicators including national security, economy/finance, public morals, daily life, social stability, government management, and resources/environment. A stable and harmonious society is the basic guarantee for the sound development of the stock market. What we concern about is whether the variations of the societal risk are related to stock market volatility. The correlations between societal risk and stock market volatility are investigated. Although there are no trading data on holidays and weekends, the risk information of no-trading days is also taken into consideration to discuss if there are any impacts on stock market volatility. Three different econometric approaches are developed to explore the relationship between them. The results show that the risk of finance/economy, social stability, and government management could cause the fluctuation of stock market. Moreover, risk information of no-trading days has an impact on the stock's volatility as well. The research demonstrates that capturing online societal risk based on public searching data is feasible and significant.

**Keywords:** Societal risk perception · Stock market volatility · Granger causality test · Multiple linear regression

# 1 Introduction

The transformation of development of economy and society in China is at a crucial stage. The contradictions and conflicts among different social strata, such as the rich and the poor, become increasingly salient and exert adverse effects on social stability, leading to a lot of societal risks. It is known that a stable and harmonious society is the basic guarantee for economy development. That is to say, external social circumstances such as nation's security and foreign relations, government policy, anti-corruption, etc. may have impacts on economic

© Springer Nature Singapore Pte Ltd. 2017
J. Chen et al. (Eds.): KSS 2017, CCIS 780, pp. 175–185, 2017.
https://doi.org/10.1007/978-981-10-6989-5_15

and financial markets. Meanwhile, higher risk may arise in the process of economy development, which poses a threat to social stability. Therefore, in order to guarantee social stability and the healthy growth of national economy, the possible risks existing in economy and finance and the causes of those risks are worth being analyzed. The stock market could be regarded as the barometer of economic development, to a certain extent, reflecting conditions of the national economy. It is of great significance to study the relationship between societal risk and stock market volatility.

In the Web 2.0 era, traditional media has been surpassed by the Internet which is more open and interactive. Internet users are not only content viewers but also content producers. Among many Internet services, search engines have been the most common tools to access information, not only meeting search requirements but also recording foci of netizens. There have been some research achievements exploiting the search data for predicting economic activities, providing a new perspective to understand the conditions of economy. The CPI was well predicted through utilizing Google search query data [1], as well as retail sales [2]. In the field of finance, search query data were successfully used for predicting dynamics of stock market volatility [3], and measuring the retail investors' attention [4,5]. The research on the correlation between the search volume and investors' attention indicated that the increase of search activity was associated with increase of trading activity [6]. Furthermore, there were studies found that an increase in search volume tended to precede stock market falls [7,8]. The search query data are closely related to stock market at information times. In this paper, we try to investigate the relationship between online societal risk acquired by Internet searching behavior and stock market volatility.

Societal risk perception is the subjective evaluation of public concerns to risk events. Traditional research on societal risk perception is studied from social psychology. The psychometric paradigm of risk perception is designed to ask the public to fill out questionnaires about acceptable risks, then assess public's attitude toward critical societal risk [9]. However, the result of this methodology is restricted by the selection of the samples, as well as the authenticity of the answers. Large-scale surveys of societal risk perception are generally expensive and time-consuming to be conducted. The researchers constructed a new framework of societal risk indicators based on word association tests [10]. Online public concerned data are mapped into respective societal risks based on societal risk indicators. Instead of asking respondents to answer questions about societal risk, the query data provided by the public actively reflect what are real concerned, and thus provide researchers a novel access to analyze societal issues.

In this paper, we concern about the question whether the variations of the societal risk have significant effects on stock market volatility. To answer the question, quantitative societal risk levels are presented and the Shanghai Composite Index is chosen to study stock market volatility. Different from previous research [3], the risk level data of holidays and weekends are also taken into consideration to discuss whether there are any impacts on stock price change or not. Three different econometric methods are developed to explore relationships between societal risk levels and Shanghai Composite Index.

This paper is organized as follows: Sect. 2 illustrates the measurement of societal risk and introduction of Shanghai Composite Index. Section 3 introduces three methods on the relationships between societal risk levels and Shanghai Composite Index. Section 4 presents results analysis. Conclusions and future work are given in Sect. 5.

## 2 Data Review

### 2.1 Measurement of Societal Risk

Due to the limitations of the traditional psychological methods, online societal risk perception is proposed to acquire public's concerns toward critical societal risk. Baidu is now the biggest Chinese search engine. Baidu hot news search words (HNSW) are based on real-time search behaviors of hundreds of millions of Internet users and released at Baidu News Portal[1], reflecting the current concerns and ongoing social topics. Therefore, HNSW can serve as the corpus to get netizens' attention to the highlighted events and provide a perspective to analyze societal risk.

HNSW and their relevant hot news with URLs at the first page of hot words search results are crawled every hour [11]. Researchers from CAS Institute of Psychology constructed a framework of societal risk indicators including 7 categories which are national security, economy/finance, public morals, daily life, social stability, government management, and resources/environment [10]. Tang has tried to map HNSW into either risk-free event or one event with risk label from these 7 risk categories, then aggregate all risky events over the whole concerns as the on-line societal risk perception [12]. The daily total risk level is obtained by computing the risky proportion of all hot search words in a day. Figure 1 shows daily total risk levels based on HNSW from January of 2012 to December of 2016 with a total number of 1827 data. The generated 5-year (2012-2016) daily societal risk data provide an overview of China's societal risk situation.

Among the 5-year data, the average risk level of 2012 is 0.554, while that of 2013 is 0.586. However, the average risk levels of 2014, 2015 and 2016 are 0.421, 0.330 and 0.27 respectively, with the annual average risk level decreasing year by year. It may show the achievements of our country in building harmonious socialist society in a way. Moreover, the societal risk levels have dropped off during the Spring Festival, as well as the Olympic Games. As the concept "Harmonious society" was proposed in 2004, there were many studies on the measurement of a harmonious society. Such the Beijing's Harmonious Society Index and Green GDP were put forward to measure the harmonious degree of a society [13]. Besides, a study named "Happiness Survey" was carried out by China Central Television (CCTV) by asking people if they were happy and what

---

[1] http://news.baidu.com/.

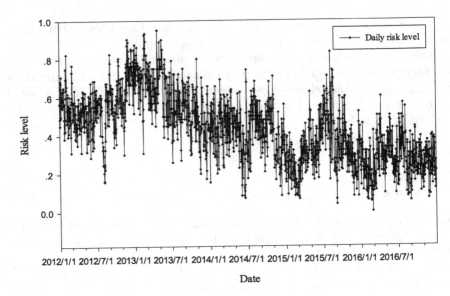

**Fig. 1.** Daily societal risk level of HNSW

happiness meant. However, these indexes are difficult to be measured and come from different government offices. The daily societal risk levels can serve as an additional way to measure the harmonious degree of a society, which truly and more efficiently reflect the social situations. In other words, the societal risk perception indicators, daily societal risk levels, could be the thermometer of Chinese society to some extent.

Furthermore, the risk level of one category is calculated as one ratio of frequency of words labeled that category over total frequency of hot words. Temporal risk levels of each risk category are acquired, which indicates specific societal risks change over time in details. Researchers had demonstrated that risk levels of each category were predictive of public moods including Happiness, Sadness, Fear, Anger and Disgust when investigating the correlation between societal risk and social moods [14]. In this paper, we study the relation of individual risk level of each category with Shanghai Composite Index.

## 2.2   Shanghai Composite Index

Shanghai Composite Index is the broadest and most influential comprehensive index across Chinese Security Markets, which can accurately reflect the overall situation of China's stock market. The Shanghai Composite Index is more commonly utilized in many domestic studies [15,16]. In this paper, we download the data from Wind financial terminal, including a total number of 1214 transactions from January 1, 2012 to December 31, 2016. Different from the daily risk level, there are no trading data on holidays and weekends, and thus leads to different sizes of those two sets of data.

The economic and financial issues as one of societal risk categories may cause the oscillations on the risk levels. We wonder if the changes on risk levels of 7 categories could affect financial and economic fluctuations. Therefore, we try to explore the relationship among the levels of 7 categories of societal risks and the Shanghai Composite Index.

## 3    Correlation Between Societal Risk Levels and Shanghai Composite Index

Baidu hot news search words, online public concerned data, have been mapped into either risk-free events or one event with a label from 7 categories. The risk level of one category is calculated as one ratio of frequency of words labeled that category over the total frequency of hot search words. In this section, we investigate whether the variations of the societal risk have a significant effect on the ups and downs of the stock price. We choose the daily closing price increment to indicate the fluctuation of stock market. Since there are no trading data on holidays and weekends, the data of no-trading days are dropped when analyzing the relation between them. However, risky events happened on no-trading days may have an impact on the future values of Shanghai Composite Index when the next work day is coming. As a result, three approaches with different data processing are developed to analyze the correlation between levels of 7 categories of societal risks and the closing price increment from January 1, 2012 to December 31, 2016.

### 3.1    Method Based on Granger Causality Test

Due to the lack of trading data on the holidays and weekends, we remove the risk level data of dates that have no transaction data, and we carry out the econometric method of Granger causality test on these two sets of data. The Granger causality test is used to measure the ability to predict the future value of a time series X utilizing prior values of another time series Y [17]. If the time series X is said to Granger-cause Y, the lagged values of X will provide statistically significant information about future values of Y. In order to avoid spurious regression problem, we first test the stationary of each variable (unit root) using the augmented Dickey-Fuller test (ADF test). The original hypothesis is the sequence has at least one unit root. Test results are as shown in Table 1, all the original series are stationary series, that is to say, all variables are suitable for Granger causality test directly.

Then we utilize vector autoregressive model (VAR) to carry out Granger causality test. The daily societal risk level is chosen as independent variable, while the closing price increment is chosen as the dependent variable to detect if societal risk perception can affect the volatility of stock price. As far as the lag orders, we just consider it delaying for 5 days. Table 2 shows the results of Granger causality correlation of the closing price increment and societal risk levels of 7 categories.

**Table 1.** Results of ADF test

| Variable | ADF t-statistics | P value | 1% level | 5% level | Results |
|---|---|---|---|---|---|
| National security | −9.73 | 0.00 | −3.96 | −3.41 | stationary |
| Economy/finance | −10.46 | 0.00 | −3.96 | −3.41 | stationary |
| Public morals | −14.34 | 0.00 | −3.96 | −3.41 | stationary |
| Daily life | −12.21 | 0.00 | −3.96 | −3.41 | stationary |
| Social stability | −9.44 | 0.00 | −3.96 | −3.41 | stationary |
| Government management | −8.51 | 0.00 | −3.96 | −3.41 | stationary |
| Resources/environment | −19.08 | 0.00 | −3.96 | −3.41 | stationary |
| Closing price increment | −14.95 | 0.00 | −3.96 | −3.41 | stationary |

**Table 2.** Granger causality correlation of the closing price increment and 7 societal risks

| Variable | Lag | | | | |
|---|---|---|---|---|---|
| | 1 day | 2 days | 3 days | 4 days | 5 days |
| National security | 0.0685 | 0.0926 | 0.1087 | 0.1756 | 0.2599 |
| Economy/finance | 0.4523 | 0.0601 | 0.1757 | **0.0074 | **0.0007 |
| Public morals | 0.2377 | 0.4075 | 0.5714 | 0.8174 | 0.6452 |
| Daily life | 0.2042 | 0.4247 | 0.4813 | 0.5079 | 0.6522 |
| Social stability | *0.0372 | 0.0696 | 0.0994 | 0.0669 | 0.0900 |
| Government management | 0.3277 | 0.2541 | 0.1861 | 0.3233 | 0.4216 |
| Resources/environment | 0.8368 | 0.3535 | 0.5559 | 0.6335 | 0.7661 |

*Correlation is significant at the 0.05 level
**Correlation is significant at the 0.01 level

According to the results of Granger causality in Table 2, we observe that economy/finance has the Granger causality relation with closing price increment lagging from 4 and 5 days while social stability lagging 1 day.

## 3.2    Method Based on Decaying Exponential Function

Different from the first approach, in order to capture the risk information of no-trading days, we define a decaying exponential function. Through setting different weight coefficients, the risk information of holidays or weekends is added to the risk level of the day just before holidays or weekends. Therefore, the risk level of the day before holidays or weekends $r_m$ is recomputed based on original risk level and given by Eq. 1.

$$r_m = \sum_{i=0}^{n} \frac{r_{m+i}}{c^{n+1-i}}. \tag{1}$$

Here, $n$ is the length of successive days with no trading data. For normal weekends, $n = 2$. $c$ is a dampening constant that gives a higher weight to the risk level of recent days. In this paper, we choose $c = 2$. The exponential function is chosen so that distant risk level affects $r_m$ without outweighing more recent risk levels.

Based on the defined decaying exponential function above, we obtain a new sequence of risk levels based on the original data. Then ADF test is conducted. We test the unit root of closing price increment and the new sequence of risk levels. The test results in Table 3 show that all of them have no unit root.

**Table 3.** Results of ADF test

| Variable | ADF t-statistics | P value | 1% level | 5% level | Results |
|---|---|---|---|---|---|
| National security | −9.69 | 0.00 | −3.96 | −3.41 | stationary |
| Economy/ finance | −10.36 | 0.00 | −3.96 | −3.41 | stationary |
| Public morals | −14.45 | 0.00 | −3.96 | −3.41 | stationary |
| Daily life | −12.25 | 0.00 | −3.96 | −3.41 | stationary |
| Social stability | −9.59 | 0.00 | −3.96 | −3.41 | stationary |
| Government management | −8.57 | 0.00 | −3.96 | −3.41 | stationary |
| Resources/ environment | −17.44 | 0.00 | −3.96 | −3.41 | stationary |
| Closing price increment | −14.95 | 0.00 | −3.96 | −3.41 | stationary |

Again, we employ Granger causality analysis. The results are shown in Table 4. As we can see from Table 4, economy/finance has the Granger causality relation with closing price increment for lagging 2, 4 and 5 days, while social stability has the Granger causality relation for lagging 4 and 5 days.

## 3.3   Method Based on Multiple Linear Regression 3

To avoid changing the original societal risk data, we apply the multiple linear regressions to analyze the relationship between the temporal data of societal risk and closing price increment, which is chosen as the dependent variable. Different lays of the societal risk level of one category are chosen as independent variables. There will generate a corresponding number of sequences to the lag number of lays for any risk category. As a result, we construct 7 regression models for each

**Table 4.** Granger causality correlation of the closing price increment and 7 societal risks

| Variable | Lag | | | | |
|---|---|---|---|---|---|
| | 1 day | 2 days | 3 days | 4 days | 5 days |
| National security | 0.2257 | 0.2848 | 0.2341 | 0.2860 | 0.3357 |
| Economy/finance | 0.6153 | *0.0190 | 0.0554 | **0.0095 | **0.0026 |
| Public morals | 0.8974 | 0.9968 | 0.9695 | 0.9109 | 0.5051 |
| Daily life | 0.3423 | 0.6086 | 0.6206 | 0.7193 | 0.8257 |
| Social stability | 0.0778 | 0.0618 | 0.0809 | *0.0145 | *0.0246 |
| Government management | 0.5790 | 0.2877 | 0.2845 | 0.4411 | 0.5636 |
| Resources/environment | 0.9005 | 0.3722 | 0.5455 | 0.6995 | 0.8168 |

*Correlation is significant at the 0.05 level
**Correlation is significant at the 0.01 level

category of societal risk with a total of 35 risk level sequences. The multiple linear regression function is defined as Eq. 2:

$$y_t = \gamma + \sum_{j=0}^{k} \alpha_j y_{t-j} + \sum_{j=0}^{k} \beta_j x_{t-j} + u_t. \tag{2}$$

Here, $y_t$ represents the closing price increment of the $t$th day, while $y_{t-j}$ indicates that of lagging $j$ trading days. $x_{t-j}$ is societal risk level of one category which lags $j$ days. $\alpha_j$ and $\beta_j$ are the regression coefficients. $u_t$ is the error term which reflects the parts of changes in $y$ that cannot be explained by $x$. To match the lag orders with the two approaches above, we set $k = 5$. The results of statistic values for either the significance of coefficients or significance of multiple linear regression models are respectively listed in Tables 5 and 6.

**Table 5.** Results of significance of coefficients

| Model | Lag | | | | |
|---|---|---|---|---|---|
| | 1 day | 2 days | 3 days | 4 days | 5 days |
| M1: National security | 0.0896 | 0.3854 | 0.9964 | 0.5955 | 0.2114 |
| M2: Economy/finance | 0.2975 | *0.0168 | 0.1648 | 0.3096 | 0.8724 |
| M3: Public morals | 0.6498 | 0.7102 | 0.8918 | 0.9043 | 0.5231 |
| M4: Daily life | 0.7196 | 0.6811 | 0.1005 | 0.4515 | 0.8274 |
| M5: Social stability | 0.4479 | *0.0116 | 0.0824 | 0.5251 | 0.2407 |
| M6: Government management | 0.9353 | 0.4999 | 0.3789 | 0.0594 | *0.0476 |
| M7: Resources/environment | 0.9264 | 0.8683 | 0.9208 | 0.3362 | 0.8298 |

*Correlation is significant at the 0.05 level

As shown in Table 6, all the regression models are significant based on F statistics. The sequences of economy/finance lagging 2 days have significant effects on closing price increment, so as those of social stability lagging 2 days, and those of government management lagging 5 days. The result is different from those of previous two approaches.

**Table 6.** Results of significance of multiple linear regression models

| Model | F-statistics | P value | D.W |
|-------|-------------|---------|--------|
| M1 | 5.5301 | 0.0000 | 2.0026 |
| M2 | 5.8288 | 0.0000 | 1.9993 |
| M3 | 5.2323 | 0.0000 | 2.0008 |
| M4 | 5.4664 | 0.0000 | 1.9992 |
| M5 | 6.3818 | 0.0000 | 1.9977 |
| M6 | 5.6184 | 0.0000 | 1.9984 |
| M7 | 5.2745 | 0.0000 | 2.0000 |

## 4 Results Analysis

Based on the above results, three different approaches demonstrate distinguishing findings on the correlation between societal risk and stock volatility. As to the first approach, the original data sequence of societal risk level is used to carry out the research, while the risk data on no-trading days are ignored without regarding their effects on stock volatility. The experimental results show that economy/finance and social stability have significant effects on closing price increment. In terms of the second approach, the decaying exponential function is adopted to capture the risk information of no-trading days by processing the original data. The results are different from the first with different delay days on the risk of economy/finance and social stability. The subcategories of economy/finance include financial problems and economic problems. As is known, the macroeconomic policy adjustment, the economic slowdown, the monetary policy, RMB exchange rate etc. will cause the fluctuation of stock prices. Besides, stable social situations is a vital prerequisite for development of finance and economy. Events that threaten social stability such as public security, crime, and major infectious diseases etc. which are the subcategories of social stability have a certain impact on stock market.

Compared to the two previous modeling approaches, not only do we consider the effects of risk information on no-trading days in the third approach, but also avoid processing the raw data. We construct regression models by generating a corresponding number of sequences to the lag number of days for any risk category. Moreover, the result has a novel finding that the risk government management has an impact on closing price increment, which shows that the raw data carry more information. The risk levels of holidays and weekends influence the closing price increment through different processing of risk information on no-trading days. In other words, the search queries data on holidays and weekends may have some effects on stock market. The subcategories of government management such as government policy and anti-corruption etc. also influence the volatility of the stock market. For instance, the economy and stock market of Lvliang city had crashed since the anti-corruption movement in Shanxi. However, the data processing of the third approach is a more complex work due to the construction of a total of 35 risk level sequences for each category.

The experimental results of the three methods show that social risk perception has a certain prediction effect on the Shanghai Composite Index. Given all that, the collective risk perception, reflect the public opinions' impacts on the stock market volatility, as well as social and economic conditions to some extent.

## 5    Conclusions

Societal risk perception is the subjective evaluation of public concerns to risk events. It reflects the public attitudes to social issues as well as government decision-making, which are the key indicators for effective social management and policy making. With the rapid growth of Internet data, an increasing number of researchers make use of user-generated contents to study social issues.

In this paper, we map on-line community concerns into respective societal risk events based on Baidu search engine by using the public searching behavior. Then we conduct the research on exploring the relationship between societal risk perception and stock market volatility. The main contributions and innovations are summarized as follows.

(1) Societal risk perception is quantitatively described. We present a reliable, scalable and timely assessment of the societal risk instead of large-scale survey which is expensive and time consuming to be conducted.
(2) The study on correlation between societal risk and stock price change is first carried out. Public searching queries are mapped into specific societal risk events. Then we analyze what kinds of risk events with risk labels from 7 risk categories have impacts on stock volatility. The results show that the risk of finance/economy, social stability and government management may cause the fluctuation of stock market.
(3) Different from previous research, the risk information of holiday is also taken into consideration to discuss the impact on stock price change based on three different methods. The results indicate that risk information of no-trading days can affect the stock volatility.

Lots of improvements are needed. The intrinsic mechanism of the relationship between societal risk and stock price fluctuation requires further research. Meanwhile, quantitative societal risk data which are more time-sensitive will be essentially used to study more social issues.

**Acknowledgments.** This research is supported by National Key Research and Development Program of China (2016YFB1000902) and National Natural Science Foundation of China (61473284 & 71371107).

# References

1. Zhang, C., Lv, B.F., Peng, G., Liu, Y., Yuan, Q.Y.: A study on correlation between web search data and CPI. In: Gaol, F. (ed.) Recent Progress in Data Engineering and Internet Technology. LNEE, vol. 157, pp. 269–274. Springer, Heidelberg (2012). doi:10.1007/978-3-642-28798-5_36
2. Choi, H., Varian, H.: Predicting the present with Google trends. Econ. Rec. **88**, 2–9 (2012)
3. Thomas, D., Jank, S.: Can internet search queries help to predict stock market volatility? Eur. Financ. Manage. **22**, 171–192 (2016)
4. Bank, M., Larch, M., Peter, G.: Google search volume and its influence on liquidity and returns of German stocks. Fin. Markets Portfolio Mgmt. **25**, 239–264 (2011)
5. Da, Z., Engelberg, J., Gao, P.: In search of attention. J. Financ. **66**, 1461–1499 (2011)
6. Takeda, F., Wakao, T.: Google search intensity and its relationship with returns and trading volume of Japanese stocks. Pac. Basin Financ. J. **27**, 1–18 (2014)
7. Curme, C., Preis, T., Stanley, H.E., Moat, H.S.: Quantifying the semantics of search behavior before stock market moves. Proc. Nat. Acad. Sci. **111**, 11600–11605 (2014)
8. Zhao, Y., Ye, Q., Li, Z.: The relationship between online attention and share prices. In: WHICEB 2014 Proceedings, pp. 462–469 (2014)
9. Xie, X.F., Xu, L.C.: The study of public risk perception. Psychol. Sci. **6**, 723–724 (2002). (in Chinese)
10. Zheng, R., Shi, K., Li, S.: The influence factors and mechanism of societal risk perception. In: Zhou, J. (ed.) Complex 2009. LNICSSITE, vol. 5, pp. 2266–2275. Springer, Heidelberg (2009). doi:10.1007/978-3-642-02469-6_104
11. Hu, Y., Tang, X.: Using support vector machine for classification of Baidu hot word. In: Wang, M. (ed.) KSEM 2013. LNCS, vol. 8041, pp. 580–590. Springer, Heidelberg (2013). doi:10.1007/978-3-642-39787-5_49
12. Tang, X.J.: Exploring on-line societal risk perception for harmonious society measurement. J. Syst. Sci. Syst. Eng. **22**, 469–486 (2013). doi:10.1007/s11518-013-5238-1
13. Tang, X.J.: Applying search words and BBS posts to societal risk perception and harmonious society measurement. In: IEEE International Conference on Systems, Man, and Cybernetics, pp. 2191–2196. IEEE Press (2013). doi:10.1109/SMC.2013.375
14. Dong, Y.H., Chen, H., Tang, X.J., Qian, W.Y., Zhou, A.Y.: Collective emotional reaction to societal risks in China. In: IEEE International Conference on Systems, Man, and Cybernetics, pp. 557–562. IEEE Press (2015). doi:10.1109/SMC.2015.108
15. Zhao, B., He, Y., Yuan, C., Hang, Y.: Stock market prediction exploiting microblog sentiment analysis. In: IEEE International Joint Conference on Neural Networks, pp. 4482–4488. IEEE Press (2016). doi:10.1109/IJCNN.2016.7727786
16. Chu, X., Wu, C., Qiu, J.: A nonlinear Granger causality test between stock returns and investor sentiment for Chinese stock market: a wavelet-based approach. Appl. Econ. **48**, 1915–1924 (2016). doi:10.1080/00036846.2015.1109048
17. Gao, T.M.: Econometric Analysis and Modeling. Tsinghua University Press, Beijing (2006). (in Chinese)

# A New Hybrid Linear-Nonlinear Model Based on Decomposition of Discrete Wavelet Transform for Time Series Forecasting

Warut Pannakkong$^{(\boxtimes)}$ and Van-Nam Huynh

School of Knowledge Science, Japan Advanced Institute of Science and Technology,
Ishikawa, Japan
{warut,huynh}@jaist.ac.jp

**Abstract.** Time series forecasting research area generally aims at improving prediction accuracy. Discrete wavelet transform (DWT) has been applied to time series for decomposing it into approximation and detail. Nevertheless, typically, the property of the approximation and the detail are presumed as either linear or nonlinear. Actually, the purpose of the DWT is not decomposing the original time series into linear and nonlinear time series. Hence, this paper develops a new hybrid model of autoregressive integrated moving average (ARIMA), artificial neural network (ANN), and the DWT without prior assumption on linear and nonlinear property of the approximation and the detail. The different Khashei and Bijari's hybrid models involving the ARIMA and the ANN are built for the approximation and the detail in order to extract their both linear and nonlinear components and fit the relationship between the components as the function instead of additive relationship. Finally, the forecasted approximation and detail are combined to obtain final forecasting. The prediction capability of the proposed model is examined with two well-known time series: the sunspot and the Canadian lynx time series. The results show that the proposed model has the best performance in all two data sets and all three measures (i.e. MSE, MAE and MAPE).

**Keywords:** Hybrid model · Time series forecasting · Autoregressive integrated moving average (ARIMA) · Artificial neural network (ANN) · Discrete wavelet transform (DWT)

## 1 Introduction

For decades, time series forecasting research area contributes several real-world applications in their prediction and decision making support [1]. This research area attempts to achieve better prediction accuracy by developing effective forecasting models. Traditionally, autoregressive integrated moving average (ARIMA) and artificial neural network (ANN) were developed and widely applied in time series forecasting. The ARIMA has an advantage in dealing with

© Springer Nature Singapore Pte Ltd. 2017
J. Chen et al. (Eds.): KSS 2017, CCIS 780, pp. 186–196, 2017.
https://doi.org/10.1007/978-981-10-6989-5_16

both stationary and non-stationary time series but it presumes that the relationship between inputs (e.g. historical time series) and outputs (e.g. future time series) is linear. On the other hand, the ANN has no such assumption. However, there is no a universal forecasting model that has the best performance in all situations. Hence, applying sole single forecasting model to time series prediction is not adequate to predict real-world time series [2].

Discrete wavelet transform (DWT), a transformation technique for signals, is adapted to transform time series into approximation (trend) and detail (noise) before further analysis. With the DWT, the prediction accuracy of the ARIMA and the ANN has been improved in many applications such as short term load [3]; electrical price [4,5]; groundwater level [6]; river discharge [7–9]; hourly flood [10]; rainfall and runoff [11,12].

Moreover, Khandelwal et al. [13] has developed the hybrid model of the ARIMA and the ANN (Zhang's model [2]) with the DWT. Nevertheless, this hybrid model considers the approximation as only a nonlinear component; in fact, there is no theoretical prove whether the approximation is linear or nonlinear. In addition, the additive relationship between linear and nonlinear components is assumed in the final forecasting step.

This study proposes a new hybrid model that can capture both linear and nonlinear components of the approximation and the detail, and has no assumption on relationship between linear and nonlinear components. Firstly, the discrete wavelet transform (DWT) is used to decompose the time series. Then, the hybrid model of ARIMA and ANN are constructed for the approximation and the detail to extract their linear and nonlinear components. Eventually, the final prediction is the combination of the predicted approximation and detail.

The rest of this paper is organized as follows. In Sect. 2, the ARIMA, the ANN, and the DWT are briefly explained. In Sect. 3, the proposed model is presented. The experiments and their results are shown and interpreted in Sect. 4. Finally, Sect. 5 provides the conclusions.

## 2   Preliminaries

### 2.1   Autoregressive Integrated Moving Average (ARIMA)

The autoregressive integrated moving average (ARIMA) is a popular forecasting model for decades due to its capability in handling both stationary and nonstationary time series [1]. However, The ARIMA assumes the relationship between predicted and historical time series as linear relationship. The ARIMA consists of three parts: autoregressive (AR), integration (I), and moving average (MA). In the situation that the time series is nonstationary, the time series is transformed by differencing in integration (I) step. The mathematical expression of the ARIMA can be written as:

$$\phi_p(B)(1-B)^d y_t = c + \theta_q(B)a_t \tag{1}$$

where $y_t$ and $a_t$ denote the time series and random error in period $t$ respectively, $\phi_p(B) = 1 - \sum_{i=1}^{p} \phi_i B^i$, $\theta_q(B) = 1 - \sum_{j=1}^{q} \theta_j B^j$, $B$ denotes the backward shift operator defined as $B^i y_t = y_{t-i}$, $\phi_i$ and $\theta_j$ denote the parameters of AR and MA respectively, $p$ and $q$ denote the orders of AR and MA respectively, and $d$ denotes the degree of differencing, and $c$ denotes the constant.

## 2.2 Artificial Neural Network (ANN)

The artificial neural network (ANN) is an artificial intelligent imitating biological neurons, and it is good at nonlinear modeling [14]. The ANN is widely used in time series forecasting because it is more flexible than the ARIMA in capturing relationship between predicted and historical time series without assumption. Typically, the ANN consists of three types of layer: input, hidden, and output layers. There are nodes in each layer. Normally, the architects choose the number of the layers and the nodes by their intuition in the problem and trial and error. Nevertheless, a feed-forward neural network that has only one hidden layer has been tested that it can be considered as a universal approximator [15]. The mathematical expression of the feed-forward neural network [16] can be written as:

$$y_t = f\left( b_h + \sum_{h=1}^{R} w_h g\left( b_{i,h} + \sum_{i=1}^{Q} w_{i,h} p_i \right) \right) \qquad (2)$$

where $y_t$ denotes the time series at period $t$, $b_{i,h}$ and $b_h$ denote the biases of hidden and output layers, $f$ and $g$ denote the transfer functions which are typically linear and nonlinear functions respectively, $w_{i,h}$ and $w_h$ denote the connection weights between the layers, $Q$ and $R$ denote the numbers of the input nodes and the hidden nodes respectively.

In this paper, the feed-forward neural network that has only one hidden layer and Levenberg-Marquardt algorithm with Bayesian regularization training algorithm [17] is applied in the experiments.

## 2.3 Discrete Wavelet Transform (DWT)

The wavelet transform is a tool for simultaneously analysis of both time and frequency of signals [18]. After the analysis, the original signal is decomposed into low frequency (approximation) and high frequency (detail) by applying low and high frequency pass filters. In case of multiple decomposition level, the approximation and the detail in the next level are the decomposition of the approximation in the previous level. In fact, there are two main categories of the wavelet transform such as continuous and discrete wavelet transforms. Nevertheless, in real-word applications, the time series are discrete and appropriate to be decomposed by the discrete wavelet transform (DWT) as:

$$\begin{aligned} y_t &= A_J(t) + \sum_{j=1}^{J} D_j(t) \\ &= \sum_{k=1}^{K} c_{J,k} \phi_{J,k}(t) + \sum_{j=1}^{J}\sum_{k=1}^{K} d_{j,k} \psi_{j,k}(t) \end{aligned} \qquad (3)$$

where $y_t$ denotes the time series in period $t$; $A_J(t)$ denotes the approximation in the highest decomposition level $(J)$; $D_j(t)$ denotes the detail in decomposition level $j$; $c_{j,k}$ and $d_{j,k}$ denote the coefficients of the approximation and detail respectively, in decomposition level $j$ and in period $k$; $\phi_{j,k}(t)$ and $\psi_{j,k}(t)$ denote low (approximation) and high (detail) pass filters respectively, in decomposition level $j$ and at period $k$; $K$ denotes the length of the time series; $J$ denotes the highest level of decomposition.

## 3 Proposed Forecasting Model

The main objective of developing the proposed model is to obtain the advantage of both the ARIMA and the ANN in fitting linear and nonlinear components from the time series without presuming the characteristic of the approximation and the detail as either linear or nonlinear. The proposed model can be divided into three steps: decomposition of the time series, capturing linear and nonlinear components, and final forecasting (Fig. 1).

In the first step, the actual time series $(y_t)$ is decomposed by the DWT with Daubechies wavelet basis function in order to obtain the approximation $(y_t^{app})$, which presents the trend, and the detail $(y_t^{det})$, which is the difference between actual value and the trend. The pattern of the detail reveals the seasonality, the white noise, etc.

Instead of applying a forecasting model to time series consisting of both trend and noise, using different forecasting models to separately predict the trend from the approximation and predict the noise (e.g. seasonality and white noise) from

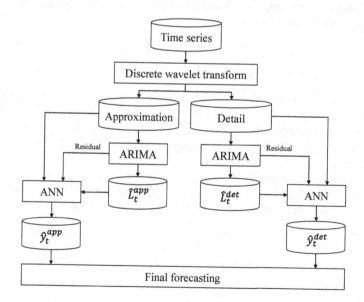

**Fig. 1.** The proposed forecasting model

the detail can provide better prediction results because each forecasting model deal with only either trend or noise, not both of them simultaneously.

In the second step, the Khashei and Bijari's hybrid model of the ARIMA and the ANN [19] is applied to both the approximation and the detail. This step contributes the new approach that does not make either linear or nonlinear assumption to the property of the approximation and the detail, and does not assume additive relationship between their linear and nonlinear components as well.

Generally, the Khashei and Bijari's model forecasts the future time series at period $t$ ($\hat{y}_t$) by using linear ($\hat{L}_t$) and nonlinear ($\hat{N}_t$) components as:

$$\hat{y}_t = f(\hat{L}_t, \hat{N}_t) \tag{4}$$

The linear component ($\hat{L}_t$) is the result of adopting the ARIMA to the actual time series ($y_t$). After that, the residual of the ARIMA ($e_t$) can be computed as:

$$e_t = y_t - \hat{L}_t \tag{5}$$

In case of the nonlinear component ($\hat{N}_t$), it can be obtained from the ANN that has the lagged values of both the time series ($y_t$) and the ARIMA residual ($e_t$) as its inputs:

$$\hat{N}_t^1 = f^1(e_{t-1}, e_{t-2}, \ldots, e_{t-n}) \tag{6}$$

$$\hat{N}_t^2 = f^2(y_{t-1}, y_{t-2}, \ldots, y_{t-m}) \tag{7}$$

where $f^1$ and $f^2$ denote the function fitted by the ANN, $n$ and $m$ is total included lagged periods.

In the proposed model, the Khashei and Bijari's model is separately built for the approximation and the detail as:

$$\hat{y}_t^{app} = f^{app}(\hat{L}_t^{app}, \hat{N}_t^{app}) \tag{8}$$

$$\hat{y}_t^{det} = f^{det}(\hat{L}_t^{det}, \hat{N}_t^{det}) \tag{9}$$

where $\hat{y}_t^{app}$ and $\hat{y}_t^{det}$ denote the forecasted approximation and detail respectively, in period $t$; $f^{app}$ and $f^{det}$ denote the function fitted by the ANN; $\hat{L}_t^{app}$ and $\hat{L}_t^{det}$ denote the linear components of the approximation and the detail respectively, in period $t$; $\hat{N}_t^{app}$ and $\hat{N}_t^{det}$ denote the nonlinear components of the approximation and the detail respectively, in period $t$.

The linear components ($\hat{L}_t^{app}$ and $\hat{L}_t^{det}$) denote the result of applying the ARIMA to $y_t^{app}$ and $y_t^{det}$ respectively. Then, the ARIMA residual of the approximation ($e_t^{app}$) and the detail ($e_t^{det}$) can be mathematically expressed as:

$$e_t^{app} = y_t^{app} - \hat{L}_t^{app} \tag{10}$$

$$e_t^{det} = y_t^{det} - \hat{L}_t^{det} \tag{11}$$

For the nonlinear components ($\hat{N}_t^{app}$ and $\hat{N}_t^{det}$), they can be produced from the ANN as:

$$\hat{N}_t^{app1} = f^{app1}(e_{t-1}^{app}, e_{t-2}^{app}, \ldots, e_{t-n_1}^{app}) \tag{12}$$

$$\hat{N}_t^{app2} = f^{app2}(y_{t-1}^{app}, y_{t-2}^{app}, \ldots, y_{t-m_1}^{app}) \tag{13}$$

$$\hat{N}_t^{det1} = f^{det1}(e_{t-1}^{det}, e_{t-2}^{det}, \ldots, e_{t-n_2}^{det}) \tag{14}$$

$$\hat{N}_t^{det2} = f^{det2}(y_{t-1}^{det}, y_{t-2}^{det}, \ldots, y_{t-m_2}^{det}) \tag{15}$$

where $f^{app1}$, $f^{app2}$, $f^{det1}$, and $f^{det2}$ denote functions fitted by the ANN; $n_1$, $n_2$, $m_1$, and $m_2$ denote total lagged periods that are identified by trial and error in the experiments.

After that, the forecasted approximation $(\hat{y}_t^{app})$ and the forecasted detail $(\hat{y}_t^{det})$ can be obtained as:

$$\hat{y}_t^{app} = f^{app}(\hat{L}_t^{app}, e_{t-1}^{app}, e_{t-2}^{app}, \ldots, e_{t-n_1}^{app}, y_{t-1}^{app}, y_{t-2}^{app}, \ldots, y_{t-m_1}^{app}) \tag{16}$$

$$\hat{y}_t^{det} = f^{det}(\hat{L}_t^{det}, e_{t-1}^{det}, e_{t-2}^{det}, \ldots, e_{t-n_2}^{det}, y_{t-1}^{det}, y_{t-2}^{det}, \ldots, y_{t-m_2}^{det}) \tag{17}$$

Finally, the final forecasting step is performed by combining of the forecasted approximation $(\hat{y}_t^{app})$ and the forecasted detail $(\hat{y}_t^{det})$ as:

$$\hat{y}_t = \hat{y}_t^{app} + \hat{y}_t^{det} \tag{18}$$

In sum, rather than applying the Khashei and Bijari's model direct to the time series, the DWT is used at first to transform the time series into the approximation (trend) and the detail (noise). Then, without assuming linear or nonlinear properties of the approximation and the detail, the Khashei and Bijari's model is adopted to both of them. After the specific Khashei and Bijari's models have been separately built to capture the trend and the noise, they would give the better forecasting result because the different Khashei and Bijari's models concentrate on only either trend or noise (not both of them simultaneously). In addition, the relationship of the linear and nonlinear components is defined as the function instead of additive relationship. Finally, the final forecasting is performed by additive combination between forecasted approximation the detail because the relationship between them is additive as well.

## 4    Experiments and Results

To assess forecasting capability of the proposed model, two well-known time series (Table 1) are used as case studies such as Wolf's sunspot (Fig. 2) and Canadian lynx (Fig. 3). The measures of forecasting performance used in this paper are mean square error (MSE), mean absolute error (MAE) and mean absolute percentage error (MAPE). The performance of the proposed model is

Table 1. Detail of time series and experiment

| Time series | Size (total, training, test) |
| --- | --- |
| Sunspot (1700–1987) | (288, 221, 67) |
| Canadian lynx (1821–1934) | (114, 100, 14) |

Amount of spot

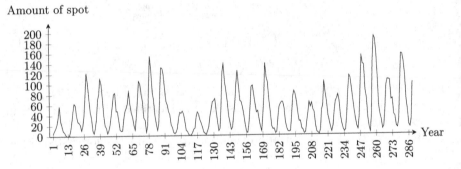

**Fig. 2.** Sunspot time series (1700–1987)

Amount of lynx

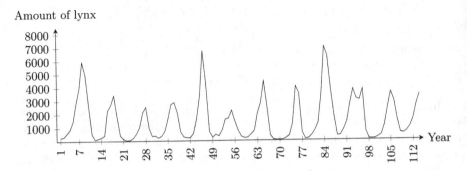

**Fig. 3.** Canadian lynx time series (1821–1934)

**Table 2.** Sunspot forecasting result

| Model | 35 year ahead | | | 67 year ahead | | |
|---|---|---|---|---|---|---|
| | MSE | MAE | MAPE | MSE | MAE | MAPE |
| ARIMA | 197.87 | 10.52 | 29.17% | 323.48 | 13.25 | 32.86% |
| ANN | 164.08 | 9.51 | 31.76% | 413.90 | 14.19 | 33.34% |
| Zhang | 156.76 | 9.63 | 30.22% | 300.88 | 12.74 | 32.08% |
| Khashei and Bijari | 127.67 | 8.60 | 22.85% | 273.15 | 12.14 | 25.31% |
| Khandelwal et al. | 144.09 | 8.21 | 20.22% | 378.00 | 12.21 | 21.95% |
| Proposed model | **121.52** | **5.51** | **16.21%** | **206.32** | **8.07** | **19.19%** |

compared with the ARIMA, the ANN, the Zhang's model, Khashei and Bijari's model, and Khandelwal et al.'s model.

For the sunspot time series, it contains 288 annual records (1700–2987). The training and test sets are 221 records (1700–1920) and 67 records (1921–1987) respectively. Firstly, the sunspot time series is decomposed by the DWT into the approximation and the detail. Secondly, the ARIMA is applied to both the approximation and the detail. The most fitted ARIMA for the approximation

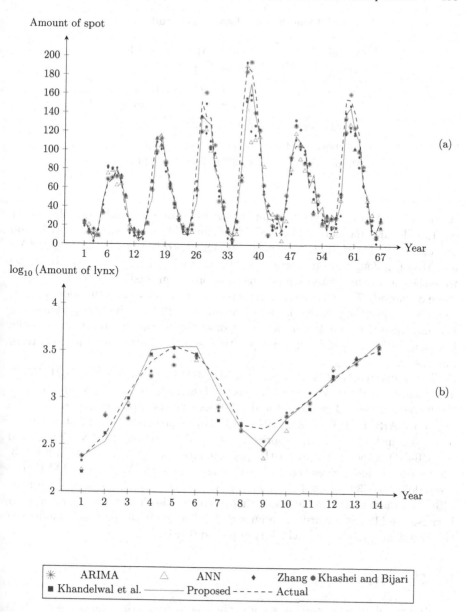

Fig. 4. Forecasted values: (a) Sunspot, (b) Canadian lynx

and the detail are ARIMA$(0, 0, 6)$ and ARIMA$(0, 0, 3)$ respectively. Thirdly, the forecasted approximation and detail are generated by the best fitted ANNs that are ANN(16-1-1) and ANN(9-10-1) respectively. Then, the final forecasting is computed from the combination of the predicted approximation and detail. After obtaining the final forecasting, the performance measures are computed for short term (35 years) and long term (67 years) horizontal predictions (Table 2).

**Table 3.** Lynx forecasting result

| Model | MSE | MAE | MAPE |
|---|---|---|---|
| ARIMA | 0.0229 | 0.1120 | 3.7062% |
| ANN | 0.0201 | 0.1165 | 4.0156% |
| Zhang | 0.0247 | 0.1083 | 3.5504% |
| Khashei and Bijari | 0.0160 | 0.0980 | 3.2381% |
| Khandelwal et al. | 0.0195 | 0.0873 | 2.9737% |
| Proposed model | **0.0071** | **0.0639** | **2.1114%** |

According to the performance comparison, the proposed model has the lowest error in all three measures. The MSE, MAE and MAPE in short term prediction are 121.52, 5.51 and 16.21% respectively. On the other hand, the MSE, MAE and MAPE in long term prediction are 206.32, 8.07 and 19.19% which are higher than the short term prediction because the long term prediction has the highest value at period 37 (see Fig. 4a) that increases the variance causing more prediction error. Nevertheless, the proposed model can still have the best performance because the other models also perform worse. Hence, the proposed model is the best forecasting model for the sunspot time series in both short and long term forecasting.

The Canadian lynx time series consists of 114 annual records (1821–1934). The training and test sets are 100 records (1821–1920) and 14 records (1921–1934) respectively. The best fitted ARIMAs of the approximation and the detail are ARIMA$(0,0,5)$ and ARIMA$(2,0,0)$ respectively. The most appropriate ANNs in forecasting the approximation and the detail are ANN(7-9-1) and ANN(6-3-1) respectively. From the performance comparison shown in Table 3, the proposed model gives the best performance in MSE, MAE and MAPE that are 0.0071, 0.0639 and 2.1114% respectively. The most improved measure is MSE. The lower MSE gives more chance to promise lower maximum of error because the MSE is sensitive to a huge error. Therefore, the proposed model has the lowest maximum error which is in period 9 (Fig. 4b).

## 5   Conclusions

In order to enhance forecasting accuracy in time series prediction, the new hybrid model of the ARIMA, the ANN, and the DWT has been proposed. The proposed model analyses the time series without assuming linear and nonlinear properties on the approximation and the detail, and defines the relationship of the linear and nonlinear components of both the approximation and the detail as the function. The prediction capability of the proposed model is examined with two well-known time series: the sunspot and the Canadian lynx time series. The results show that the proposed model has the best performance in all two data sets and all three measures (i.e. MSE, MAE and MAPE). The improved performance implies benefit of hybridization of the ARIMA, the ANN, and the DWT in capturing the linear and nonlinear components of the approximation and the detail without prior assumption on their properties.

The limitation in the experiment is that the level of decomposition is one. For future works, the impact of different decomposition levels will be considered. Moreover, the statistical test will be performed to measure the significant level of performance improvement.

# References

1. De Gooijer, J.G., Hyndman, R.J.: 25 years of time series forecasting. Int. J. Forecast. **22**(3), 443–473 (2006)
2. Zhang, G.P.: Time series forecasting using a hybrid ARIMA and neural network model. Neurocomputing **50**, 159–175 (2003)
3. Fard, A.K., Akbari-Zadeh, M.R.: A hybrid method based on wavelet, ANN and ARIMA model for short-term load forecasting. J. Exp. Theor. Artif. Intell. **26**(2), 167–182 (2014)
4. Conejo, A.J., Plazas, M.A., Espinola, R., Molina, A.B.: Day-ahead electricity price forecasting using the wavelet transform and ARIMA models. IEEE Trans. Power Syst. **20**(2), 1035–1042 (2005)
5. Tan, Z., Zhang, J., Wang, J., Xu, J.: Day-ahead electricity price forecasting using wavelet transform combined with ARIMA and GARCH models. Appl. Energ. **87**(11), 3606–3610 (2010)
6. Adamowski, J., Chan, H.F.: A wavelet neural network conjunction model for groundwater level forecasting. J. Hydrol. **407**(1), 28–40 (2011)
7. Zhou, H.C., Peng, Y., Liang, G.H.: The research of monthly discharge predictor-corrector model based on wavelet decomposition. Water Resour. Manage. **22**(2), 217–227 (2008)
8. Wei, S., Zuo, D., Song, J.: Improving prediction accuracy of river discharge time series using a wavelet-NAR artificial neural network. J. Hydroinformatics **14**(4), 974–991 (2012)
9. Adamowski, J., Sun, K.: Development of a coupled wavelet transform and neural network method for flow forecasting of non-perennial rivers in semi-arid watersheds. J. Hydrol. **390**(1), 85–91 (2010)
10. Tiwari, M.K., Chatterjee, C.: Development of an accurate and reliable hourly flood forecasting model using wavelet-bootstrap-ANN (WBANN) hybrid approach. J. Hydrol. **394**(3), 458–470 (2010)
11. Nourani, V., Komasi, M., Mano, A.: A multivariate ANN-wavelet approach for rainfallruno modeling. Water Resour. Manage. **23**(14), 2877–2894 (2009)
12. Partal, T., Kisi, Ö.: Wavelet and neuro-fuzzy conjunction model for precipitation forecasting. J. Hydrol. **342**(1), 199–212 (2007)
13. Khandelwal, I., Adhikari, R., Verma, G.: Time series forecasting using hybrid ARIMA and ANN models based on DWT decomposition. Procedia Comput. Sci. **48**, 173–179 (2015)
14. Zhang, G., Patuwo, B.E., Hu, M.Y.: Forecasting with artificial neural networks: the state of the art. Int. J. Forecast. **14**(1), 35–62 (1998)
15. Hecht-Nielsen, R.: Theory of the backpropagation neural network. In: International Joint Conference on Neural Networks, IJCNN 1989, pp. 593–605 (1989)

196     W. Pannakkong and V.-N. Huynh

16. Dayhoff, J.A.: Neural Network Architectures: An Introduction. MIT press, Cambridge (1995)
17. MacKay, D.J.: A practical bayesian framework for backpropagation networks. Neural comput. **4**(3), 448–472 (1992)
18. Fliege, N.J.: Multirate Digital Signal Processing: Multirate Systems, Filter Banks. Wavelets. John Willy & Sons, Chichester (1994)
19. Khashei, M., Bijari, M.: A novel hybridization of artificial neural networks and ARIMA models for time series forecasting. Appl. Soft Comput. **11**(2), 2664–2675 (2011)

# The Distribution Semantics of Extended Argumentation

Nguyen Duy Hung[(✉)]

Sirindhorn International Institute of Technology, Pathum Thani, Thailand
hung.nd.siit@gmail.com

**Abstract.** The distribution semantics is a de facto approach for integrating logic programming with probability theory, and recently has been applied for the standard abstract argumentation framework. In this paper, we define the distribution semantics for extended argumentation frameworks, and moreover derive inference procedures from existing proof procedures of such extended argumentation frameworks. While doing so we focus on extended argumentation frameworks with attacks on attacks and the inductive defense semantics thereof. However our results can be easily obtained for other extended frameworks and semantics.

**Keywords:** Distribution semantics · Argumentation · Procedures

## 1 Introduction

In his seminal work [17], noticing that definite logic programs augmented with probabilistic elements can emulate probabilistic graphical models such as Bayesian networks, Sato defined a probabilistic logic program as a pair of a set of ground probabilistic facts $F$ and a definite program $\Pi$ whose heads do not occur in $F$. The probability measure on the interpretations of $F$ is then extended to produce a probability measure of LHMs (Least Herbrand Models) of definite programs $\{\Pi \cup F' \mid F' \subseteq F\}$. However, when $\Pi$ is extended to allow negations, LHM may not exist and hence some generalization of LHM has to be used. In particular, the state-of-the-art Probabilistic Logic Programming (PLP) languages (e.g. [6,16,18]) generalize Sato's distribution semantics to normal logic programs using either the well-founded semantics or the answer set semantics.

It is worth noting that when negations are present, there can be many attacking arguments constructable from the program, and the question is then which arguments should be *acceptable* - a central topic in argumentation studies [4]. In his landmark paper [3], Dung defines an abstract argumentation framework (AF) as a pair $(Arg, Att)$ where $Arg$ is a set of arguments and $Att \subseteq Arg \times Arg$ is a set of attacks. He then introduces several argumentation semantics all relying on the same natural notion of argument acceptability, namely an argument $X$ is acceptable wrt a set $S$ of arguments (aka $S$ defends $X$) iff $S$ attacks every argument attacking $X$. For example, wrt the AF framework in Fig. 1(a)[1], $B$ and

---

[1] Arguments and attacks are shown as nodes and directed edges respectively.

© Springer Nature Singapore Pte Ltd. 2017
. Chen et al. (Eds.): KSS 2017, CCIS 780, pp. 197–211, 2017.
https://doi.org/10.1007/978-981-10-6989-5_17

198    N.D. Hung

$B_1$ are acceptable wrt $\{B, B_1\}$. On the relationships between argumentation and logic programming, Dung shows that different semantics of argumentation frameworks such as grounded extensions and stable extensions correspond to the well-founded and answer set semantics of normal logic programs [3]. Since [3], AF becomes a unified reasoning formalism in AI, and also extended in several directions to address different shortcomings. Recently, Probabilistic Argumentation frameworks (PAF) [5,8,11,12,15,19] extends AF with classical probability theory. Intuitively, if a PAF framework uses the distribution semantics[2], then it will be viewed as a probabilistic distribution of AF frameworks representing different "possible worlds". An argument may or may not occur in a possible world, as exemplified by Fig. 1(b) which shows an PAF framework with three possible worlds $\{\omega_0, \omega_1, \omega_2\}$[3]. The probability that an argument $X$ is acceptable can be obtained by marginalizing the joint probability distribution of possible worlds and the acceptability of $X$.

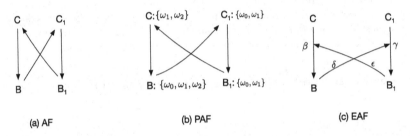

**Fig. 1.** Different argumentation frameworks

The main purpose of this paper is to investigate the distribution semantics for Extended Argumentation frameworks (EAFs [1,2,7,9,14]) where attacks, as arguments, can be attacked. An attack can come from an argument (as in [7,9]), or an attack (as in [2]), or a set of them (as in [14]). In particular, we show that it is possible to define the distribution semantics for any extended abstract argumentation framework and moreover, its inference procedures can be derived automatically from the proof procedures of the extended argumentation framework. While doing so we focus on EAFs with only attacks coming from arguments [7,9] (for illustration, see the EAF framework in Fig. 1-c[4]) and the inductive defense semantics for these EAFs introduced in [9]. However our results can be easily obtained for other extended argumentation frameworks and semantics. To the best of our knowledge, this paper is the first proposal for Probabilistic Extended Argumentation framework (PEAF). However focusing on PEAF semantics and inference procedures, we only briefly discuss applications.

---

[2] Not all PAF proposals use the distribution semantics (detailed in the paper body).
[3] Each argument is annotated with possible worlds it occurs.
[4] An attack $(C, B)$ where $B$ is an argument is shown by an arrow $C \rightarrow B$. An attack $(B, \gamma)$ where $\gamma$ is an attack is shown by an arrow from $B$ to the arrow showing $\gamma$.

The rest of the paper is structured as follows: Sect. 2 recalls AF, PAF, EAF and also EAF proof procedures; Sect. 3 presents our model of PEAF; Sect. 4 translates EAF proof procedures into PEAF inference procedures; Sect. 5 presents the procedural form of the resulted PEAF inference procedures and demonstrates a Prolog-based implementation[5]; Sect. 6 concludes. For the lack of space, we prove the technical results only sketchily.

## 2   Background

### 2.1   Argumentation Frameworks

An Abstract Argumentation framework (AF) is a pair $(Arg, Att)$ where $Arg$ is a set of arguments, $Att \subseteq Arg \times Arg$ and $(A, B) \in Att$ means that $A$ attacks $B$ [3]. $S \subseteq Arg$ attacks $A \in Arg$ iff $(B, A) \in Att$ for some $B \in S$. $S$ defends $A$ (aka $A$ is acceptable wrt to $S$) iff $S$ attacks every argument attacking $A$. $S$ is *conflict-free* iff $S$ does not attack itself; *admissible* iff $S$ is conflict-free and defends each argument in $S$; *complete* iff $S$ is admissible and contains all arguments that $S$ defends; a *preferred* extension iff $S$ is a maximal (wrt set inclusion) complete set; the *grounded* extension iff $S$ is the least complete set. An argument $A$ is accepted under semantics *sem*, denoted AF $\mathcal{F} \vdash_{sem} A$, iff $A$ is in a *sem* extension. In this paper, we restrict ourselves to *sem* $\in \{pr, gr\}$[6].

Probabilistic Argumentation (PAF) [5,8,11,12,15,19] extends AF with classical probability theory. The earliest known proposal [5] defines a PAF framework by a triple $(\mathcal{F}, \mathcal{W}, P)$ where $\mathcal{F} = (Arg, Att)$ is a standard AF framework, $\mathcal{W}$ is a set of possible worlds such that each $\omega \in \mathcal{W}$ defines a subset of arguments $Arg_\omega \subseteq Arg$, and $P : \mathcal{W} \rightarrow [0, 1]$ is a probability distribution over $\mathcal{W}$ (i.e. $\sum_{\omega \in \mathcal{W}} P(\omega) = 1$). The probability that an argument $A$ is acceptable under a semantics *sem* is obtained by marginalizing the joint probability distribution of possible worlds and the acceptability of $A$. Concretely,

$$Prob_{sem}(A) = \sum_{\omega \in \mathcal{W}} P(\omega).P(\mathcal{F}_\omega \vdash_{sem} A \mid \omega) = \sum_{\omega \in \mathcal{W}: \mathcal{F}_\omega \vdash_{sem} A} P(\omega)$$

where $\mathcal{F}_\omega$ denotes the AF framework $(Arg_\omega, Att \cap (Arg_\omega \times Arg_\omega))$.

*Example 1.* Figure 2 shows AF $\mathcal{F}_\omega$, $\omega \in \{\omega_0, \omega_1, \omega_2\}$ of the PAF framework in Fig. 1(b). Since $\mathcal{F}_{\omega_0} \vdash_{gr} B_1$ and $\mathcal{F}_{\omega_1} \vdash_{pr} B_1$, $Prob_{gr}(B_1) = P(\omega_0)$ while $Prob_{pr}(B_1) = P(\omega_0) + P(\omega_1)$.

It is worth noting that the PAF proposals in [12,15] also use the above distribution semantics but the ones in [8,11,19] do not[7].

---

[5] Download link: http://ict.siit.tu.ac.th/~hung/peafengine.

[6] preferred/grounded.

[7] The PAF proposals of [8,11,19] define their semantics in terms of *rational conditions* on Probabilistic Distribution Function (PDF) $f : Arg \rightarrow [0, 1]$, for $f(A)$ to represent some value of argument $A$, which may not relate to the acceptability of $A$.

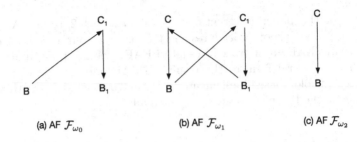

**Fig. 2.** AF frameworks $\mathcal{F}_{\omega_0}, \mathcal{F}_{\omega_1}, \mathcal{F}_{\omega_2}$.

In fact an even earlier research line to extend AF is known as Extended Argumentation (EAF) where attacks, as arguments, can be attacked [1,2,7,14]. An attack can come from an argument, or an attack (as in [2]), or a set of them (as in [14]). In this paper we consider only attacks coming from arguments, so following [7] we define an EAF framework as a pair $(Arg, Att)$ where $Arg$ is a set of arguments and $Att$ is a set of attacks s.t. $Att = \bigcup_{i=0}^{\infty} Att^i$, where $Att^0 \subseteq Arg \times Arg$ and $Att^{i+1} \subseteq Arg \times Att^i$.

*Example 2.* Figure 1(c) shows an EAF framework $(Arg, Att)$ where $Arg = \{B, C, B_1, C_1\}$ and $Att = Att^0 \cup Att^1$ with $Att^0 = \{\beta, \gamma\}$ and $Att^1 = \{\delta, \epsilon\}$.

One of intuitive uses of EAF is to model preference-based argumentation. For illustration let's borrow the following example from [13].

*Example 3.* Consider two persons **P** and **Q** arguing about weather forecast:

**P:** "Today will be dry in London since the BBC forecast sunshine" $(A)$.
**Q:** "Today will be wet in London since CNN forecast rain" $(B)$.
**P:** "But the BBC are more trustworthy than CNN" $(C)$.
**Q:** "However, statistically CNN are more accurate forecasters than BBC" $(C')$.
**Q:** "And basing a comparison on statistics is more rigorous and rational than basing a comparison on your instincts about their relative trustworthiness" $(E)$.

$A$ and $B$ claim contradictory conclusions and hence attack each other. $C$ and $C'$ are arguments that express preferences for $A$ over $B$ and $B$ over $A$ respectively. $C$ hence attacks the attack from $B$ against $A$. Similarly $C'$ attacks the attack from $A$ against $B$. These preferences are contradictory, so $C$ and $C'$ attack each other. At last $E$ states that $C'$ is preferred to $C$.

The above debate can be represented by the EAF in Fig. 3 where $AR = \{A, B, C, C', E\}$ and $Att = Att^0 \cup Att^1$ with

- $Att^0 = \{\alpha, \beta, \epsilon, \zeta\}$, $\alpha = (A, B), \beta = (B, A), \epsilon = (C', C)$ and $\zeta = (C, C')$
- $Att^1 = \{\gamma, \delta, \eta\}$, $\gamma = (C', \alpha), \delta = (C, \beta)$ and $\eta = (E, \zeta)$.

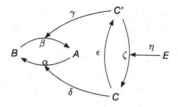

**Fig. 3.** EAF for preference-based argumentation

Also many semantics have been proposed for EAF (e.g. [1,7,9,13]). Here we are interested in the i-defense semantics [9], which is recalled below.

For $\alpha = (A, X) \in Att$, $src(\alpha)$ and $targ(\alpha)$ refers to $A$ and $X$ respectively. For $X \in Arg \cup Att$, $Attack_X \triangleq \{\alpha \in Att \mid targ(\alpha) = X\}$. For $S \subseteq Arg$,

- $S$ *i-defends* (inductively defends) $\alpha \in Att$ iff $S$ i-defends $\alpha$ within some steps. $S$ i-defends $\alpha$ within 0 steps iff there is no $\beta \in Att$ such that $targ(\beta) = \alpha$. $S$ i-defends $\alpha$ within $(k+1)$ steps iff $S$ i-defends $\alpha$ within k steps or for each $\beta \in Att$ where $targ(\beta) = \alpha$, there is $\gamma \in Att$ such that $S$ i-defends $\gamma$ within k steps, $src(\gamma) \in S$ and $targ(\gamma) \in \{\beta, src(\beta)\}$.
- $S$ i-defends $A \in Arg$ iff for each $\beta \in Att$ where $targ(\beta) = A$, there is $\gamma \in Att$ such that $S$ i-defends $\gamma$, $src(\gamma) \in S$ and $targ(\gamma) \in \{\beta, src(\beta)\}$.
- $S$ is *i-conflict free* iff $S$ does not i-defends any $(A, B) \in Att$ where $A, B \in S$.
- $S$ is *i-admissible* iff $S$ is i-conflict free and i-defends every argument in $S$.
- $S$ is an *i-preferred* extension iff $S$ is a maximally i-admissible set; an *i-complete* extension iff $S$ is i-admissible and contains all arguments that it i-defends; the *i-grounded* extension iff $S$ is the least i-complete extension. Finally, argument $A$ is accepted under semantics $sem \in \{ipr, igr\}$[8], denoted $\mathcal{F} \vdash_{sem} A$, iff $A$ is in a $sem$ extension.

*Example 4* (Continue Example 2). $\{B, B_1, C, C_1\}$ is the only i-preferred extension and the i-grounded extension is $\{C, C_1\}$.

## 2.2 EAF Proof Procedures

The i-acceptability of an argument can be evaluated by simulating two-party disputes, where a proponent starts by putting forward the argument, then alternates with an opponent in attacking each other's previous arguments and the associated attacks. Formally in [9] a dispute is represented by a dispute derivation in which states are tuples $t_i = \langle P_i, O_i, SP_i, SO_i \rangle$. $P_i \subseteq Arg \cup Att$ is a set of arguments and attacks presented by the proponent that have not been defended by the proponent. $SP_i \subseteq Arg$ is the set of all arguments presented by the proponent. $O_i \subseteq Att$ is a set of attacks of the opponent against arguments presented by the proponent in previous steps that are not counter-attacked yet by the proponent. $SO_i$ contains attacks by the opponent that have been counter-attacked by the proponent.

---

[8] i-preferred/i-grounded.

**Definition 1.** *Given a selection function, a dispute derivation for an argument $A$ is a sequence $\langle P_0, O_0, SP_0, SO_0 \rangle \ldots \langle P_n, O_n, SP_n, SO_n \rangle$ where*

1. $P_i \subseteq Arg \cup Att;\ SP_i \subseteq Arg,$ and $O_i, SO_i \subseteq Att.$
2. $P_0 = SP_0 = \{A\}$ and $O_0 = SO_0 = P_n = O_n = \emptyset.$
3. At step $i$, let $X$ be an element selected from either $P_i$ or $O_i$

   (a) If $X \in P_i$ then: $\begin{array}{ll} P_{i+1} = P_i \setminus \{X\} & O_{i+1} = O_i \cup Attack_X \\ SP_{i+1} = SP_i & SO_{i+1} = SO_i \end{array}$

   (b) If $X \in O_i$, then there exists some attack $\alpha$ such that

      i. if $\alpha \in src(X) \in SP_i$ then $\alpha \in Attack_X \setminus (SO_i \cup O_i),$

      ii. otherwise $\alpha \in (Attack_{src(X)} \cup Attack_X) \setminus (SO_i \cup O_i)^9,$

      and: $P_{i+1} = P_i \cup \{src(\alpha), \alpha\}$ if $src(\alpha) \notin SP_i$, otherwise $P_{i+1} = P_i \cup \{\alpha\}^{10}$

      $O_{i+1} = O_i \setminus \{X\} \quad SP_{i+1} = SP_i \cup \{src(\alpha)\} \quad SO_{i+1} = SO_i \cup \{X\}$

As shown in [9], for a bounded EAF[11], dispute derivations represent sound and complete proofs for i-preferred semantics because: (1) if $\langle P_0, O_0, SP_0, SO_0 \rangle \ldots \langle P_n, O_n, SP_n, SO_n \rangle$ is a dispute derivation for argument $A$, then $SP_n$ is i-admissible and contains $A$; (2) if an argument $A$ belongs to an i-admissible set $S$ of arguments in a finite argumentation framework, then for any selection function there is a dispute derivation for $A$, whose component $SP_n$ of the final tuple is a subset of $S$.[12]

Continue Example 4, a dispute derivation for argument $B_1$ is shown below.

| i | $P_i$ | $O_i$ | $SP_i$ | $SO_i$ | Rule (of Definition 1) used | Remark |
|---|---|---|---|---|---|---|
| 0 | $\{\underline{B_1}\}$ | $\{\}$ | $\{B_1\}$ | $\{\}$ | Rule 2 | Proponent starts dispute |
| 1 | $\{\}$ | $\{\underline{\gamma}\}$ | $\{B_1\}$ | $\{\}$ | 3(a): $Attack_{B_1} = \{\gamma\}$ | Opponent attacks $B_1$ with $\gamma$ |
| 2 | $\{B, \underline{\delta}\}$ | $\{\}$ | $\{B_1, B\}$ | $\{\gamma\}$ | 3(b)ii: $Attack_{C_1} \cup Attack_\gamma = \{\delta\}$ | Proponent counters by $\delta$ |
| 3 | $\{\underline{B}\}$ | $\{\}$ | $\{B_1, B\}$ | $\{\gamma\}$ | 3(a): $Attack_\delta = \{\}$ | There are no attacks against $\delta$ |
| 4 | $\{\}$ | $\{\underline{\beta}\}$ | $\{B_1, B\}$ | $\{\gamma\}$ | 3(a): $Attack_B = \{\beta\}$ | Opponent attacks $B$ with $\beta$ |
| 5 | $\{\underline{\epsilon}\}$ | $\{\}$ | $\{B_1, B\}$ | $\{\gamma, \beta\}$ | 3(b)ii: $Attack_C \cup Attack_\beta = \{\epsilon\}$ | Proponent attacks $\beta$ with $\epsilon$ |
| 6 | $\{\}$ | $\{\}$ | $\{B_1, B\}$ | $\{\gamma, \beta\}$ | 3(a): $Attack_\epsilon = \{\}$ | Proponent wins |

# 3   Probabilistic Extended Argumentation

In this section we define Probabilistic Extended Argumentation frameworks (PEAFs). In accordance with the distribution semantics, a PEAF framework defines a probabilistic distribution of EAF frameworks.

---

[9] Since $X \in Att$, the proponent can attack $src(X)$ or $X$. He should not attack $src(X)$ if this is an argument he moved previously.

[10] If $src(\alpha) \in SP_i$, then the proponent does not need to re-defend $src(\alpha)$.

[11] An EAF $(Arg, Att)$ is bounded if for each $X \in Arg \cup Att$, $Attack_X$ is finite.

[12] To compute i-grounded semantics, dispute derivations have to be equipped with slightly different filtering mechanisms which we do not explore here.

**Definition 2.** *A Probabilistic Extended Argumentation (PEAF) framework $\mathcal{P}$ is a triple $\mathcal{P} = (\mathcal{F}, \mathcal{W}, P)$ where $\mathcal{F} = (Arg, Att)$ is an EAF framework, $\mathcal{W}$ is a set of possible worlds such that each $\omega \in \mathcal{W}$ defines a subset of arguments $Arg_\omega \subseteq Arg$, and $P : \mathcal{W} \to [0,1]$ is a probability distribution over $\mathcal{W}$.*

As illustrated by Example 5, we can visualize a PEAF framework $\mathcal{P} = (\mathcal{F}, \mathcal{W}, P)$ by annotating each node $A \in Arg$ of the graph of EAF $\mathcal{F}$ with the set of possible worlds in which $A$ occurs, i.e. $\{\omega \in \mathcal{W} \mid A \in Arg_\omega\}$. From now on, we refer to this set by $W_A$.

*Example 5.* Figure 4(a) shows a PEAF framework $\mathcal{P} = (\mathcal{F}, \mathcal{W}, P)$ where $\mathcal{F}$ is the EAF defined in Example 2; $\mathcal{W} = \{\omega_0, \omega_1, \omega_2\}$ where $Arg_{\omega_0} = \{B, B_1, C_1\}$, $Arg_{\omega_1} = \{B, C, B_1, C_1\}$ and $Arg_{\omega_2} = \{B, C\}$.

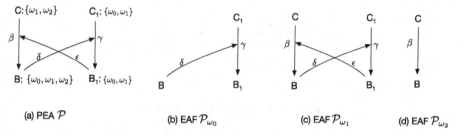

(a) PEA $\mathcal{P}$     (b) EAF $\mathcal{P}_{\omega_0}$     (c) EAF $\mathcal{P}_{\omega_1}$     (d) EAF $\mathcal{P}_{\omega_2}$

**Fig. 4.** An PEAF framework

Each possible world $\omega$ identifies an EAF framework $\mathcal{P}_\omega$ obtained by restricting EAF $\mathcal{F}$ to the set of arguments that occur in $\omega$.

**Definition 3.** *For each $\omega \in \mathcal{W}$ of a PEAF framework $\mathcal{P} = (\mathcal{F}, \mathcal{W}, P)$, let $\mathcal{P}_\omega$ denote the EAF framework $(Arg_\omega, Att_\omega)$ with $Att_\omega = \bigcup_{i=0}^{\infty} Att_\omega^i$ where $Att_\omega^i$ is defined inductively as follows.*

- $Att_\omega^0 = (Arg_\omega \times Arg_\omega) \cap Att^0$
- *For each $i > 0$, $Att_\omega^{i+1} = (Arg_\omega \times Att_\omega^i) \cap Att^{i+1}$*

Continue Example 5, Fig. 4(b–d) shows EAF $\mathcal{P}_\omega$ for each $\omega \in \mathcal{W}$.

The semantics of PEAF semantics are defined as follows.

**Definition 4.** *Wrt a PEAF framework $\mathcal{P} = (\mathcal{F}, \mathcal{W}, P)$, the probability of the acceptability of argument $A$ under semantics sem, denoted $Prob_{sem}(A)$, is obtained by marginalizing the joint probability distribution of possible worlds and the acceptability of $A$. Concretely,*

$$Prob_{sem}(A) = \sum_{\omega \in \mathcal{W}} P(\omega).P(\mathcal{P}_\omega \vdash_{sem} A \mid \omega) = \sum_{\omega \in \mathcal{W} : \mathcal{P}_\omega \vdash_{sem} A} P(\omega)$$

204     N.D. Hung

*Example 6* (Continue Example 5). It is easy to see that $\mathcal{P}_{\omega_0} \vdash_{igr} B_1$ and $\mathcal{P}_{\omega_1} \vdash_{ipr} B_1$, and hence $Prob_{igr}(B_1) = P(\omega_0)$ while $Prob_{ipr}(B_1) = P(\omega_0) + P(\omega_1)$.

To see practical uses of PEAF, let's twist the debate in Example 3 slightly.

*Example 7.* Suppose that **P** continues the debate with an utterance $E'$ intended to be a counter-argument for $E$, but **Q** does not fully see $E'$ as an argument proper (e.g. **P** may arrogantly say "my instincts are as good as facts."). Now, if we represent the debate by an EAF from **P**'s point of view, the EAF shall contain argument $E'$. However, if we follow **Q**'s point of view, the EAF does not contain $E'$. Hence, from the point of view of a neutral third party, the debate can be represented by the PEAF shown in Fig. 5.

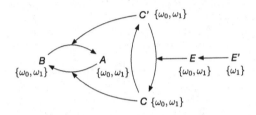

**Fig. 5.** Another PEAF framework

## 4   PEAF Inference Procedures

Inference procedures for PEAF compute $Prob_{sem}(A)$ for a given an argument $A$ wrt a PEAF framework[13]. This section aims at automatically deriving such procedures from EAF proof procedures, which are supposedly given. First let us consider a naive approach below.

```
result = 0;
foreach ω ∈ W do
    if P_ω ⊢_sem A   /* invoke EAF proof procedure*/ then
        result = result + P(ω);
    end
end
```

The major advantage of the above approach is that it does not need to make any surgery to a given EAF proof procedure (and hence it can be fully automated). However, since EAF proof procedure is invoked afresh for different possible worlds, the common parts of different dispute derivations (for convenience, let's call them partial dispute derivations) are repeatedly constructed.

*Example 8* (Continue Example 6). The following tables show two dispute derivations for $B_1$ (in $\omega_0$ and $\omega_1$ respectively) sharing a common part, which is a partial dispute derivation of four steps (step 0 to step 3).

---

[13] In this section we always refer to an arbitrary but fixed PEAF framework $\mathcal{P} = (\mathcal{F}, \mathcal{W}, P)$ with $\mathcal{F} = (Arg, Att)$ if not explicitly stated otherwise.

| i | $P_i$ | $O_i$ | $SP_i$ | $SO_i$ |
|---|-------|-------|--------|--------|
| 0 | $\{B_1\}$ | $\{\}$ | $\{B_1\}$ | $\{\}$ |
| 1 | $\{\}$ | $\{\gamma\}$ | $\{B_1\}$ | $\{\}$ |
| 2 | $\{B,\delta\}$ | $\{\}$ | $\{B_1,B\}$ | $\{\gamma\}$ |
| 3 | $\{B\}$ | $\{\}$ | $\{B_1,B\}$ | $\{\gamma\}$ |
| 4 | $\{\}$ | $\{\}$ | $\{B_1,B\}$ | $\{\gamma\}$ |

| i | $P_i$ | $O_i$ | $SP_i$ | $SO_i$ |
|---|-------|-------|--------|--------|
| 0 | $\{B_1\}$ | $\{\}$ | $\{B_1\}$ | $\{\}$ |
| 1 | $\{\}$ | $\{\gamma\}$ | $\{B_1\}$ | $\{\}$ |
| 2 | $\{B,\delta\}$ | $\{\}$ | $\{B_1,B\}$ | $\{\gamma\}$ |
| 3 | $\{B\}$ | $\{\}$ | $\{B_1,B\}$ | $\{\gamma\}$ |
| 4 | $\{\}$ | $\{\beta\}$ | $\{B_1,B\}$ | $\{\gamma\}$ |
| 5 | $\{\varepsilon\}$ | $\{\}$ | $\{B_1,B\}$ | $\{\gamma,\beta\}$ |
| 6 | $\{\}$ | $\{\}$ | $\{B_1,B\}$ | $\{\gamma,\beta\}$ |

A dispute derivation wrt EAF $\mathcal{P}_{\omega_0}$          A dispute derivation wrt EAF $\mathcal{P}_{\omega_1}$

**Data:** A dispute derivation state $t = \langle P, O, SP, SO \rangle$; a selection function $sl$.
**Result:** $Follow_{\mathcal{F}}(t, sl)$
**if** $sl$ selects $X \in P$ **then**
$\quad \mid \quad t' = \langle P \setminus \{X\}, O \cup Attack_X, SP, SO \rangle$;
$\quad \mid \quad$ **return** $\{t'\}$;
**end**
**if** $sl$ selects $X \in O$ **then**
$\quad$ **if** $src(X) \in SP$ **then**
$\quad\quad \mid \quad PossibleMoves = Attack_X \setminus (SO_i \cup O_i)$;
$\quad$ **else**
$\quad\quad \mid \quad PossibleMoves = (Attack_{src(X)} \cup Attack_X) \setminus (SO_i \cup O_i)$;
$\quad$ **end**
$\quad Result = \{\}$;
$\quad$ **foreach** $\alpha \in PossibleMoves$ **do**
$\quad\quad$ **if** $src(\alpha) \notin SP$ **then**
$\quad\quad\quad \mid \quad P' = P \cup \{src(\alpha), \alpha\}$;
$\quad\quad$ **else**
$\quad\quad\quad \mid \quad P' = P \cup \{\alpha\}$;
$\quad\quad$ **end**
$\quad\quad O' = O \setminus \{X\}$;
$\quad\quad SP' = SP \cup \{src(\alpha)\}$;
$\quad\quad SO' = SO \cup \{X\}$;
$\quad\quad t' = \langle P', O', SP', SO' \rangle$;
$\quad\quad Result = Result \cup \{t'\}$;
$\quad$ **end**
$\quad$ **return** $Result$;
**end**

**Algorithm 1.** $Follow_{\mathcal{F}}(t, sl)$

To address the above problem, common partial dispute derivations need to be reused across possible worlds. As in [10], this can be done by annotating each partial dispute derivation with a set of possible worlds in which it is valid. When it is extended, possible worlds in which it is not longer valid are removed. For example, the partial dispute derivation in Example 8 can be annotated by $\{\omega_0, \omega_1\}$. The extension at step 4 yields two partial dispute derivations: one is annotated by $\{\omega_0\}$ (i.e. $\omega_1$ is removed) and the other is still annotated by $\{\omega_0, \omega_1\}$. To avoid any surgery to given EAF proof procedures (so that the approach can be fully automated), the annotations should not be incorporated into dispute derivation states. However, we can assume that these states are readable and so are state transitions. Thus we define a function $Follow_{\mathcal{F}}(\_,\_)$ which can be considered as the "API" that EAF proof procedures provide.

**Definition 5.** *Let $\mathcal{F}$ be an EAF framework. For a dispute derivation state $t$ and a selection function $sl$, $Follow_{\mathcal{F}}(t, sl)$ denotes the set of states that can immediately follow $t$ in a dispute derivation using $sl$.*

The implementation of this function is up to EAF proof procedures and PEAF inference procedures should not be concerned with that. For example, Algorithm 1, which follows directly from Definition 1, implements the function for i-preferred semantics. PEAF inference procedures call $Follow_{\mathcal{F}}(t, sl)$, examine each returned state $t' \in Follow_{\mathcal{F}}(t, sl)$ and probably also the cause of state transition. In the following definition, which defines our PEAF inference procedures abstractly, notation $t \xrightarrow[O:Attack_X]{P:X} t'$ says that the transition $t \to t'$ is caused by the opponent's movement of $Attack_X$ to attack $X \in Arg \cup Att$ moved previously by the proponent. Similarly $t \xrightarrow[P:\alpha]{O:X} t'$ says that the proponent moves $\alpha$ to attack $X$ moved previously by the opponent.

**Definition 6.** *Given a selection function, a **world derivation** for an argument $A$ is a sequence $T_0, T_1, \ldots, T_n$ where*

1. *$T_i$ is a set of pairs of the form $(t, W)$ where $t$ is a dispute derivation state and $W \subseteq \mathcal{W}$ is a subset of possible worlds.*
2. *At each step $i$, a pair $(t, W)$ is selected from $T_i$, and $T_{i+1} = T_i \setminus \{(t, W)\} \cup \Delta T$ where*
   *(a) If $W = \emptyset$ then $\Delta T = \emptyset$,*
   *(b) Otherwise, an element $X$ is selected from the $P$ or $O$ component of $t$, and*
      *i. If $X \in P$ then $\Delta T = \{(t', W)\}$ where $t' \in Follow_{\mathcal{F}}(t, sl)$.[14]*
      *ii. If $X \in O$ then $\Delta T = \{(\langle P, O\setminus\{X\}, SP, SO\rangle, W\setminus W_{src(X)})\} \cup \{(t', W \cap W_{src(\alpha)}) \mid t' \in Follow_{\mathcal{F}}(t, sl) \text{ and } t \xrightarrow[P:\alpha]{O:X} t'\}.$*

The intuition behind Definition 6 is as follows.

- Rule 2(a): the current partial dispute derivation is valid in no possible worlds, and hence it should be abandoned.
- Rule 2(b)i: the current partial dispute derivation is extended by one state transition $t \xrightarrow[O:Attack_X]{P:X} t'$.
- Rule 2(b)ii: the current partial dispute derivation is extended by either ways:
  - By a state transition $t \xrightarrow[P:\alpha]{O:X} t'$ where $t' \in Follow_{\mathcal{F}}(t, sl)$: Since the proponent is using an argument $src(\alpha)$, possible words that do not contain this argument should be removed[15].

---

[14] In this case $Follow(t, sl) = \{t'\}$ and $t \xrightarrow[O:Attack_X]{P:X} t'$.

[15] Note that $W_{src(\alpha)}$ is the set of possible worlds containing $src(\alpha)$.

- By removing possible worlds that contain argument $src(X)$: In the remaining possible worlds, the proponent does not have to counter-attack $X$ because $X$ does not occur.

**Definition 7.** *A world derivation $d$ for an argument $A$ is a world derivation $T_0, T_1, \ldots, T_n$ where $T_0$ consists of only one pair $((\{A\}, \emptyset, \{A\}, \emptyset), W_A)$ and $T_n$ consists of only pairs of the form $(\langle \emptyset, \emptyset, \_, \_ \rangle, \_)$. The world set derived by $d$, denoted $\mathcal{W}_d$, is $\bigcup\limits_{(\_, W) \in T_n} W$.*

*Example 9* (Continue Example 5). A world derivation for $B_1$ with derived world set $\{\omega_0, \omega_1\}$ is given in the following table. It is easy to see that steps $T_0, \ldots, T_4$ constructs the common partial dispute derivation in Example 8, while steps $T_4, \ldots, T_7$ extends this into two dispute derivations for $B_1$.

| | $(t, W)$ | | Remark |
|---|---|---|---|
| | $t$ | $W$ | |
| $T_0$ | $\langle \{\underline{B_1}\}, \emptyset, \{B_1\}, \emptyset \rangle$ | $\{\omega_0, \omega_1\}$ | |
| $T_1$ | $\langle \emptyset, \{\underline{\gamma}\}, \{B_1\}, \emptyset \rangle$ | $\{\omega_0, \omega_1\}$ | Use Definition 6, rule 2(b)i: $Attack_{B_1} = \{\gamma\}$ |
| $T_2$ | $\langle \emptyset, \emptyset, \{B_1\}, \emptyset \rangle$ | $\emptyset$ | rule 2(b)ii: $W_{src(\gamma)} = \{\omega_0, \omega_1\}$ |
| | $\langle \{B, \delta\}, \emptyset, \{B_1, B\}, \{\gamma\} \rangle$ | $\{\omega_0, \omega_1\}$ | $Attack_{src(\gamma)} \cup Attack_\gamma = \{\delta\}$; $W_{src(\delta)} = \{\omega_0, \omega_1, \omega_2\}$ |
| $T_3$ | $\langle \{B, \underline{\delta}\}, \emptyset, \{B_1, B\}, \{\gamma\} \rangle$ | $\{\omega_0, \omega_1\}$ | rule 2(a) |
| $T_4$ | $\langle \{\underline{B}\}, \emptyset, \{B_1, B\}, \{\gamma\} \rangle$ | $\{\omega_0, \omega_1\}$ | rule 2(b)i: $Attack_\delta = \emptyset$ |
| $T_5$ | $\langle \{\}, \{\underline{\beta}\}, \{B_1, B\}, \{\gamma\} \rangle$ | $\{\omega_0, \omega_1\}$ | rule 2(b)i: $Attack_B = \{\beta\}$ |
| $T_6$ | $\langle \emptyset, \emptyset, \{B_1, B\}, \{\gamma\} \rangle$ | $\{\omega_0\}$ | rule 2(b)ii: $W_{src(\beta)} = \{\omega_1, \omega_2\}$ |
| | $\langle \{\underline{\epsilon}\}, \emptyset, \{B_1, B\}, \{\gamma, \beta\} \rangle$ | $\{\omega_0, \omega_1\}$ | $Attack_{src(\beta)} \cup Attack_\beta = \{\epsilon\}$; $W_{src(\epsilon)} = \{\omega_0, \omega_1\}$ |
| $T_7$ | $\langle \emptyset, \emptyset, \{B_1, B\}, \{\gamma\} \rangle$ | $\{\omega_0\}$ | |
| | $\langle \emptyset, \emptyset, \{B_1, B\}, \{\gamma, \beta\} \rangle$ | $\{\omega_0, \omega_1\}$ | rule 2(b)i: $Attack_\epsilon = \emptyset$ |

**Lemma 1.** *Assume the soundness of dispute derivations for EAF semantics sem (i.e. if there is a dispute derivation for an argument $A$ wrt EAF $\mathcal{F}$ then $\mathcal{F} \vdash_{sem} A$). If $\omega \in \mathcal{W}_d$ where $d$ is a world derivation for an argument $A$ in PEAF framework $\mathcal{P}$, then $\mathcal{P}_\omega \vdash_{sem} A$.*

*Proof* (Sketch). If $\omega \in \mathcal{W}_d$ then we can extract from $d$ a dispute derivation for $A$ wrt EAF $\mathcal{P}_\omega$. From the assumed soundness of dispute derivations, $\mathcal{P}_\omega \vdash_{sem} A$.

**Lemma 2.** *Assume the completeness of dispute derivations for EAF semantics sem (i.e. for any selection function, there is a dispute derivation for an argument $A$ wrt EAF $\mathcal{F}$ if $\mathcal{F} \vdash_{sem} A$). If $\mathcal{P}_\omega \vdash_{sem} A$ then for any world derivation $d$ for $A$, $\omega \in \mathcal{W}_d$.*

*Proof* (Sketch). If $\mathcal{P}_\omega \vdash_{sem} A$, then from the assumed completeness of dispute derivations, for any selection function there exists a dispute derivation for $A$ wrt EAF $\mathcal{P}_\omega$. Hence $\omega \in \mathcal{W}_d$.

The above lemmas ensure the correctness of the following theorem.

**Theorem 1.** *Assume the soundness and completeness of dispute derivations for EAF semantics sem. If $\mathcal{W}_d$ is the world set derived by a world derivation d for an argument A, then $Prob_{sem}(A) = \sum\limits_{\omega \in \mathcal{W}_d} P(\omega)$.*

Continue Example 9, Theorem 1 says that $Prob_{ipr}(B_1) = P(\omega_0) + P(\omega_1)$.

## 5    Algorithmic Issues and Implementation

The procedural form of our PEAF inference procedures is given by Algorithm 3 below, which uses Algorithm 2 to compute $T_n$, the last element of a world derivation $T_0, \ldots, T_n$ for a given argument $A$.

> **Data:** An argument $A$ in a PEAF framework $\mathcal{P} = (\mathcal{F}, \mathcal{W}, P)$; a selection
>     function $sl$.
> **Result:** $T_n$, the last element of a world derivation $T_0, \ldots, T_n$ for $A$.
> $T = \{((\{A\}, \emptyset, \{A\}, \emptyset), W_A)\}$;
> **while** *a pair* $(t, W)$ *with* $t = \langle P, O, SP, SO \rangle$ *s.t. if* $P \cup O = \emptyset$ *then* $W = \emptyset$ *can be*
> *selected from* $T$ **do**
> > $T = T \setminus \{(t, W)\}$;
> > **if** $W \neq \emptyset$ **then**
> > > Let $X$ be an element selected from $P$ or $O$ component of $t$;
> > > **if** $X \in P$ **then**
> > > > Let $Follow_{\mathcal{F}}(t, sl) = \{t'\}$;
> > > > $T = T \cup \{(t', W)\}$;
> > >
> > > **else**
> > > > $T = T \cup \{((\langle P, O \setminus \{X\}, SP, SO \rangle, W \setminus W_{src(X)})\}$;
> > > > **foreach** $t' \in Follow_{\mathcal{F}}(t, sl)$ **do**
> > > > > Let $t \xrightarrow[P:\alpha]{O:X} t'$;
> > > > > $T = T \cup \{(t', W \cap W_{src(\alpha)})\}$;
> > > >
> > > > **end**
> > >
> > > **end**
> >
> > **end**
>
> **end**
> **return** $T$;

Algorithm 2. worldDerivation(A)

> **Data:** An argument $A$ in a PEAF framework $\mathcal{P} = (\mathcal{F}, \mathcal{W}, P)$.
> **Result:** $Prob_{sem}(A)$.
> $T = worldDerivation(A)$; /* use algorithm 2*/
> $\mathcal{W}_d = \bigcup\limits_{(\_, W) \in T} W$;
> **return** $\sum\limits_{\omega \in \mathcal{W}_d} P(\omega)$;

Algorithm 3. $Prob_{sem}(A)$

In the following, we demonstrate a Prolog-based implementation of the above algorithms. As illustrated by the following code listing, users need to specify PEAF frameworks using a tiny syntax consisting of several predicates: arg(., [.]) declares, for each argument $A$, the set of possible worlds $W(A)$ that contain $A$; att(.,.) declares an attack; and p(.,.) assigns probabilities to possible words.

```
arg(c,  [w1,w2]).
arg(b,  [w0,w1,w2]).
arg(c1, [w0,w1]).
arg(b1,[w0,w1]).

att(c,b,beta).
att(c1,b1,gamma).
att(b,gamma,delta).
att(b1,beta,eps).

p(w0,0.1).
p(w1,0.2).
p(w2,0.7).
```

**Code listing 1:** Specifying the PEAF framework in Example 5.

Note that a PEAF specification turns out to be a valid Prolog program. The screen shot in Fig. 6 shows how users could load the specification in the above code listing, then call our PEAF procedures to compute $Prob_{ipr}(.)$ using Prolog predicate probipr/1. More examples and demonstrations are included in our implementation downloadable from http://ict.siit.tu.ac.th/~hung/peafengine.

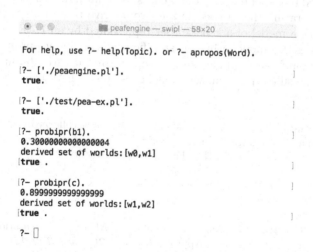

**Fig. 6.** Calling PEAF procedures

# 6  Conclusions and Related Works

The distribution semantics has become a de facto approach for integrating logic programming with probability theory, and is recently applied for the standard abstract argumentation framework. In this paper we investigate the distribution semantics for Extended Argumentation according to which a Probabilistic Extended Argumentation framework (PEAF) defines a probabilistic distribution of EAF frameworks representing different possible worlds. We show that PEAF inference procedures can be derived automatically from the proof procedures of EAFs. To improve efficiency, common partial results of different invocations of EAF proof procedures are reused. This is done by annotating such partial results with "possible worlds" in which the results are valid. To avoid a surgery to EAF proof procedures, the annotations are not defined as part of their states. While doing so we focus on EAFs with only attacks coming from arguments [7,9] and the inductive defense semantics for these EAFs introduced in [9]. However our results can be easily obtained for other extended argumentation frameworks and semantics. To the best of our knowledge, this paper is the first proposal for Probabilistic Extended Argumentation framework (PEAF). However we have focused on only PEAF semantics and inference procedures. So to close let us briefly discuss potential applications of PEAF. An important issue in extending abstract argumentation is to offer natural or comprehensible interpretations to the extended features. For example, in Modgil's extended argumentation framework [13], attacks on attacks represent preferences between conflicting arguments. The subclass of stratified EAF frameworks presented in [9] generalizes Modgil's extended argumentation framework, aiming at applications including but not limited to reasoning with preferences. Imagine a recommendation system predicting the rating that a specific user would give to a product. The system may start with a model of a generic user represented by an EAF framework. It then customizes this model using the specific user's preference profile. Often the specific user's preference profile can not be determined with absolute certainty, and hence the resulted personalized model could be a PEAF framework, but not an EAF framework.

**Acknowledgment.** This work was funded by Center of Excellence in Intelligent Informatics, Speech and Language Technology and Service Innovation; and Intelligent Informatics and Service Innovation, SIIT, Thammasat University.

# References

1. Baroni, P., Cerutti, F., Giacomin, M., Guida, G.: AFRA: argumentation framework with recursive attacks. Int. J. Approx. Reason. **52**(1), 19–37 (2011)
2. Barringer, H., Gabbay, D., Woods, J.: Temporal dynamics of support and attack networks: from argumentation to zoology. In: Hutter, D., Stephan, W. (eds.) Mechanizing Mathematical Reasoning. LNCS, vol. 2605, pp. 59–98. Springer, Heidelberg (2005). doi:10.1007/978-3-540-32254-2_5

3. Dung, P.M.: On the acceptability of arguments and its fundamental role in non-monotonic reasoning, logic programming and n-person games. Artif. Intell. **77**(2), 321–357 (1995)
4. Dung, P.M., Son, T.C., Thang, P.M.: Argumentation-based semantics for logic programs with first-order formulae. In: Baldoni, M., Chopra, A.K., Son, T.C., Hirayama, K., Torroni, P. (eds.) PRIMA 2016. LNCS (LNAI), vol. 9862, pp. 43–60. Springer, Cham (2016). doi:10.1007/978-3-319-44832-9_3
5. Dung, P.M., Thang, P.M.: Towards (probabilistic) argumentation for jury-based dispute resolution. In: COMMA 2010, pp. 171–182 (2010)
6. Fierens, D., Van Den Broeck, G., Renkens, J., Shterionov, D., Gutmann, B., Thon, I., Janssens, G., De Raedt, L.: Inference and learning in probabilistic logic programs using weighted boolean formulas. Theory Pract. Logic Program. **15**(3), 358–401 (2015)
7. Gabbay, D.M.: Semantics for higher level attacks in extended argumentation frames part 1: overview. Stud. Logica **93**(2–3), 357–381 (2009)
8. Gabbay, D.M., Rodrigues, O.: Probabilistic argumentation. An equational approach. In: CoRR (2015)
9. Hanh, D.D., Dung, P.M., Hung, N.D., Thang, P.M.: Inductive defense for sceptical semantics of extended argumentation. J. Logic Comput. **21**(2), 307–349 (2011)
10. Hung, N.D.: A generalization of probabilistic argumentation with dempster-shafer theory. In: Kern-Isberner, G., Fürnkranz, J., Thimm, M. (eds.) KI 2017: Advances in Artificial Intelligence. LNCS, vol. 10505, pp. 155–169. Springer, Cham (2017). doi:10.1007/978-3-319-67190-1_12
11. Hunter, A.: A probabilistic approach to modelling uncertain logical arguments. Int. J. Approx. Reason. **54**(1), 47–81 (2013)
12. Li, H., Oren, N., Norman, T.J.: Probabilistic argumentation frameworks. In: Modgil, S., Oren, N., Toni, F. (eds.) TAFA 2011. LNCS, vol. 7132, pp. 1–16. Springer, Heidelberg (2012). doi:10.1007/978-3-642-29184-5_1
13. Modgil, S.: Reasoning about preferences in argumentation frameworks. Artif. Intell. **173**(9–10), 901–934 (2009)
14. Nielsen, S.H., Parsons, S.: A generalization of dung's abstract framework for argumentation: arguing with sets of attacking arguments. In: Maudet, N., Parsons, S., Rahwan, I. (eds.) ArgMAS 2006. LNCS, vol. 4766, pp. 54–73. Springer, Heidelberg (2007). doi:10.1007/978-3-540-75526-5_4
15. Polberg, S., Doder, D.: Probabilistic abstract dialectical frameworks. In: Fermé, E., Leite, J. (eds.) JELIA 2014. LNCS (LNAI), vol. 8761, pp. 591–599. Springer, Cham (2014). doi:10.1007/978-3-319-11558-0_42
16. Poole, D.: The independent choice logic and beyond. In: De Raedt, L., Frasconi, P., Kersting, K., Muggleton, S. (eds.) Probabilistic Inductive Logic Programming. LNCS, vol. 4911, pp. 222–243. Springer, Heidelberg (2008). doi:10.1007/978-3-540-78652-8_8
17. Sato, T.: A statistical learning method for logic programs with distribution semantics. In: ICLP 1995, pp. 715–729 (1995)
18. Sato, T., Kameya, Y.: New advances in logic-based probabilistic modeling by PRISM. In: De Raedt, L., Frasconi, P., Kersting, K., Muggleton, S. (eds.) Probabilistic Inductive Logic Programming. LNCS, vol. 4911, pp. 118–155. Springer, Heidelberg (2008). doi:10.1007/978-3-540-78652-8_5
19. Thimm, M.: A probabilistic semantics for abstract argumentation. In: ECAI, vol. 242, pp. 750–755. ISO Press (2012)

# An Ontology-Based Knowledge Framework for Software Testing

Shanmuganathan Vasanthapriyan[1,2](✉), Jing Tian[1], and Jianwen Xiang[1](✉)

[1] School of Computer Science and Technology, Wuhan University of Technology,
Wuhan, People's Republic of China
priyan@appsc.sab.ac.lk
[2] Department of Computing and Information Systems,
Sabaragamuwa University of Sri Lanka, Belihuloya, Sri Lanka

**Abstract.** Software development is conceptually a complex, knowledge intensive and a collaborative activity, which mainly depends on knowledge and experience of the software developers. Effective software development relies on the knowledge collaboration where each and every software engineer shares his or her knowledge or acquires knowledge from others. Software testing which is a sub area of software engineering is related to various activities such as test planning, test case design, test implementation, test execution and test result analysis and they are all essential. Given great importance to knowledge for software testing, and the potential benefits of managing software testing knowledge, an ontological approach to represent the necessary software testing knowledge within the software testers context was developed. Using this approach, software testing ontology to include information needs identified for the software testing activities was designed. Competency questions (contextualized information) were used to determine the scope of the ontology and used to identify the contents of the ontology because contextualized information fulfills the expressiveness and reasoning requirements of the software testing ontology. SPARQL query was used to query the competency questions. A web based KM Portal was developed using semantic web technologies for knowledge representation. Software testers can annotate their testing knowledge with the support of ISTQB and IEEE 829-2008 terms. Both ontology experts and non-experts evaluated the developed ontology. We believe our software testing ontology can support other software organizations to improve the sharing of knowledge and learning practices.

**Keywords:** Software testing ontology · Software testing knowledge · Ontology-based knowledge sharing · Knowledge management system

## 1 Introduction

Software testing is a sub area of software engineering which is also a knowledge intensive and collaborative activity [1,2]. Knowledge can be applied to different

© Springer Nature Singapore Pte Ltd. 2017
J. Chen et al. (Eds.): KSS 2017, CCIS 780, pp. 212–226, 2017.
https://doi.org/10.1007/978-981-10-6989-5_18

testing tasks and purposes. Since software development is an error prone task, in order to achieve quality software products, Validation and Verification should be carried throughout the development [3]. Therefore, software testers have to work with all the other software practitioners who are working in the development activities. As software testers, they would be familiar with the several software testing methods and considerably aware of the software development models and need information relevant to their context. For instance, software testers may either need assistance on a test case design information relevant to the similar project which was handled previously for testing purposes or to design a test case. Moreover, this information would also have a greater impact on their decision-making process. Our previous study [4] results revealed some of the issues such as outdated knowledge in the repositories, unstructured internal documents, varied formats and less accessing facilities and lack of targeted delivery methods. Hence, efficient mechanisms for capturing, representing, reusing, and sharing the software testing knowledge involved are sorely needed. Indeed, the software companies need to foster unstructured knowledge creation culture, good organizational learning and knowledge sharing culture and determining the organization's knowledge for software testing activities. Therefore, Knowledge Management(KM) practices needed to be incorporated into software development [2].

According to Gruber [5], an ontology is a formal, explicit specification of a shared conceptualization. Ontology provides a structured view of domain knowledge and acts as a repository of concepts in the domain.

Given great importance for knowledge for software testing, and the potential benefits of managing software testing knowledge, using semantic web technologies, Ontology-based knowledge framework is developed. The context has also been decided to confine the study to a particular Sri Lankan software development company. The key reasons are based on the geographical location of the researcher, practicality and ease of access to those software development companies and comparability of research data due to companies' same jurisdiction, same economic and regulatory regimes governing their operation. Further, we briefly explain each type of the high-level concepts based on IEEE 829-2008 [6], also known as the 829 Standard for Software and System Test Documentation and ISTQB (International Software Testing Qualifications Board) [7]. Even though the standard specifies the procedures of software testing, we have also included what industries are stipulating in their practice.

The remainder of this paper is organized as follows. Succinct analysis of related research is presented in the second section. Section 3 discusses software testing ontology in detail. Development of knowledge framework to manage software testing knowledge is discussed in Sect. 4. Section 5 discusses the evaluation of the ontology developed under two points-of-view: domain experts and non-experts. Finally, Sect. 6 concludes the paper and presents directions for future work.

## 2   Related Work

We evaluated the existing software testing ontologies mentioned in the literature and there are still problems related to the establishment of an explicit common conceptualization regarding this domain. The five ontologies [8–12] described as just OWL artifacts, thus are not enough for the purposes of applying ontologies to manage software knowledge. Arnicans [13] and Cai [11] presented ontologies as taxonomies thus they cannot be considered for ontologies to manage software testing knowledge.

Eventhough STOWS [14], OntoTest [15], TaaS [16] and ROoST [17] proposes software testing ontologies, they have not included much on the following (a) conceptual design (b) domain coverage (c) reusing of ontologies (d) implementation of international standard (e) implementation of relationships, axioms (f) vocabulary standards (g) evaluation methodologies of ontologies and (h) implemented as KMS, while designing software testing ontology. Importantly, none of them have implemented any knowledge management system to manage software testing knowledge. Also, these ontologies are not specific enough to satisfy the software testers need for timely information in context. Further, some of these characteristics are difficult to evaluate, since there was not much information about them in the literature presenting those corresponding ontologies.

Having discovered this research gap we have focused on our attention to develop a software testing ontology to represent information needs according to software testers context. That is, we intend to develop an ontology-based knowledge framework to manage software testing-related knowledge. This would assist the software testers in the software companies to manage software testing knowledge.

## 3   Designing of Software Testing Ontology

There are many approaches existing to organize software testing knowledge and information such as relational data models, hypertext documents etc. To represent the domain knowledge in user context, we needed an approach which relies on the expressive features to represent appropriate description of the context. Importantly, ontology represents a richer knowledge than a relational data model and provides reasoning support to infer new knowledge. For such reasons, ontology based approaches are better suited to represent contextualized knowledge that can be used to find a response to queries with more expressive relationships within a specified context in software testing domain.

### 3.1   Ontology Engineering Approach

In literature, many methodologies have been proposed until now to build an ontology [18–20], but we considered the Grüninger and Fox methodology [21], which considered a formal approach to design ontology as well as providing a framework for evaluating the adequacy of the developed ontology. This methodology focuses on building ontology based on first-order logic (FOL) by providing strong semantics. In our scenario, we introduce Description Logic (DL) which is a decidable fragment of FOL since we are designing with OWL2 Web Ontology Language [22] for semantic. The widely-used Protégé system (http://protege.stanford.edu) has recently been extended with support for the additional constructs provided by OWL2. We describe "context specific" [23] to the software testers belonging to a leading software company in Sri Lanka and the approach which will be used to design the ontology to provide context-specific information and knowledge to software testers. To identify software testers' context clearly, we have extracted domain specific knowledge of software testing ontologies from existing literature and interviewed the software testing experts belonging to a particular company.

## 3.2 Ontology Components

At the first instance, high-level ontology concepts, their properties and their relationships should be identified and it is shown in Fig. 1. The basic high level ontology concepts are identified as Test Environment, Testing Activities, Static Testing Techniques, Test Design Techniques, Test Objects, Test Level, Testing Artifacts and Organizational Team. For example, the concept Organizational Team having the properties of TeamID, TeamName, Team Size and Team Type. Secondly, sub classes of the high-level ontology concepts, their properties and relationships are also defined. For example, Test Environment has Human Resource, Software Tools and Hardware as it's sub classes. These sub classes are related to their superclass by is_a relation. During the modeling of software testing domain ontology, some special types of axioms such as Instantiation, Assertion, Subsumption, Domain, Range and Disjointness are included. The classes have been created in Protégé 5.1 can be described by necessary and sufficient conditions as illustrated in Table 1. The Fig. 2 illustrates the Artifacts Concept and it's relationships in our ontology. Both Existential restrictions ('someValuesFrom' or 'some' restrictions) and Universal restrictions ('allValuesFrom' restrictions, or 'only' restrictions) have been used in designed ontology.

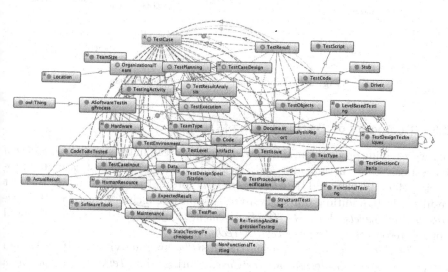

**Fig. 1.** Part of software testing ontology

**Table 1.** Necessary and sufficient conditions written in Protégé 5.1

| Testing activities | Axioms written in Protégé 5.1 for equivalent classes |
|---|---|
| TestCaseDesign | (isGeneratesTestCase only TestCase) and (isGeneratesTestCase min 1 TestCase) |
| TestExecution | hasExecutesTestCases only (TestCase and (hasTestCaseExpectedResult some ExpectedResult)) |
| TestPlanning | (hasCreatesTestPlan some TestPlan) and (hasCreatesTestPlan only TestPlan) |
| TestResultAnalysis | hasEvaluatesTestCases some (TestCase and (hasTestActualResult some ActualResult)) |

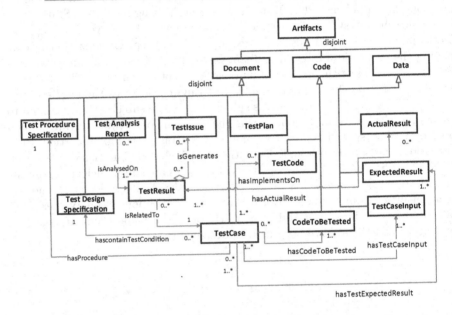

**Fig. 2.** Artifacts class and it's relationships

### 3.3   Taxonomy of Testing Concepts

Software testing process contains *TestPlanning* for planning tests, *TestCaseDesign* for test case construction, and *TestExecution* for execution of test cases and producing *ActualResult*, and *TestResultAnalysis* for analysis and evaluate the test results [24]. In addition, our software testing process includes *Test Design Techniques, Test Levels, Artifacts, TestEnviorment* (includes hardware, software and Human resources) and *Static Testing Techniques*.

A *TestPlan* is produced during the *TestPlanning* activity. TestCaseDesign activity targets to design a *TestCase*. During the *TestCaseDesign*, if a

*TestDesignTechniques* is used then the following axiom can be defined for *Test-CaseDesign* to design any *TestCase* as follows.

$\forall$ *tc:TestCase, tdt:TestDesignTechnique, tcd:TestCaseDesign, hasDesignAccordingTo(tc,tdt)* $\wedge$ *isGeneratesTestCase(tcd,tc)* $\longrightarrow$ *isAdopts (tcd,tdt)*

Several Artifacts have been used to derive test cases in software testing which describes the functionalities, architecture, and design of software. Such Artifacts are used as *TestCaseDesignInput* during the software *TestCaseDesign* activity and the output is test cases. Test cases can be documented as described in the IEEE 829-2008 documentation. The document that describes the steps to be taken in running a set of tests (and specifies the executable order of the tests) is called a test procedure in IEEE 829. Besides, Test case contains a set of input values (*TestCaseInput*), execution preconditions, expected results (*Expected Result*) and execution post conditions, developed for a particular objective or test condition, such as to exercise a particular program path or to verify compliance with a specific requirement.

That is, a test case targets to test a *Code To be Tested*. *Code To be Tested* can be any programs, modules, and the whole system code. Furthermore, *Test Code* is a portion of code that is to be run for executing a given set of test cases, contain three subtypes such as *Test Script, Driver* and *Stub*.

*Test Execution* activity executes any Test case. Test Execution requires as input the *Test Code* to be run and the Code To Be Tested. Notably, both *Test Code* and *Code To Be Tested* are needed for Test Execution activity. This can be illustrated in an axiom as follows.

$\forall$ *te:TextExecution, tc:TestCase hasExecutesTestCases(te,tc)* $\longrightarrow$ ($\exists$ *tcode: TestCode, CodeTobeTest: CodeToBeTested) uses (te,tcode)* $\wedge$ *isContainedOf (tcode,tc)* $\wedge$ *uses (te,CodeTobeTest)* $\wedge$ *hasCodeTobeTested(tc,tcode)*

The output of the *Test Execution* activity is the *Test Result*. Each of the *Test Result* is particularly related to a TestCase. *Test Result* may be related to an *Actual Result*, for a particular *TestCaseInput* and *ExpectedResult*. *Test Execution* also can be elaborated necessary and sufficient axioms in Protégé 5.1 as follows.

*hasExecutesTestCases only (TestCase and (hasTestCaseExpectedResult some ExpectedResult))*

A test execution can run and achieve a result (*ActualResult*), but it can also fail, and generate a *TestIssue*. Thus, a Test Result contains either an *ActualResult*, or *TestIssue* or both. This can be expressed using the following axiom.

$\forall$ *TestR:TestResult* $\longrightarrow$ ($\exists$ *ActResult:ActualResult, Issue: TestIssue) (ActualResult(ActResult)* $\vee$ *(TestIssue(Issue))* $\wedge$ *isGenarates(ActResult,Issue)*

Further, *TestIssueReport*, could report this event in detail, which requires investigation. Finally, during a Test Result Analysis, *TestResults* are analyzed and a *TestAnalysisReport* is produced.

## 4   Development of Knowledge Management Portal

The Semantic Web is an extension of the current Web, in which Web resources are given machine-understandable descriptions of data, programs, and infrastructure, better enabling computers and people to work in co-operation. From this point of view, ontologies and semantic web technologies have received more attention and been gradually used in the knowledge representation, that have been together introduced in this study. At the same time, ontology and semantic web technologies provide powerful reasoning capabilities. In this section, we describe the designing of knowledge framework to manage software testing knowledge, which is built upon the Java J2EE distributed component environment.

Our knowledge framework consists of five layers: Experience Sharing, Ontology, Storage, Reasoning Engine, and Knowledge Retrieval Layer and they are shown in Fig. 3 [25]. The process is summarized as follows. Through the Experience Sharing layer, software testers can annotate their testing knowledge with an aid of software testing variable cloud. The software testing variable cloud

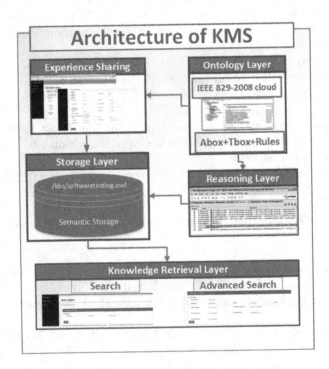

**Fig. 3.** Architecture of knowledge framework

basically contains the vocabularies mentioned in IEEE 829-2008 [6] and ISTQB [7]. Even though the standard specifies the procedures of software testing, we have also included what companies are stipulating in their practice. Through the Experience Sharing interface of the Experience Sharing layer, software testers can annotate their testing knowledge with an aid of software testing variable cloud. Software testing variable cloud is already defined in the ontology layer. The semantic data is expressed in triple structures according to the concepts and relationships of software testing ontology. After the semantic data is stored in the storage layer, the reasoning engine creates reasoning data by means of a reasoning process based on defined ontology, domain rules, and semantic data. Finally, the Knowledge Retrieval Layer retrieves relevant information related to the testing knowledge from the storage layer. The main functions of the Knowledge Retrieval Layer are basic search and Advanced search.

## 4.1   Ontology Layer

The ontology layer has our ontology which includes domain rules, axioms etc. Using the Protégé Ontology Editor 5.1, these concepts and their relationships were described in three elements of OWL: class, property, and individual. The Ontology layer is constructed on OWL DL, which includes a terminology box (TBox), an assertion box (ABox) and a rule base. The ontology layer has our ontology which includes domain rules, axioms etc. Using the Protégé Ontology Editor 5.1, these concepts and their relationships were partly described in Sect. 3.

## 4.2   Experience Sharing Layer

Through the Experience Sharing layer, software testers can annotate their testing knowledge with the support of ISTQB and IEEE 829-2008 terms. We have used three parts such as General Information, Test Case and Test Result. General Information is about Project details. TestCase part presents about the necessary information to create TestCase. Test Result details are presented in TestResult part. Once knowledge is shared, such knowledge is transformed by the semantic data generator into the semantic data in a machine-understandable format of triple structure.

## 4.3   Storage Layer

We used Triple-store, which stores RDF triples and are queried using SPARQL. There are different types of triple-stores available such as Jena SDB, Jena TDB, OWLIM, Sesame and others [23]. Among them, Jena TDB has been selected to be used in this study because it is a component of Jena for RDF storage and query. Importantly, it supports the full range of Jena APIs and can be used as a high performance of RDF store on a single machine. Importantly, it supports the full range of Jena APIs and can be used as a high performance of RDF store on a single machine.

### 4.4  Reasoning Layer

The Semantic Web Rule Language (SWRL) [26] is based on a combination of OWL with the Rule Markup Language which provides inference capabilities from existing OWL ontology. Rules in SWRL reason about OWL instances in terms of OWL classes and properties. Importantly, such rules express more complex relationships and restrictions between concepts. Software Testing rules were generated with Protégé SWRL Editor that is a plugin in Protégé 5.1 environment and with the support of the Jess Rule Engine.

### 4.5  Knowledge Retrieval Layer

To show how our ontology can be used to share software testing knowledge collected from software testers, the Knowledge Retrieval Layer includes two functionalities that use Semantic Web technologies: (1) basic search, and (2) Advanced Search. SPARQL has been used as the query language to retrieve software testing knowledge from the semantic data storage. Since the storage contains a set of knowledge in the triple structure, a SPARQL query is expressed as a list of conditions in a triple structure that is similar to the structure of knowledge in the storage. Besides, Using SPARQL, one can express complex graph pattern queries on RDF graphs. We have used Jena, which is a Semantic Web framework that is used to manipulate and store RDF graphs in storage layer. Besides, ARQ (http://jena.apache.org/documentation/query/) is a query engine of Jena that is used in this scenario for querying ontologies by SPARQL.

The basic search provides a simple triple pattern matching service, which is one of the most frequently used functions for searching documents in the Semantic Web. Besides, Advanced Search Option has been incorporated into knowledge retrieval to support more precise and effective knowledge. The search function subsequently retrieves relevant software testing knowledge from the stored semantic data. Results of the SPARQL queries are displayed in the left part of the screen.

## 5  Verification and Validation of Software Testing Ontology

The quality of an ontology should be verified and validated before it is used in practice to avoid defects [23]. Further, such verification process will prevent contain ontology with anomalies or pitfalls, inconsistent incorrect and redundant information [23]. Besides, ontology validation is important to guarantee that what is built meets the application requirements. The developed ontology is evaluated in three methods such as internal (using reasoners and OOPS!), ontology expert method and non-expert methods.

In literature, various number of semantic ontology reasoners have been studied and implemented to validate the ontologies. Also, each reasoner has used a unique method and a strategy in order to derive an accurate result. In this study,

**Table 2.** Pitfall description and solution proposed

| Pitfall | Dimension | Description | Solution proposed |
|---|---|---|---|
| Missing annotation (256 cases \| Minor) | Human understanding | Ontology terms lack annotation properties that could improve the understanding of the ontology | Included the ontology annotation properties |
| Missing domain or range in properties (7 cases \| Important) | Modelling issues, human understanding | Relationships and (or) attributes without domain or range are included | Restored missing domains |
| Missing inverse relationships (14 cases \| Minor) | Modelling issues, human understanding | When a relation has non-inverse relationship defined | Included missing inverse relationships |
| Defining wrong inverse relationships (6 cases \| Critical) | Logical consistency, modelling issues | Relationships are defined as inverse relations when they are not necessarily inverse | They were removed to improve the expressiveness |

Protégé 5.1 inbuilt reasoners such as FaCT++ 1.6.5 and HermiT 1.3.8.413 were used to check the internal consistency and inferences.

OOPS! (http://oops.linkeddata.es/) is an online ontology evaluator that has been used for our context to detect potential pitfalls that could lead to modelling errors, before ontology has been deployed in the end-user application. This method evaluates human understanding, logical consistency, modelling issues, ontology language specification, real world representation and semantic applications from the developed ontology [23]. Table 2 summarizes the pitfalls encountered, along with a brief description and the way each one of them was handled. There were three levels such as Critical, Important and Minor. Critical level is very crucial and it must be corrected in order to avoid ontology inconsistency. To make ontology nicer both Minor and Important cases were corrected.

Expert evaluation activity is performed by two senior university academic lecturers, who are expertised in ontology area and also have a good understanding of software testing. These experts are neither authors nor belonging to our research team. The main objectives of expert evaluation are (a) whether the software testing ontology meets its' requirements, standards, representation of concepts, relationships among them (b) coverage of the software testing domain and (c) checking for internal consistencies. They took 21 days for evaluation.

The report summarizes the following suggestions and improvements that were highlighted by the ontology experts. All such suggestions, comments were taken into consideration and implemented in the software testing ontology.

- +: Manchester syntax was used.
- +: Taxonomy: It follows is-a Relationship.
- +: OntoGraph: rdfs:subClassOf from a single node to 5.

- +: CQs provide clear idea of the Ontology structure.
- +: Complete software testing process.
- +: No Inconsistency with FaCT++.
- −: URIs could be shortened.
- −: Rewrite English grammar for some CQs.

## 5.1  Ontology Non-expert Evaluation

To carry out our first industrial application based evaluation from software testing experts, the developed knowledge framework was hosted locally inside one software company in Sri Lanka. A separate questionnaire was designed in English with five-point Likert-type scale to capture respondents' self-reported attitudes where respondents had to make their level of agreement such as; Strongly Agree, Agree, No Idea, Disagree and Strongly Disagree. Scores 5, 4, 3, 2, and 1 were assigned respectively for the above-mentioned categories. The profiles and demographics of the participants (Employed Group, work experience, job description, and qualification) were questioned first and continued with questions focused to check whether developed ontology was able to (a) express software testing knowledge (b) support software testing knowledge framework (c) support software testing knowledge retrieval and (d) User Satisfaction. We limited the time period to ten (10) days to collect the questionnaire data. Importantly, with the full considerable support of the management, all software testers in both groups were notified in detail in advance through an email. In addition, a prior training session was conducted through recorded video to make software testers familiar with the knowledge framework. Total of 21 (Group A (12) and Group B (9)) software testing professionals participated in this study to measure the level of user satisfaction with hosted knowledge framework. This was performed by two software testing groups working on a similar information system project for comparison and in depth understanding.

All of the software testers participated believed that the standards have been maintained, but a few have included some extra comments, such as inclusion of some vocabularies, parameters in Test Level and Test Types. Moreover, when asked about their satisfaction on the factors related to user manipulation, personalization and knowledge community, both groups have mostly responded neither satisfied nor unsatisfied, or satisfied. Notably, a very few have been very much satisfied with the user manipulation or knowledge community of the KM.

When asked to evaluate the content of the KM, the large majority (66.67%) of the respondents from group A believed (Strongly Agree or Agree) that the provided content in Web is correct. In contrast, only 44.44% of the respondents from group B believed (Strongly Agree or Agree) that the provided content in Web is correct. All of the software testers participated believed that the standards have been maintained, but a few have included some extra comments, such as inclusion of some vocabularies, parameters in Test Level and Test Types.

When asked about their satisfaction on the factors related to knowledge framework and Retrieval experiences, the following information was revealed. More precisely, if we focused on the search/retrieve functionalities to retrieve software testing knowledge, we found that from group A (41.66%) of the respondents believed (Strongly Agree or Agree) that the search or retrieval functionalities are adequate while from group B this value is 55.56%. Further, when asked to evaluate the effect of the knowledge framework functionalities, (66.67%) of the respondents from group B agreed (Agree) with the existing KM. On the other hand, 41.66% respondents from group A believed (Strongly Agree or Agree) that the knowledge framework functionalities are adequate.

Moreover, when asked about their satisfaction on the factors related to user manipulation, personalization and knowledge community, both groups have mostly responded neither satisfied nor unsatisfied, or satisfied. Notably, a very few have been very much satisfied with the user manipulation or knowledge community of the KM.

Importantly, one discrepancy exposed in this survey study is that the KM does not support to upload files such as image, audio and even images with high quality.

To conclude, overall 80.95% of the respondents would like to recommend the use of such KM among the software testers for knowledge framework and knowledge retrieval.

# 6   Discussion and Conclusion

Software testing is a knowledge intensive and a collaborative activity, which mainly depends on knowledge and experience of the software testers. Knowledge management in software companies aims to create an environment for continuous knowledge framework and creation to remain competitive. Therefore in this research, great importance is given to knowledge for software testing, and the potential benefits of managing software testing knowledge, an ontological approach to represent the necessary software testing knowledge within the software testers' context was developed. Using this approach, software testing ontology to include information needs to be identified for software testing activities to be designed. Competency questions (contextualized information) were used to determine the scope of the ontology and used to identify the contents of the ontology because contextualized information fulfils the expressiveness and reasoning requirements of the software testing ontology. SPARQL query was used to query the competency questions.

The ontology-based knowledge framework is introduced into the semantic representation of software testing knowledge. Such ontology-based knowledge framework would support and provide knowledge sharing and retrieval of software testing knowledge [27]. Moreover, this framework would encourage the software testers to involve in sharing knowledge with others within their software company.

Mostly, quality of an ontology depends on its validity. Hence, by establishing the validity of the developed software testing ontology, quality can be ascertained. We used reasoners inbuilt with the Protégé 5.1 to check the logical inconsistency. The web-based tool OOPS! was also used to detect potential pitfalls in the developed software testing ontology. Based on Ontology experts responses, comments, and suggestions the ontology was redeveloped. Finally, application-based evaluation, that is ontology non-expert evaluation provided the information where they would like more information in this application. The results from the industrial experimental investigation show that, it is adequate to support knowledge framework and reuse, allowing:

- knowledge representation and organization;
- distributed knowledge inference and retrieval;
- management of organizational knowledge on software testing knowledge. Their feedback was considered for future refinement of the software testing ontology.

In this research, we developed a pioneering prototype ontology based knowledge framework to manage software testing knowledge in Sri Lankan software company. The following limitations and difficulties have been identified.

- Even though the KM was hosted locally in a software organization, to provide a richer software testing knowledge through this knowledge framework supported by semantic Web, project documents from previous projects and tests should be included into semantic storage.
- At present, the KM Web focuses on the issue of knowledge framework and provides a simple way for knowledge searching. Still it has been employed with many participants to evaluate and validate it.

We believe our software testing ontology can support other software companies to improve the sharing of knowledge and learning practices. In the future work, the reasoning engine with Query-enhanced Web Rule Language (SQWRL) [28] will be incorporated into knowledge searching to support more precise and effective knowledge framework.

**Acknowledgement.** We thank Virtusa software company (www.virtusa.com/) and its testing team members who were actively facilitating this research since beginning. We also thank both Academic Staffs who provided their immense support through expert evaluation. This work was partially supported by the National Natural Science Foundation of China (Grant No. 61672398), the Key Natural Science Foundation of Hubei Province of China (Grant No. 2015CFA069), and the Applied Fundamental Research of Wuhan (Grant No. 20160101010004).

# References

1. Rus, I., Lindvall, M.: Knowledge management in software engineering. IEEE Softw. **19**(3), 26–35 (2002)
2. Vasanthapriyan, S., Tian, J., Xiang, J.: A survey on knowledge management in software engineering. In: 2015 IEEE International Conference on Software Quality, Reliability and Security-Companion (QRS-C), pp. 237–244. IEEE (2015)
3. Santos, V., Goldman, A., De Souza, C.R.: Fostering effective inter-team knowledge sharing in Agile software development. Empir. Softw. Eng. **20**(4), 1006–1051 (2015)

4. Vasanthapriyan, S., Tian, J., Xiong, S., Xiang, J.: Knowledge synthesis in software industries - a survey in Sri Lanka. Knowl. Manage. Res. Pract. **15**(3), 413–430 (2017)
5. Gruber, T.R.: Toward principles for the design of ontologies used for knowledge sharing? Int. J. Hum.-Comput. Stud. **43**(5–6), 907–928 (1995)
6. Software and Systems Engineering Committee: IEEE standard for software and system test documentation. IEEE Computer Society, Fredericksburg, VA, USA (2008)
7. Graham, D., Van Veenendaal, E., Evans, I.: Foundations of Software Testing: ISTQB Certification. Cengage Learning EMEA, London (2008)
8. Guo, S., Tong, W., Zhang, J., Liu, Z.: An application of ontology to test case reuse. In: 2011 International Conference on Mechatronic Science, Electric Engineering and Computer (MEC), pp. 775–778. IEEE (2011)
9. Nasser, V.H., Du, W., MacIsaac, D.: Knowledge-based software test generation. In: Software Engineering and Knowledge Engineering, pp. 312–317 (2009)
10. Ryu, H., Ryu, D.K., Baik, J.: A strategic test process improvement approach using an ontological description for MND-TMM. In: Seventh IEEE/ACIS International Conference on Computer and Information Science, ICIS 2008, 561–566. IEEE (2008)
11. Cai, L., Tong, W., Liu, Z., Zhang, J.: Test case reuse based on ontology. In: 15th IEEE Pacific Rim International Symposium on Dependable Computing, PRDC 2009, pp. 103–108. IEEE (2009)
12. Anandaraj, A., Kalaivani, P., Rameshkumar, V.: Development of ontology-based intelligent system for software testing. arXiv preprint arXiv:1302.5215 (2013)
13. Arnicans, G., Romans, D., Straujums, U.: Semi-automatic generation of a software testing lightweight ontology from a glossary based on the ONTO6 methodology. In: DB&IS, pp. 263–276 (2012)
14. Zhang, Y., Zhu, H.: Ontology for service oriented testing of web services. In: IEEE International Symposium on Service-Oriented System Engineering, SOSE 2008, pp. 129–134. IEEE (2008)
15. Barbosa, E.F., Nakagawa, E.Y., Maldonado, J.C.: Towards the establishment of an ontology of software testing. In: SEKE, pp. 522–525 (2006)
16. Yu, L., Zhang, L., Xiang, H., Su, Y., Zhao, W., Zhu, J.: A framework of testing as a service. In: International Conference on Management and Service Science, MASS 2009, pp. 1–4. IEEE (2009)
17. Souza, E.F., Falbo, R., Vijaykumar, N.L.: Using ontology patterns for building a reference software testing ontology. In: 2013 17th IEEE International Enterprise Distributed Object Computing Conference Workshops (EDOCW), pp. 21–30. IEEE (2013)
18. Fernández-López, M., Gómez-Pérez, A., Juristo, N.: Methontology: from ontological art towards ontological engineering (1997)
19. Sure, Y., Staab, S., Studer, R.: On-to-knowledge methodology (OTKM). In: Staab, S., Studer, R. (eds.) Handbook on Ontologies. INFOSYS, pp. 117–132. Springer, Heidelberg (2004). doi:10.1007/978-3-540-24750-0_6
20. Noy, N.F., McGuinness, D.L., et al.: Ontology development 101: a guide to creating your first ontology (2001)
21. Grüninger, M., Fox, M.S.: Methodology for the design and evaluation of ontologies (1995)
22. W3C OWL Working Group: OWL2 Web Ontology Language: Document Overview. W3C Recommendation (2009). http://www.w3.org/TR/owl2-overview/

23. Vasanthapriyan, S., Tian, J., Zhao, D., Xiong, S., Xiang, J.: An ontology-based knowledge sharing portal for software testing. In: 2017 IEEE International Conference on Software Quality, Reliability and Security-Companion (QRS-C), pp. 237–244. IEEE (2017)
24. Myers, G.J., Sandler, C., Badgett, T.: The Art of Software Testing. Wiley, Hoboken (2011)
25. Vasanthapriyan, S., Tian, J., Zhao, D., Xiong, S., Xiang, J.: An ontology-based knowledge management system for software testing. In: The Twenty-Ninth International Conference on Software Engineering and Knowledge Engineering (SEKE), pp. 522–525 (2017)
26. Horrocks, I., Patel-Schneider, P.F., Boley, H., Tabet, S., Grosof, B., Dean, M., et al.: SWRL: a semantic web rule language combining OWL and RuleML. W3C Memb. Submiss. **21**, 79 (2004)
27. Yoo, D., No, S.: Ontology-based economics knowledge sharing system. Expert Syst. Appl. **41**(4), 1331–1341 (2014)
28. Zhang, Y., Luo, X., Zhao, Y., Zhang, H.C.: An ontology-based knowledge framework for engineering material selection. Adv. Eng. Inform. **29**(4), 985–1000 (2015)

# The Effect of Task Allocation Strategy on Knowledge Intensive Team Performance Based on Computational Experiment

Shaoni Wang[✉], Yanzhong Dang, and Jiangning Wu

Institute of System Engineering, Dalian University of Technology,
Dalian 116024, China
shaoni_wang@163.com

**Abstract.** The purpose of this study is to research the task allocation problem of the knowledge intensive team (abbreviated as KIT), which is different from the traditional task assignment. We built a KIT system model, designed task allocation strategies and team performance measurement scale, based on complex adaptive system (abbreviated as CAS) theory with regarding the knowledge requirement of tasks as a primer mover, additionally, took into consideration that knowledge exchange behaviors and processes would be contingent when different team members deal with different tasks. The computational experimental method was used to analyze how different allocation strategies impact KIT performance. The experimental results show that different allocation strategies variously influence KIT performance when the team members, team structures, and tasks to be assigned are different. We would be appreciated to help the decision maker, before the real tasks are executed, to apply the computational experiment method proposed in this paper to carry out the task allocation to provide with decision support.

**Keywords:** Knowledge intensive team · Task allocation strategies · Team performance · Computational experiment

## 1 Introduction

Knowledge intensive team (KIT) [1–4] can solve complex and vague problems and enhance adaptability and innovation ability of enterprises. It has become an important organizational form of enterprises and has gradually become the academic focus.

KIT generally deals with knowledge intensive tasks, which require strong expertise, independent thinking and collaborating, rather than routine tasks with fixed workflow and strong structure [5]. The level of knowledge, autonomous behavior, and interaction among team members, as well as task difficulty, make the KIT a complex adaptive system (CAS) with openness and autonomy [6–8]. Task allocation scheme basically refers to assigning tasks to team members according to certain strategies, and different task allocation strategies (abbreviated as TASs) have different effects on operation process and team performance. Accordingly, the TASs are important managerial tools to affect team performance. The task allocation problem is essentially about how to assign tasks appropriately to achieve the best performance of a team. How can we

© Springer Nature Singapore Pte Ltd. 2017
Chen et al. (Eds.): KSS 2017, CCIS 780, pp. 227–241, 2017.
https://doi.org/10.1007/978-981-10-6989-5_19

determine a TAS by various team conditions, the team can optimally accomplish tasks and improve the overall performance? Finding a manner to that is the purpose of this study.

However, due to complex interaction and knowledge exchange [9, 10] between team members of a CAS during the execution of tasks, it is likely that the initial "optimal assignment" may not produce the best overall team performance. In reality, the KIT task allocation problem significantly differs from the classical task assignment problem. The essential difference is that classical assignment considers team members are "all-powerful" who can deal with any tasks regardless of costs, with no interaction between coworkers, and also, team member's ability will not change after tasks are assumed. While in this KIT task allocation study, team members are not "all-powerful" and their abilities are heterogeneous; not any of them can solve any tasks due to insufficient ability; however, team members are able to improve abilities during the process of interacting with others. So, it can be observed evidently that the task assignment problem being researched here is not a classic assignment problem, and the traditional task allocation methods have great limitations on solving the problem raised in this paper. Comparatively speaking, the KIT task allocation methods are more in line with the real team task allocation.

The existing task allocation studies could be divided into two categories: the traditional assignment problem research and task allocation by using the multi-agent system (MAS) [11, 12]. The classical assignment problem mainly concentrates on assignment algorithms. It was first discovered in the 1940s, and it basically uses linear programming in operational research to solve the problem of using existing resources to complete fixed tasks [13, 14]. With the continuous increasing constraints imposed on task allocation, the cost of solving the problem got higher. Inspired by Sagar [15], a plethora of intelligent optimization algorithms [16, 17] emerged in new task allocation problems. After the 90s, with the rapid development of agent technology, the task allocation problem gradually shift to focus MAS. Compared with the traditional task allocation problem, the task allocation in MAS is closer to reality and more humane. Loads of agent-based task allocation methods, including extended contract network protocol [18], self-organization [19, 20], etc. have greatly promoted the development of task allocation theory. Recently, the task allocation problem has gradually changed from static to dynamic. Sighn et al. have designed algorithms to dynamically measure the individual utility and team utility of task allocation [12]. Chinese researchers also proposed a centralized dynamic task assignment model based on the Markoff decision process, which is suitable for small scale systems to achieve task balance [21]. With the increasing complexity of management problem, the computational organization theory [22] and computational experiment method [23] came into being. The computational experiment can reproduce the "happened", "happening" and "may happen in future" phenomenon through the scenario modeling, analyze the occurrence of the phenomenon to find its inherent laws, to better guide the practice [24].

Although current researches in China and abroad are mature, also some have been applied in practice, but they differ from the problem of this study. Features of this study are: (1) Knowledge exchange among team members is an important prerequisite for a team to complete tasks efficiently. We study the impact of different TASs on team performance considering the knowledge exchange as a core. (2) Different TASs generate different ways of knowledge exchange, and members dissimilarly make adaptive adjustments, which will differently impact team performance. This series of processes is complexly adaptive. (3) The "non-exclusive" knowledge resource significantly differentiates from the traditional "exclusive" resources such as material, capital and so on. Knowledge resources are not consumed in time of using whereas they may be innovated and proliferated due to flowing.

On the whole, the emergence and development of computational experiments, especially agent-based experiments, provide a new perspective for CAS research and a new way of thinking about task allocation for KIT. Based on the theory and viewpoints of CAS, this paper utilizes MAS technologies and tools to model the KIT in the real world as knowledge-intensive MAS models according to the characteristics of KIT. The model is used to construct the artificial team and to carry out the dynamic computational experiment, obtain the research data, analyze the influence of the task allocation strategy on the team performance, so as to provide decision support for the best allocation strategy for the realistic KIT.

# 2   KIT System Model

## 2.1   Team Model

**Interpersonal Communication Network.** Members' power could be integrated into teams which fully rely on interaction and reciprocal collaboration to complete tasks. Meanwhile, the interaction depends on interpersonal communication network [25–27] (abbreviated as ICN). The nodes in ICN represent members, and each node has equal status in ICN, the edges in ICN express connections among members, which serve as knowledge flow channels, and members are capable of getting required knowledge through this channel. The team present connections number (abbreviated as PCN) is defined as the total number of connections in the team currently. The member present connections number (abbreviated as $PCN(a_i)$) is defined as the number of connections currently between member $a_i$ and the others.

In this paper, we resort to the team present interpersonal communication network connectivity coefficient (abbreviated as ICNCC) to express the connection status among entire team members. The ICNCC defines as the ratio of PCN and the number of connections with full connectivity, which accounts for the total connections number (abbreviated as TCN) in the team. Calculate according to Eq. 1.

$$ICNCC = \frac{PCN}{TCN},\tag{1}$$

Where $TCN = C_m^2 = \frac{m(m-1)}{2}$, $m$ is the number of team members. $ICNCC \in [0, 1]$, $ICNCC = 0$ shows there is no edge in ICN and no communication channel between members; $0 < ICNCC < 1$ indicates certain edges exists, which means part of members can communicate with others, the larger the ICNCC is, the more edges exist, the more members could interact. $ICNCC = 1$ means ICN is fully connected, which means members can communicate with all other members. There are several different connections types under the same ICNCC, which affects team performance.

**Team Multi-agent System Model.** We take a team as a CAS, the team member as well as the node in ICN defined as an agent. We assume that $A = \{a_1, a_2, \cdots, a_i, \cdots, a_m\}$ is a team composed of m agents, $a_i$ represents the $i$ th team member. Whether agent $a_i$ can interact or not depends on whether there is an edge between the nodes in ICN, if there is one, they can interact and communicate with each other, or else they cannot.

**Knowledge Space.** Knowledge is an indispensable resource for KIT to perform tasks, if members' knowledge fulfils needs of tasks, the members can complete the tasks, otherwise, cannot. We defined a knowledge space $K = \{k_1, k_2, \cdots, k_l \cdots, k_s\}$ to formally represent and measure the amount of knowledge, $k_l$ represents one kind of specialized knowledge or specific skill.

## 2.2   Member Model

Each agent has attributes and abilities or behaviors to communicate with other agents, and it can be expressed as $a_i = \{attribute, behavior\_mechanism\}$, $i = 1, 2, \cdots, m$, the *attribute* value of $a_i$ is distinct, the *behavior_mechanism* refers to what each agent follows during interacting. We exclusively focus on knowledge attribute.

**The Knowledge Attribute.** Each agent $a_i$ has the certain professional knowledge and specific skills as prerequisites to complete tasks. We call both knowledge and ability as knowledge resource, which is represented as $K(a_i) = \left[ak_i^1, ak_i^2, \cdots, ak_i^l, \cdots ak_i^s\right]$, $ak_i^l \in [0, 1]$, $i = 1, 2, \cdots, m$ indicates how well $a_i$ knows $l$th dimension knowledge, $ak_i^l = 0$ represents $a_i$ knows nothing about $l$th dimension knowledge, while $ak_i^l = 1$ means $a_i$ absolutely knows it. Different members have different knowledge.

**The Behavior Mechanism of an Agent.** After $a_i$ receiving a task, it inspects the knowledge itself to judge whether its knowledge meets the task's requirement or not, if it does then complete the task directly, otherwise $a_i$ produces demand for knowledge and motivation of knowledge acquisition, further generate knowledge acquisition behavior. Knowledge can be taught by themselves or by interacting, here we only delve the latter form. The $a_i$ chooses to communicate and carries on knowledge interaction, if acquired knowledge has reached the task's need, then to complete the task, otherwise, $a_i$ continues to choose another agent to interact. When $a_i$ fully interacts with all other agents who are connected with $a_i$ in ICN, if $a_i$ still cannot achieve the task, $a_i$ gives up the received task. The behavior mechanism of agent $a_i$ is shown in Fig. 1.

There are three rules in the agent behavior mechanism:

1. if agent_knowledge $\geq$ task_knowledge then do Task execute;
2. if agent_knowledge < task_knowledge then do Agent_choose;
3. if agent_knowledge < task_knowledge then do Continue?
   if Continue? = Yes then do 2 else do Task_end.

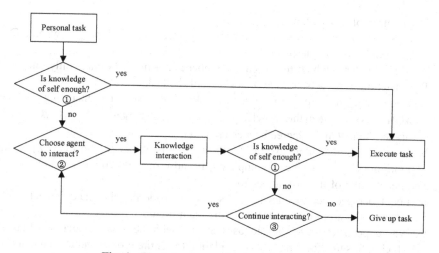

**Fig. 1.** The behavior mechanism of the agent $a_i$

The selection rules of the agent $a_i$:

1. Agent $a_i$ choose one candidate randomly according to the connection_list which is a list of candidate agents who is connected with $a_i$ directly.
2. if more than one wants to interact with agent $a_i$;
   then agent $a_i$ preferentially choose the one with the smaller number.

## 2.3  Task Model

We suppose team project P is a collection of tasks, shown as $P = \{t_1, t_2, \cdots, t_j, \cdots, t_n\}$, $t_j$ is the $j$ th task. We only consider the need for knowledge to complete tasks here. The knowledge of $t_j$ is defined as $K(t_j) = \left[ tk_j^1, tk_j^2, \cdots, tk_j^l \cdots, tk_j^s \right]$, $j = 1, 2, \cdots, n$, $tk_j^l \in [0, 1]$, which represents the minimum amount of the $l$th dimension knowledge to complete $t_j$, it also indicates the minimum mastery of the $l$th knowledge of agent. $tk_j^l = 0$ indicates $a_i$ do not need to grasp the $l$th dimension knowledge to complete $t_j$, while $tk_j^l = 1$ means $a_i$ need to fully master $l$th dimension knowledge to complete $t_j$.

# 3  Task Allocation Strategy

## 3.1  TAS Design

Tasks can be distributed in varieties of ways, which is based on TASs, such as random election. Here, we attempt to design several TASs from following three perspectives according to actual needs or situation of a team or some rules, specifically as follows:

1. The number of acceptable tasks:
   a. One_task, which means each member only accept one task despite whether the task could be completed or not.
   b. Multiple_tasks, which means each member is not limited to take only one task.
2. The matching type between the members and the tasks:
   a. Knowledge_matching, to allocate tasks following the matching degree $M_{ij}(ak_i, tk_j)$ between the knowledge of $a_i$ and $t_j$. The bigger the $M_{ij}(ak_i, tk_j)$ is, the higher the matching degree between $a_i$ and $t_j$ is.
   b. Connections_matching, to consider $PCN(a_i)$ in *ICN*. The greater $PCN(a_i)$ is, the more possibly $a_i$ is able to complete $t_j$ because of more communication.
3. The perspective of the task allocation:
   a. Local_perspective, to allocate tasks only considering the matching situation between the arbitrary one task and the team members.
   b. Whole_perspective, to allocate tasks accords with the overall matching of tasks, which refers to allocating tasks considering the entire project and team members.

### 3.2  Allocation Strategies

We constitute a strategic space to form eight different TASs, as Figs. 2 and 3:

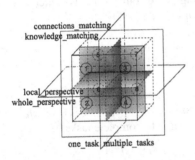

| matching type | acceptable task number | local_perspective | whole_perspective |
|---|---|---|---|
| knowledge_ matching | one_task | ① | ② |
| | multiple_tasks | ③ | ④ |
| connections_ matching | one_task | ⑤ | ⑥ |
| | multiple_tasks | ⑦ | ⑧ |

**Fig. 2.**  Task allocation strategy space

**Fig. 3.**  Eight task allocation strategies

Apart from the eight TASs above, we provide a random strategy as computational experimental comparison basis, which is 0-RS in this paper. The nine TASs are as follows:

0-RS: Randomly select a member to receive task, which is random selection TAS.
1-LOK: In case of considering only one task and each member only accept one task, meanwhile to allocate tasks in the form of knowledge_matching, which is called local_one_knowledge TAS.
2-WOK: In case of considering the entire tasks set and each member only accept one task, meanwhile to allocate tasks in the form of knowledge_matching, which is called whole_one_knowledge TAS.
3-LMK: In case of considering only one task and each member can accept multiple tasks, meanwhile to allocate tasks in the form of knowledge_matching, which is called local_multiple_knowledge TAS.

4-WMK: In case of considering the entire task set and each member can accept multiple tasks, meanwhile to allocate tasks in the form of knowledge_matching, which is called whole_multiple_knowledge TAS.

5-LOC: In case of considering only one task and each member can only accept one task, meanwhile to allocate tasks in the form of connections_matching, which is called local_one_connection TAS.

6-WOC: In case of considering the entire tasks set and each member can only accept one task, meanwhile to allocate tasks in the form of connections_matching, which is called whole_one_connection TAS.

7-LMC: In case of considering only one task and each member can accept multiple tasks, meanwhile to allocate task in the form of connections_matching, which is called local_multiple_connection TAS.

8-WMC: In case of considering the entire tasks sets and each member can accept multiple tasks, meanwhile to allocate task in the form of connections_matching, which is called local_multiple_connection TAS.

## 4 Metric Definition and Experimental Settings

### 4.1 Metric Definition

Here we propose some metrics to analyze the whole system of task allocation.

- Knowledge Matching Degree. We calculate the knowledge matching degree $M_{ij}(ak_i, tk_j)$ between $a_i$ and $t_j$ by Eq. 2,

$$M_{ij}(ak_i, tk_j) = \frac{|ak_i| \times \cos\langle ak_i, tk_j \rangle}{|tk_j|} = \frac{ak_i \times tk_j}{|tk_j|^2} (i = 1, 2, \cdots, m; j = 1, 2, \cdots, n)$$

(2)

Table 1. The matching table

|       | $t_1$    | $t_2$    | $\cdots$ | $t_j$    | $\cdots$ | $t_n$    |
|-------|----------|----------|----------|----------|----------|----------|
| $a_1$ | $M_{11}$ | $M_{12}$ | $\cdots$ | $M_{1j}$ | $\cdots$ | $M_{1n}$ |
| $\vdots$ | $\vdots$ | $\vdots$ | $\ddots$ | $\vdots$ | $\vdots$ | $\vdots$ |
| $a_i$ | $M_{i1}$ | $M_{i2}$ | $\cdots$ | $M_{ij}$ | $\cdots$ | $M_{in}$ |
| $\vdots$ | $\vdots$ | $\vdots$ | $\cdots$ | $\vdots$ | $\cdots$ | $\vdots$ |
| $a_m$ | $M_{m1}$ | $M_{m2}$ | $\cdots$ | $M_{mj}$ | $\cdots$ | $M_{mn}$ |

To calculate matching degree in accord with the Eq. 2, formulating a matching table shown as Table 1, $M_{ij}$ represents matching degree between $a_i$ and $t_j$.

- Competence Threshold (abbreviated as *CT*), which is on behalf of psychological expectations of decision makers, the decision makers think that only when $M_{ij}(ak_i, tk_j) \geq CT$, $a_i$ is competent for $t_j$, otherwise not. Different decision makers have different psychological expectations, hence they have different *CT* values, a part thereof are conservative distributors who suppose only a few members could complete the tasks, they have larger *CT*; while some are positive distributors who suppose most of the members are able to complete the tasks, they have smaller *CT*.

- Sociability Threshold (abbreviated as *ST*), which is also a psychological expectation of a decision maker, decision makers think only when $PCN(a_i)$ is greater than *ST*, that is, $PCN(a_i) \geq CT$, the member $a_i$ is competent for the task $t_j$, otherwise not. Different psychological expectations lead to different *ST* value, *ST* is related to $PCN(a_i)$. Parameter $\alpha$ is a measure of decision makers' subjective expectation when the one who thinks larger connections of $a_i$ contribute to a larger possibility of competence for $t_j$ has a larger $\alpha$.

$$ST = \alpha Max(PCN(a_i)) \ (i = 1, 2, \cdots, m) \tag{3}$$

- Competence of Team Project $C(P)$ refers to how entire tasks of team project match team members, assumed as $C(P) = \{C(t_1), C(t_2), \cdots, C(t_j), \cdots, C(t_n)\}$, where $C(t_j)$ represents the number of members who are competent for $t_j$. The smaller the $C(t_j)$ is, the fewer members are competent for $t_j$, when $C(t_j) = 0$, it shows none of the members are qualified for $t_j$. To preferentially allocate $t_j$ attached smaller $C(t_j)$, $C(t_j) = 0$ means no one is competent, then to allocate $t_j$ at last.

- Completion Rate, which is the ratio of the number of tasks completed $N^{completed}$ and the total number of tasks $N$, shown as Eq. 4.

$$R = \frac{N^{completed}}{N} \tag{4}$$

- Time Cost of Unit Knowledge Learning, expressed as *tc*, which indicates how long the $a_i$ learns one unit of knowledge.

- Team Project Time, expressed as $t^{team}$, which demonstrates the maximum time used to complete the task allocated to each member, the time of member expressed as $t(a_i)$, all members simultaneously receive tasks and deal with them immediately.

$$t^{team} = Max(t(a_i)) \ i = 1, 2, \cdots, m, \tag{5}$$

where $t(a_i)$ indicates the total time for $a_i$ from the very beginning of receiving to the end of executing the task, including five parts, shown in Eq. 6.

$$t(a_i) = t_{ask}(a_i) + t_{wai}(a_i) + t_{int}(a_i) + t_{res}(a_i) + t_{exe}(a_i) \ i = 1, 2, \cdots, m, \tag{6}$$

where $t_{ask}(a_i)$ means the time cost for $a_i$ sending a request and asking interaction object, $t_{wai}(a_i)$ means the time $a_i$ waiting for interaction object, $t_{int}(a_i)$ means the time $a_i$ interacting, $t_{int}(a_i)$ relating to knowledge difference and *tc*, shown as Eq. 7, $t_{res}(a_i)$

means the time $a_i$ responding other members' request, $t_{exe}(a_i)$ means the time $a_i$ executing the task, when $a_i$ acquired enough knowledge. Where $ak^l_{int}$ means how interaction object masters the $l$th dimension knowledge.

$$t_{int} = \begin{cases} \left|ak^l_{int} - ak^l_i\right| * tc, & \left|ak^l_{int}\right| \leq \left|tk^l_j\right| \\ \left|tk^l_j - ak^l_i\right| * tc, & \left|ak^l_{int}\right| \geq \left|tk^l_j\right| \end{cases} \tag{7}$$

- Total Task Completion Time $t^{sum}$, which indicates the total time of team completing the entire tasks set, it's the total time for all members completing their tasks, as shown in Eq. 8.

$$t^{sum} = \sum_{i=1}^{m} t(a_i) \quad i = 1, 2, \cdots, m \tag{8}$$

## 4.2  Experimental Settings

We take advantage of the model proposed above to simulate a KIT performing tasks. The specific experimental variables and parameters are set as the Table 2:

**Table 2.** The form of experimental parameters setting rules

| Variables/parameters | Implication | Range/value | Setting rules |
|---|---|---|---|
| $s$ | Knowledge dimension number | 10 | Constant |
| $m$ | The scale of the *KIT* | 10 | Constant |
| $P$ | The number of team projects | 100 | Constant |
| $n$ | Tasks number of each project | 9 | Constant |
| $ak^s_i$ | Knowledge of team member | 0–1 | Random value |
| $tk^s_j$ | Knowledge of team task | 0–1 | Random value |
| $CT$ | The competence threshold | 0.7 | Constant |
| $\alpha$ | Decision makers' expectation | 0.7 | Constant |
| $ICNCC$ | The connectivity coefficient | $0, 0.1, \cdots, 1$ | Discrete value |
| $Q$ | Connections types under same *ICNCC* | 20 | Constant |
| $O$ | Interaction types under same *ICNCC* | 10 | Constant |
| $t_{ask}(a_i)$ | The asking time of members | 1 | Constant |
| $tc$ | Time to study one unit knowledge | 100 | Constant |
| $t_{exe}(a_i)$ | The executing time members | 10 | Constant |

## 5  Experimental Results and Analysis

### 5.1  Experimental Results and Output Data

Run KIT computational experiment according to the Table 2, obtain 1,980,000 items data. Each item represents a possibility for KIT to deal with tasks. The data obtained by a complete experiment, which illuminates task execution and results of a team with specific TAS through certain ICNCC, structure and selecting type, shown in Table 3.

**Table 3.** The form of the output data

| Experiment number | Project number | ICNCC | Structure number | Selecting number | TAS | R | $t^{team}$ | $t^{sum}$ |
|---|---|---|---|---|---|---|---|---|
| 1 | 23 | 0.4 | 2 | 2 | WMC | 0.7 | 399 | 908 |
| ⋮ | ⋮ | ⋮ | ⋮ | ⋮ | ⋮ | ⋮ | ⋮ | ⋮ |
| 1980000 | ... | ... | ... | ... | ... | ... | ... | ... |

## 5.2    Analysis of the Effect of TASs on Task Completion Rate

**Analysis of TASs under the One_Task and Multiple_Tasks.** Under the given parameters, the above TASs are divided into two groups: the knowledge_matching group, shown as Fig. 4(a), and the connections matching group, shown in Fig. 4(b). The comparison of the best TAS in Fig. 4(a), (b) and the 0-RS is shown in Fig. 5. The horizontal axis indicates the progressively increasing ICNCC, and each curve reveals that task completion rate $R$ varies with the change of ICNCC.

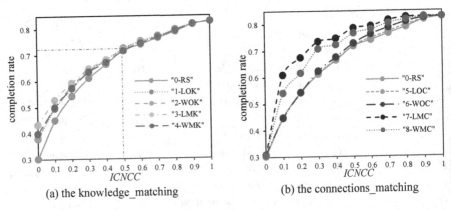

(a) the knowledge_matching          (b) the connections_matching

**Fig. 4.** The influence of TASs on task completion rate $R$

In Fig. 4(a) and (b), $R$ of nine TASs are rising along with the increasing ICNCC. When $ICNCC = 1$, it means the team ICN is fully connective, $R$ of all TASs are same under the nine strategies. When $ICNCC < 1$, which is more realistic and it means not all members are able to exchange knowledge, different TASs contributes to different $R$. In pace with the increasing ICNCC, the difference of $R$ under TASs is decreasing; besides, it proves that although members cannot achieve their task, they interact more adequately with others under a circumstance of bigger ICNCC to raise $R$ of the team.

Moreover, we find the multiple_tasks TASs (3-LMK, 4-WMK, 7-LMC, 8-WMC) has greater $R$ than the one_task TASs (1-LOK, 2-WOK, 5-LOC, 6-WOC).

Under the knowledge_matching TASs, $R$ of TAS 3-LMK and 4-WMK are better in comparison with 1-LOK and 2-WOK (shown as Fig. 4(a)), which results from the former two TASs assign tasks to competent members, no matter if they burden lots of works. To compare 3-LMK and 4-WMK, we find $R$ of the former is higher, and the four knowledge_matching TASs have bigger $R$ compare to 0-RS.

In conclusion, under certain ICNCC, the TAS 3-LMK is optimum to accounting for the highest $R$. When ICNCC increases to an absolute value (Fig. 4(a), ICNCC = 0.5), the four knowledge_matching TASs' $R$ becomes very close to each other.

There exists similar appearance under connections_matching TASs in Fig. 4(b), the multiple_tasks TASs' $R$ are bigger, while the causes in Fig. 4(b) differ in Fig. 4(a), we do not take agent knowledge into account through connections_matching but merely count on the channels connected. The 7-LMC and 8-WMC prioritize members owning plenty of connections without thinking about member taking too many tasks, the priority ensures the probability that members will complete the task, thus achieve a high $R$ against the TASs 5-LOC and 6-WOC, besides, it can be seen, 7-LMC's $R$ is better than 8-WMC's. Exceptionally, when $ICNCC = 0$, no any connections in ICN, the four connections_matching TASs' allocation rules is same with the 0-RS, which brings about the same $R$ under those TASs.

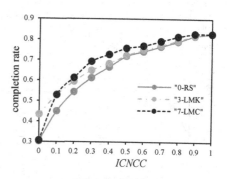

**Fig. 5.** The influence of three TASs on completion rate

In Fig. 5, when $ICNCC = 0$, the TAS 3-LMK's $R$ is high than the 7-LMC's and 0-RS's, while $ICNCC > 0$, the 7-LMC has bigger $R$ compare to the other two TASs, furthermore, the disparity diminishes along with the incremental ICNCC.

We can obtain the following inspirations: when there is a low ICNCC in team, directly choosing multiple_tasks TASs lead to a high $R$, and when ICNCC reaches to certain value (Fig. 4(a), $ICNCC = 0.5$), choosing any knowledge_matching TASs results in an approximately same $R$. When in early stage of team formation, which means ICNCC is small, whatever the TAS is, internal communication should be advocated in order to enlarge ICNCC in team, consequently, to improve team $R$.

**Analysis of TASs of the Knowledge_Matching and the Connections_Matching.** The experimental results of the one_task group TASs (1-LOK, 2-WOK, 5-LOC, 6-WOC) and the multiple_tasks group TASs (3-LMK, 4-WMK, 7-LMC, 8-WMC) are respectively shown in Fig. 6(a) and (b).

In Fig. 6(a), when $ICNCC = 0$, the knowledge_matching TASs 1-LOK and 2-WOK, obviously are superior in terms of $R$ to the connections_matching TASs, 5-LOC and 6-WOC; while $ICNCC < 1$, the superiority of the former two TASs shrinks with ICNCC increasing, when ICNCC approaches to one, $R$ of the four TASs reach the highest equal point. It indicates when in an early stage of a team (small ICNCC), select knowledge_matching TASs to guarantee a higher $R$, however when $ICNCC > 0.5$ (Fig. 6(a)), $R$ of knowledge_matching or connections_matching TASs are radically consistent.

In Fig. 6(b), when $ICNCC = 0$, the knowledge_matching TASs 3-LMK and 4-WMK's $R$ are better than two connections_matching TASs 7-LMC and 8-WMC, while $ICNCC > 0$, $R$ of the latter two TASs rapidly grow to greater than the former two, and the advantage decreases when ICNCC increases.

The implications are: When $ICNCC = 0$, to select the superior knowledge_matching TASs no matter the one_task or multiple_tasks, when $ICNCC > 0$, to select the multiple_tasks TASs to get higher $R$.

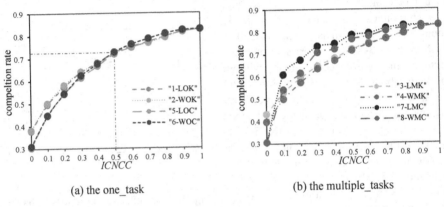

(a) the one_task                    (b) the multiple_tasks

**Fig. 6.** The influence of TASs under the constraint of one_task and multiple_tasks

### 5.3    Analysis of the Effect of TASs on Completion Time

Different TASs lead to different knowledge exchange and interaction status, which also affects the time of team performing tasks. The analysis below is to find best practice.

**Analysis of TASs under the Knowledge_Matching Method.** The Fig. 7(a) and (b) separately shows the change trend of $t^{team}$ and $t^{sum}$ with the incremental ICNCC. In Fig. 7(a), when $ICNCC = 0$, $t^{team}$ is broadly same under four knowledge_matching TASs and 0-RS, with the rising ICNCC, $t^{team}$ of the four knowledge_matching TASs is obviously less than of the 0-RS, and 2-WOK has the smallest $t^{team}$ among those TASs, which owing to 2-WOK could always find the most appropriate member for a task, without additional communication time, it ensures everyone gets their own suitable tasks, thus making $t^{team}$ the shortest.

In Fig. 7(b), the TAS 0-RS' $t^{sum}$ is much more than four knowledge_matching TASs, 2-WOK's $t^{sum}$ is evidently smaller and it remains at a low level although the other TASs' $t^{sum}$ gradually increase in pace with ICNCC, because when ICNCC rises, the interaction channels enlarge and interaction time elongate, accordingly $t^{sum}$ becomes longer. While the 2-WOK always manage to allocate task to the most suitable member, without any additional interaction time.

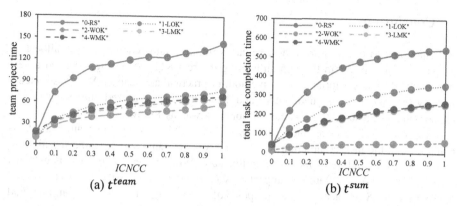

**Fig. 7.** The influence of the knowledge matching strategies on $t^{team}$ and $t^{sum}$

The obtained illumination is: the 2-WOK is of optimum when in time oriented team.

**Analysis of the TASs under the One_Task Constraint.** See in Fig. 8(a) and (b), when $ICNCC = 0$, $t^{team}$ and $t^{sum}$ of TASs are approximately equal. In addition, compare the knowledge_matching TASs 1-LOK and 2-WOK with connections_matching TASs 5-LOC and 6-WOC, both $t^{team}$ and $t^{sum}$ of the former two TASs are clearly less than of the latter two. As a result of discovering competent members accurately, the knowledge_matching helps to reduce members' waiting time and interaction time. While connections_matching TASs focus on more connections and allocate with no considering members' knowledge reserves. When $ICNCC > 0$, the TAS 2-WOK has lowest $t^{team}$ and $t^{sum}$.

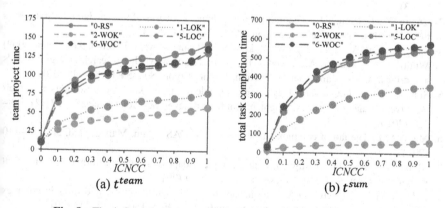

**Fig. 8.** The influence of the single task strategies on the $t^{team}$ and $t^{sum}$

Inspirations are acquired as follows: when a team is confined to one_task condition and merely concentrating on completion time, choose the TAS 2-WOK, while a team is restricted in both one_task condition and connections_matching, there is little difference of choosing 5-LOC or 6-WOC.

## 6  Conclusions

We attempted to put forward a new research thinking and scheme on task allocation decision-making problem through simulating the real KIT, accordingly shaped a new view for such issues, supplied a new method for practical application. We constructed a KIT system model, designed TASs and team metrics, analyzed the experimental phenomena and results, the conclusion obtained was as follows:

The team performance under different TASs differs. When the team ICNCC is equal to one, the task completion rate of all TASs are same and reaching the highest, which means the failure of the TASs in this case. While in reality, ICNCC is between zero and one, TASs differently impact task completion rates, so we can choose the appropriate TAS to achieve the purpose of improving task completion rate and reducing completion time.

We have been carried out a series of work on the KIT task allocation, what the next step work is that we will consider the influence of the task allocation involved factors such as the differentiated team members, the team structure type, additionally, we would like adopt empirical study to verify the experimental results of aforementioned TASs.

**Acknowledgement.** This work is partly supported by the National Natural Science Foundation of China under Grant No. 71471028.

## References

1. Yu, Y., Hao, J.X., Dong, X.Y., et al.: A multilevel model for effects of social capital and knowledge sharing in knowledge-intensive work teams. Int. J. Inf. Manag. **33**(5), 780–790 (2013)
2. Chung, Y., Jackson, S.E.: The internal and external networks of knowledge-intensive teams the role of task routineness. J. Manag. **39**(2), 442–468 (2013)
3. Faraj, S., Yan, A.: Boundary work in knowledge teams. J. Appl. Psychol. **94**(3), 604 (2009)
4. Wildman, J.L., Thayer, A.L., Pavlas, D., et al.: Team knowledge research emerging trends and critical needs. Hum. Factors: J. Hum. Factors Ergon. Soc. **54**(1), 84–111 (2012)
5. Sun, R., Chen, G.Q.: Knowledge work, knowledge team, knowledge workers and their effective management tactics-enlightenment from drucker. Sci. Sci. Manag. S. T. **31**(2), 189–195 (2010)
6. Ma, A.M.J.: The tao of complex adaptive systems (CAS). Chin. Manag. Stud. **5**(1), 94–110 (2011)
7. Mittal, S.: Emergence in stigmergic and complex adaptive systems: a formal discrete event systems perspective. Cogn. Syst. Res. **21**(1), 22–39 (2013)
8. Polacek, G.A., Gianetto, D.A., Khashanah, K., et al.: On principles and rules in complex adaptive systems: a financial system case study. Syst. Eng. **15**(4), 433–447 (2012)
9. Argote, L., Ingram, P., Levine, J.M., et al.: Knowledge transfer in organizations: learning from the experience of others. Organ. Behav. Hum. Decis. Process. **82**(1), 1–8 (2000)
10. Levine, S.S., Prietula, M.J.: How knowledge transfer impacts performance: a multilevel model of benefits and liabilities. Soc. Sci. Electron. Publishing **23**(23), 1748–1766 (2012)

11. Weerdt, M.M., Zhang, Y., Klos, T.: Multiagent Task Allocation in Social Networks. Kluwer Academic Publishers, Netherlands (2012)
12. Singh, A.J., Dalapati, P., Dutta, A.: Multi agent based dynamic task allocation. In: Jezic, G., Kusek, M., Lovrek, I., Howlett, R.J., Jain, L.C. (eds.) Agent and Multi-agent Systems: Technologies and Applications. AISC, vol. 296, pp. 171–182. Springer, Cham (2014). doi:10.1007/978-3-319-07650-8_18
13. Wu, S.B., Liu, M.T.: Assignment of tasks and resources for distributed processing. In: Proceedings of Distributed Computing, Twenty-First IEEE Computer Society International Conference, COMPCON 1980, pp. 655–662 (1980)
14. Garey, M.R., Graham, R.L., Johnson, D.S.: Performance guarantees for scheduling algorithms. Oper. Res. 26(1), 3–21 (1978)
15. Sagar, G., Sarje, A.K., Ahmed, K.U.: Task allocation techniques for distributed computing systems - a review. J. Microcomput. Appl. 12(2), 97–105 (1989)
16. Kim, Y.C., Hong, Y.S.: A task allocation using a genetic algorithm in multicomputer systems. In: Computer, Communication, Control and Power Engineering Proceedings, TENCON 1993 (1993)
17. Protzel, P.W.: Artificial neural network for real-time task allocation in fault-tolerant, distributed-processing system. In: Parallel Processing in Neural Systems and Computers. North-Holland (1990)
18. Andersson, M.R., Sandholm, T.W., Andersson, M.R., et al.: Contract types for optimal task allocation: II experimental results. In: Aaai Spring Symposium (1997)
19. Malville, E., Bourdon, F.: Task allocation: a group self-design approach. In: International Conference on Multi agent Systems. IEEE Computer Society (1998)
20. Wang, X., Yang, J.: Self-adaptive agent model for task allocation in a manufacturing system. Comput. Integr. Manuf. Syst. 7(8), 17–58 (2001)
21. Ma, Q.Y.: Research on dynamic task allocation based on MAS. Huazhong University of Science and Technology (2006)
22. Zhang, J., Li, X.W.: Artificial societies—agent based social simulation. Syst. Eng. 23(1), 13–20 (2005)
23. Sheng, Z.H., Zhang, W.: Computational experiments in management science and research. J. Manag. Sci. China 14(5), 1–10 (2011)
24. Sheng, Z.H.: Case Studies of Computational Experiment in Social Science. Shanghai Sanlian Bookstore, Shanghai (2009)
25. Robbins, S.P.: Organizational Behavior. People's University Publication House, Beijing (2005)
26. Gainforth, H.L., Latimercheung, A.E., Athanasopoulos, P., et al.: The role of interpersonal communication in the process of knowledge mobilization within a community-based organization: a network analysis. Implementation Sci. 9(1), 1–8 (2014)
27. Caimo, A., Lomi, A.: Knowledge sharing in organizations: a bayesian analysis of the role of reciprocity and formal structure. J. Manag. 41(2), 665–691 (2014)

# A Kind of Investor-Friendly Dual-Trigger Contingent Convertible Bond

Wenhua Wang and Xuezhi Qin[✉]

Faculty of Management & Economics, Dalian University of Technology,
Dalian, People's Republic of China
qinxz@dlut.edu.cn

**Abstract.** If financial systemic crisis occurred, one of the effective counter-measures is to issue contingent securities like contingent convertible bonds (CoCos). In this paper, we present a new kind of CoCos which is of investor-friendly dual-trigger property, and it is called "Contingent Convertible bond after Converted" which can be put back at a discount price or converted into CoCos prior to an imminent financial systemic risk. We provided the design rule of this bond and a closed-form pricing formula under some assumptions, and this kind of bond is likely to be more powerful in loss absorbing capacity. Consequently, it is necessary to restrain investors' option to put the CoCoCo back in order to keep loss absorbing capacity more powerful, meaning to limit the discount ratio $\alpha$ less than 1.

**Keywords:** Contingent capital · CoCos · Investor-friendly · Dual-trigger

## 1 Introduction

If some systemically important financial institutions (SIFIs) or big banks become bankrupt unexpectedly, it might threaten the solvency of other symbiotic financial institutions due to a domino effect. Some financial institutions are in trouble to raise new capital from the market for suffering a financial systemic crisis. Simultaneously, government regulators are reluctant to indulge SIFIs to fail and have to provide them capital. However, these kinds of emergency-type government interventions, or extensive amount of implicit guarantees, are controversial for keeping the trouble financial institutions viable with the taxpayers' money.

Contingent convertible bonds are financial hybrid securities, initially proposed by Flannery [1], often denoted as CoCos, which automatically transfuse capital into the banking system in times of financial distress [2, 3]. Theoretically, CoCos can strengthen loss absorption capacities of SIFIs [4, 5]. The mixed features of both debt and equity can promptly recapitalize and effectively avoid bankruptcy without the need of a public bail-out [5].

In this paper, we design a dual-trigger Contingent Convertible bond after Converted denoted as CoCoCo, which is likely to be a type of "Investor-Friendly" puttable debt instrument. Theoretically, CoCoCo similar to other contingent convertible bonds is of

J. Chen et al. (Eds.): KSS 2017, CCIS 780, pp. 242–249, 2017.
https://doi.org/10.1007/978-981-10-6989-5_20

more power in loss absorbing capacity. It is so-called emotionally Contingent Convertible bond after Converted, because it can be put back at a discount price or converted into Contingent Convertible bond (CoCo) prior to an imminent financial systemic risk. In other words, when the financial system stress exceeds the predetermined threshold, the first trigger time, the CoCoCo holders have the right to put the bond back at a discount price $\alpha N_b (0 \leq \alpha \leq 1)$, if they hold a pessimistic view about the prospect of economy and the CoCoCo-issuer. This debt put-back rule provides a way to permit issuing bank to recapitalize prior to an imminent financial systemic risk which maintains the issuing bank capital adequate. On the contrary, the CoCoCo holders may convert this bond into CoCo if they are optimistic about economy and the CoCoCo-issuer. Subsequently, the CoCo may be automatically converted into common equity when the issuer is facing debt distress, the second trigger time. This paper shares similar views to the proposals of Hilscher and Raviv [2], Flannery [5], and Spiegelleer and Schoutens [6], in which contingent capitals have potential to avoid bank bailouts in the financial crisis when banks cannot keep capital adequate. Meanwhile, the main difference between our proposal with other dual-trigger contingent capitals, proposed by McDonald [3], Squam Lake working group [7] and Allen and Tang [8], is that the CoCoCo in our paper provides an opportunity for the investors to put it back prior to probably being converted into common equity, but other dual-trigger contingent capitals do not. Therefore, the CoCoCo in this paper would be a type of "Investor-Friendly" bond.

## 2 The Contingent Convertible Bond After Converted

### 2.1 The Basic Characteristics of the CoCoCo

A systemic risk trigger is employed as the first conversion trigger similar to the proposal by Allen and Tang [8]. It can be used to warn the CoCoCo-issuer to recapitalize before a financial crisis at the first trigger time. In the meantime, the contingent capital by our proposal permits the investors to put the bond back at a discount price or convert it into CoCo. So it is vital to take an appropriate indicator variable (IndV) to indicate the imminent financial crisis. For instance, the credit-to-GDP gap will be a suitable indicator variable, which is the gap between the ratio of credit-to-GDP and its long-term backward-looking trend.

Meanwhile, the CoCoCo-issuer's stock price will be employed as the indicator variable of the second conversion trigger which resembles the McDonald [3] proposal. When the stock price falls below the second predetermined threshold, the CoCo converted from the CoCoCo is automatically converted into common equity. Hence, the conversion decision only pays close attention to the CoCoCo-issuer's financial performance without regulatory intervention. It is once again to emphasize that the CoCoCo will never be converted into common equity or be put back no matter whether the second trigger event becomes reality or not, if the first barrier is never triggered during the whole maturity of the CoCoCo.

## 2.2 Value Transfer and Conversion Price

In our proposal, the CoCoCo holders are able to "break the journey" to put back the CoCoCo at the first trigger moment $\tau_1$ when the first barrier is triggered. And the transfer value from the bondholders to equity holders is $(1-\alpha)N_b$. On the contrary, if the bondholders decide to convert the CoCoCo into CoCo at the moment $\tau_1$, no value is transferred, namely, the face value of the CoCo is also equal to the face value of the CoCoCo. When the CoCoCo-issuer suffers financial distress within financial systemic crisis meaning that the stock price falls down to the second predetermined threshold $S^*$ at moment $\tau_2$, CoCo is mandatorily converted into common stock of $m$ share with conversion price $S_{\tau_2}(S_{\tau_2} = S^*)$. In order to avoid the multiple equilibriums due to using market trigger (Sundaresan and Wang [10]), we assume that there is no value transfer when the CoCo is converted into common equity. If the first trigger event does not become reality during the whole maturity of the CoCoCo, the bondholders will be of payoff $N_b$ at maturity $T$, analogously, if the CoCo will never be converted into common equity, then the bondholders will also be of payoff $N_b$ at maturity $T$.

# 3 A Closed-Form Pricing Formula for Contingent Convertible Bond After Converted

## 3.1 Pricing Zero-Coupon CoCoCo

A closed-form pricing formula for a zero-coupon CoCoCo will be proposed with the conversion principle in Sect. 2 under the default-free condition. It is assumed that a big bank issues a zero-coupon CoCoCo with the face value of $N_b$ and the maturity $T$. When the first predetermined threshold $IndV^*$ for indicating an imminent financial crisis is triggered at the moment $\tau_1$, the CoCoCo can be put back at a discount price and will be paid out $\alpha N_b$ to the bondholders at the maturity $T$. Alternatively, the CoCoCo can be converted into the CoCo along with the face value of $N_b$ and the maturity $(T - \tau_1)$ which is mandatorily converted into common stock of $m$ share at the second trigger moment $\tau_2$. The final payoff of the default free zero-coupon CoCoCo, $V_T^{CoCoCo}$, is given by the following equations,

$$
V_T^{CoCoCo} = \begin{cases} N_b & \text{if } \tau_1 > T \\ \alpha N_b & \text{if } \tau_1 \leq T \text{ and put back} \\ V_T^{coco} & \text{if } \tau_1 \leq T \text{ and converted into CoCo} \end{cases}
$$

$$
V_T^{coco} = \begin{cases} N_b & \text{if } \tau_1 \leq T < \tau_2 \\ mS_T & \text{if } \tau_1 \leq \tau_2 \leq T \end{cases}
\tag{1}
$$

where $\tau_1$ is the conversion moment of the first trigger event, and $\tau_2$ is the second conversion moment. Denote $1_{\{\tau_1 \leq T\}}$ as the indicator function of first trigger event, now we rewrite $V_T^{CoCoCo}$ as follows.

$$V_T^{CoCoCo} = N_b + \max\left(V_T^{coco} - \alpha N_b,\, 0\right)1_{\{\tau_1 \le T\}} - (1-\alpha)N_b 1_{\{\tau_1 \le T\}} \qquad (2)$$

So $V_T^{CoCoCo}$ can be decomposed into three components—a zero-coupon common corporate bond with face value $N_b$ and the maturity $T$, and a call option on the underlying CoCo, $\max\left(V_T^{CoCoCo} - \alpha N_b, 0\right)1_{\{\tau_1 \le T\}}$ and the contingent loss $(1-\alpha)N_b 1_{\{\tau_1 \le T\}}$.

## 3.2   An Option Replication Method for Pricing Zero-Coupon CoCoCo

We assume that a default free zero-coupon contingent capital combination (CCC, for simply) contains a long position of a dual-trigger CoCo and a long position of knock-in Put (KIP). The CoCo sharing the same dual triggers with the CoCoCo is converted into common equity only when both the first threshold and the second threshold are triggered in order. And the knock-and-in Put (KIP) is on the underlying stock price starting at $\tau_1$ and expiring at $T$ with the strike price $\frac{\alpha N_b}{m}$, which will be knocked-in if the stock price falls down to the second predetermined threshold $S^*$ within the maturity. The final payoff of the contingent capital combination, $V_T^{CCC}$, is given by the following equation.

$$V_T^{CCC} = \begin{cases} N_b & \text{if } \tau_1 > T \\ V_T^{coco} & \text{if } \tau_1 \le T \\ \alpha N_b & \text{if } \tau_1 \le \tau_2 \le T \text{ and KIP was exercised} \end{cases} \qquad (3)$$

$$V_T^{coco} = \begin{cases} N_b & \text{if } \tau_1 \le T < \tau_2 \\ mS_T & \text{if } \tau_1 \le \tau_2 \le T \end{cases}$$

Analyzing CoCoCo and Eq. (1), when the first barrier is triggered, the investors elect to put the CoCoCo back at moment $\tau_1$ if and only if the risk neutral estimated value of new issued common equity to the investors is not higher than $\alpha N_b$. So the option for CoCoCo holders to put the CoCoCo back is equivalent to the knock-and-in Put (KIP) in the model (3). Through further analyzing Eqs. (1) and (3) and according to the principle of no arbitrage pricing, we can price the CoCoCo by pricing the contingent capital combination as follows,

$$V^{CoCoCo} = V_T^{CCC}$$
$$= N_b 1_{\{\tau_1 > T\}} + \left[V^{co\ cos} + KIP\right]1_{\{\tau_1 \le T\}} \qquad (4)$$

Thanks to Spiegeleer and Schoutens [6] equity derivative approach, the zero-coupon CoCo can be decomposed into two components, i.e. a zero-coupon common corporate bond with face value of $N_c$ and a knock-in Forward (KIF) that is a combination of a long position of a knock-in Call and a short position of a knock-in Put which are of the same barrier $S^*$ and the same strike price $S^*$ and $S^* = S_{\tau_2}$. Namely,

$$V^{coco} = N_c + KIF \tag{5}$$

From Eqs. (4) and (5), then $V^{CoCoCo}$ can be rewritten as

$$
\begin{aligned}
V^{CoCoCo} &= N_b 1_{\{\tau_1 > T\}} + [N_c + KIF_{\tau_1} + KIP_{\tau_1}] 1_{\{\tau_1 \le T\}} \\
&= N_b + (KIF_{\tau_1} + KIP_{\tau_1}) 1_{\{\tau_1 \le T\}} + (N_c - N_b) 1_{\{\tau_1 \le T\}}
\end{aligned} \tag{6}
$$

where $N_c$ is the face value of the CoCo. We give a further assumption that there are no value transfers at conversion moment $\tau_1$ and $\tau_2$, it means $N_b = N_c$, therefore Eq. (6) can be rewritten as

$$V^{CoCoCo} = N_b + (KIF_{\tau_1} + KIP_{\tau_1}) 1_{\{\tau_1 \le T\}} \tag{7}$$

### 3.3    A Closed-Form Pricing Formula for Zero-Coupon CoCoCo

The first trigger event may occur at any time within the bond's maturity. We naively assume that the occurrence of the first trigger event follows a Poisson process. Therefore, the probability of the first trigger event occurrence is equal to $\lambda_\theta dt$ in the time period $[t, t + dt]$ which is never touched the barrier before time $t$. And $\lambda_\theta$ is a default intensity parameter in the "reduced-form" model of Jarrow-Turnbull [9]. Then we can obtain the probability which the bond is never put back or converted into CoCo during the next $(T - t)$ years by

$$Q(t, T) = \exp\left(-\int_t^T \lambda_\theta d\theta\right) \tag{8}$$

Under the above assumptions, the value of the CoCoCo in Eq. (7) can be expressed as

$$
\begin{aligned}
V_{t_0}^{CoCoCo} &= N_b \exp(-r_f T) + \int_{t_0}^T P_\tau KIF_\tau (-dQ(t_0, \tau)) \\
&\quad + \int_{t_0}^T P_\tau KIP_\tau (-dQ(t_0, \tau))
\end{aligned} \tag{9}
$$

where $P_\tau$ is the risk-free discount factor, risk free rate is $r_f$, starting time $t_0$, expiring at $T$, and all the parameters are the same in the context except some special explanations. Certainly, the expression of $KIF_{\tau_1}$ can be found in the paper of Spiefeleer and Schoutens [6], and the expression of $KIP_{\tau_1}$ comes from Hull [11] when $\tau_1$ is fixed.

## 3.4   A Numerical Example

We can estimate the loss absorbing capacity by estimating the expected value of new issued equity in the following example. In the meantime, we also estimate the "Investors-Friendly" property of the CoCoCo by estimating the expected return of the bondholders to put the CoCoCo back. We assume that the values of the face value of the CoCoCo, maturity and risk-free interest rate are $N_b = \$1000$, $T = 30$ and $r_f = 0.01$, respectively, and the values of the stock price, conversion price, conversion share and volatility are $S_0 = 100$, $S^* = 50$, $m = 20$ and $\sigma = 0.1$, respectively. Additionally, the default intensity parameter $\lambda$ is assumed to be $0.8t$ and the discount ratio $\alpha$ is from 0 to 1. There are no value transfers, except for putting the CoCoCo back at a discount price where the transfer value from the investors to the issuing bank is $(1 - \alpha)N_b$. The second threshold of stock price is $S^* = 50$. Figure 1 shows us the CoCoCo's loss absorbing capacity with the discount ratio, and Fig. 2 the expected return of the bondholders to put the CoCoCo back with the discount ratio.

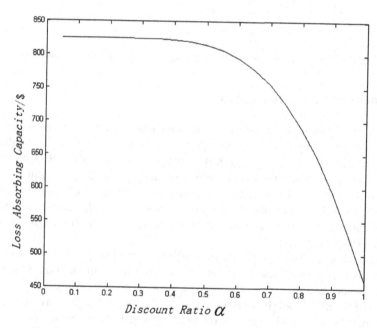

**Fig. 1.** CoCoCo's loss absorbing capacity with the discount ratio

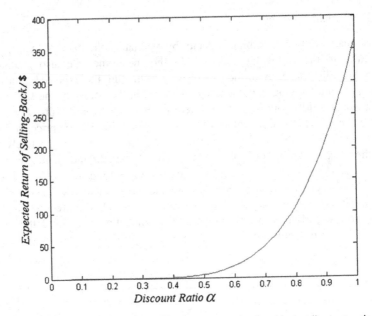

**Fig. 2.**  The expected return of the CoCoCo put-back with the discount ratio

## 4   Conclusion and Discussion

In Figs. 1 and 2, the expected return rises with the discount ratio α increases, while the loss absorbing capacity has an opposite performance. In other words, as the discount ratio α gradually increases, the "Investors-Friendly" property of the CoCoCo is getting stronger. On the contrary, the loss absorbing capacity of the CoCoCo will be getting weak. Consequently, it is necessary to restrain investors' option to put the CoCoCo back in order to keep loss absorbing capacity more powerful, meaning to limit the discount ratio α less than 1. From Figs. 1 and 2, it seems to be better to set α = 0.70−0.80, intuitively.

Compared with other proposals of contingent capitals [3, 5–8] in which the bondholders will get worthless for the mandatory convertible mechanism if the trouble CoCoCo-issuer becomes insolvent, the CoCoCo in this paper is a type of "Investors-Friendly" bond. It is because that the CoCoCo offers the investors a guarantee to put the bond back at a discount price by which the investors obtain at least $\alpha N_b$. This shiny favorable advantage may attract more potential investors to invest in this bond, such as risk-averse investors.

**Acknowledgement.** This research is jointly supported by Chinese Natural Science Foundation (Granted No. 71171032) and The Central University Basic Scientific Research Business Expenses (Granted No. DUT17RW210).

# References

1. Flannery, M.J.: No pain, no gain effecting market discipline via 'reverse convertible debentures'. In: Scott, H.S. (eds.) Capital Adequacy Beyond Basel: Banking, Securities, and Insurance. Oxford University Press, Oxford (2005)
2. Hilscher, J., Raviv, A.: Bank stability and market discipline: the effect of contingent capital on risk taking and default probability. J. Corp. Financ. **29**, 542–560 (2014). Social science electronic publishing
3. McDonald, R.L.: Contingent capital with a dual price trigger. J. Financ. Stab. **9**, 230–241 (2013)
4. Corcuera, J.M., Spiegelerr, J.D., Fajardo, J., Jonsson, H., Schoutens, W.: A close form pricing formulas for coupon cancellable CoCos. J. Banking Financ. **42**, 339–351 (2014)
5. Flannery, M.J.: Stabilizing large financial institutions with contingent capital certificates. Q. J. Financ. **06**(02), 1650006 (2016)
6. Spiegeleer, J.D., Schoutens, W.: Pricing contingent convertibles: a derivatives approach. J. Deriv. **20**(2), 27–36 (2011). Social science electronic publishing
7. Squam Lake Working Group: an expedited resolution mechanism for distressed financial firms: regulatory hybrid securities. Council on foreign relations (2009)
8. Allen, L., Tang, Y.: What's the contingency? a proposal for bank contingent capital triggered by systemic risk. J. Financ. Stab. **26**, 1–14 (2016)
9. Jarrow, R.A., Turnbull, S.M.: Pricing derivatives on financial securities subject to credit risk. J. Am. Financ. Assoc. **50**(1), 53–85 (1995)
0. Sundaresan, S., Wang, Z.: On the design of contingent capital with a market trigger. J. Financ. **70**(2), 881–920 (2015)
1. Hull, J.C.: Options, Futures, and Others Derivatives. 8th Edition. Simplified Chinese edition copyright 2012 by China Machine Press (2012)

# Analysis on Influencing Factors on Cultivation of Graduate Innovation Ability Based on System Dynamics

Bing Xiao[1,2]([✉]) and Vira Chankong[2]

[1] School of Computer Science, Guangdong Polytechnic Normal University, Guangzhou 510665, Guangdong, China
bingxiao3@outlook.com
[2] Case School of Engineering, Case Western Reserve University, Cleveland, OH 44106, USA
vira@case.edu

**Abstract.** The purpose of this paper is to clarify and determine the influencing factors on cultivation of graduate innovation ability. It is of great theoretical and practical significance to the graduate educational quality improvement and to the social and economic development as well as improvement of international competence. The present research applies the systematic thinking to propose the utilization of government-customer-industry-university-institute (GCIUI) practical cultivation mode to discuss the influencing factors on cultivation of graduate innovation ability. In the research, the systematic essence of cultivation of graduate innovation ability is firstly defined, and the systematic structural model for graduate innovation ability cultivation is constructed; then, the impacts which graduate enterprise, GCIUI practical cultivation mode and innovative environment have on graduate innovation ability are analyzed; finally, the System Dynamics software is applied to establish the System Dynamics flow chart for influencing factors on graduate innovation ability, and the dynamic modeling and simulation analysis are carried out. Simulation results show that, the graduate innovation ability improves as the government and university profit distribution ratios increase, and reduces as institute and customer profit distribution ratios increase. However, the influence from enterprise is lower compared to other entities. The influences of other parameters on graduate innovation ability rank from knowledge protection, enrollment examination, teacher-student relationship, knowledge system, research projects and funds, incentive mechanism, construction of research platforms, technological exchange and cooperation to trust mechanism. Consequently, the formulation of each relevant policy should be systematically considered from graduate entity, cultivation of graduate innovation ability system structure and cultivation of graduate innovation ability environment.

**Keywords:** Innovation ability · Government-Customer-Industry-University-Institute (GCIUI) practical cultivation mode · Influencing factors · System Dynamics

© Springer Nature Singapore Pte Ltd. 2017
J. Chen et al. (Eds.): KSS 2017, CCIS 780, pp. 250–266, 2017.
https://doi.org/10.1007/978-981-10-6989-5_21

# 1    Introduction

As the major source of innovative talents, graduates are the potential main force for improving comprehensive national power and international competence. On March 29th, 2013, it is specifically required in the Opinions about Deepening Graduate Educational Reform issued by the Ministry of Education, National Development and Reform Commission and the Ministry of Finance that graduate education needs to highlight the cultivation of innovation ability and practical ability [1]. Currently, it is a common sense that the graduates are subject to insufficient innovation ability [2]. It is a vital project for many universities and scientific research institutes in the new era to improve cultivation of graduate innovation ability, and the discussion on graduate innovation ability has become the research focus of many scholars. It is found by referring to relevant Chinese and foreign literature that the analysis on influencing factors on cultivation of graduate innovation ability is mainly focused on research opportunity [3], graduate enterprise and social capital [4], environmental factors [5] and instructors [6]. In addition, some scholars have analyzed the constraint factors on cultivation of graduate innovation ability from the perspectives of ideas and culture, system and mechanism, teaching and practices, resources and conditions [7]. These studies start from one or several perspectives, and the research methods mainly adopt qualitative analysis, which neglect the systematic, dynamic and complex cultivation of graduate innovation ability. Therefore, it is very necessary to analyze the influencing factors on cultivation of graduate innovation ability from a systematic perspective.

# 2    Systematic Content and Structural Model of Graduate Innovation Ability Cultivation

The academic community has not reached a precise definition on the concept of graduate innovation ability. Apart from the reason that different researchers have different understandings about it from different viewpoints, another more vital reason is that the concept of graduate innovation ability is developing and its essence is changing. In the cultivation system of graduate innovation ability, the core innovation entities are the graduates. However, the innovation ability of graduates cultivated from different universities differs, and the concept, channel, method and innovation environment of graduates cultivated from the same discipline of different universities differ. To gain an insight into the influencing factors on cultivation of graduate innovation ability, it is necessary to analyze the cultivation of graduate innovation ability from a systematic perspective, specify the objective, element, structure, function and environment of this system and then analyze the influencing factors of this system.

## 2.1    Systematic Essence of Cultivation of Graduate Innovation Ability

The systematic essence is to analyze the essence system of the system studied from a systematic perspective. In this research, a coordinated government-customer-industry-university-institute system on cultivation of graduate innovation ability is proposed. In other words, in the graduate cultivation process, driven by customer demand, government and graduates, a government and graduates, a government-customer-industry-university-institute practical innovation system is constructed with the objective of improving graduate innovation ability in combination of industry, university and institute. In this system, according to the graduate enterprise, under the guarantee of governmental macroscopic environment and policy, enterprises, customers and institutes are led to comprehensively participate into graduate cultivation, and the guiding role of government, customer, enterprise and institute in graduate enrollment, formulation of cultivation standard, teaching reform, coordinated innovation and social service is given to full play. In addition, the professional instructor team and joint cultivation base combining universities, institutes and enterprises are constructed. Through the systematic research training on graduates, the graduates are supported in participating in cutting-edge and high-level research practices, academic exchanges and international cooperation to broaden their academic horizons, inspire their innovative thinking and provide them with new ideas, theories, methods and innovation capabilities with economic, social and ecological value in various practical fields.

In the systematic theory, it is assumed that any system has its objective and includes element and sub-system. In addition, any system has its specific structure and function. The element is its most basic unit. The elements in cultivation system of graduate innovation ability include graduates, instructors, innovation foundation, innovation opportunities, innovation policies, innovation conditions and innovation achievements. Furthermore, the innovation foundation includes knowledge system; the innovation opportunity includes social relation, social network and joint goal; the innovation policies include incentive mechanism, trust mechanism and knowledge protection; the innovation conditions include innovation funds, research platform, practical base, innovation equipment and technologies; the innovation achievements include papers, patents and monographs. The objective and function of cultivation system of graduate innovation ability are to improve the innovation and social service ability of graduates; apart from common-sense political, economic, cultural and technological environments, the environment of this system mainly refers to the customer demand environment, the governmental policy support environment, funds guarantee environment, industry-university-institute cooperation and practical environment among enterprise, research institute and university sharing close ties with graduate innovation. From the perspective of graduate cultivation, this system can be divided into three sub-systems, including enrollment sub-system, cultivation sub-system (including another three sub-systems: teaching sub-system, innovation ability improvement sub-system and management sub-system) and feedback sub-system. The enrollment sub-system includes enrollment demand planning and enrollment examination; the teaching sub-system includes theoretical teaching, practical teaching and teaching quality assurance; the innovation ability improvement sub-system includes technological exchange and cooperation, participation in instructor projects, project declaration industry-university-institute cooperation and construction of research platform; the management sub-system includes teaching management, administrative management and academic management; the feedback sub-system includes innovation demand analysis sub-system, innovation demand prediction sub-system and graduate innovation ability evaluation sub-system.

## 2.2    Structural Model of Cultivation of Graduate Innovation Ability

According to the relation among entity, structure, function and environment [8, 9], the entity, structure and environment decide the system function together, and the function of this system is to cultivate the graduate innovation ability. Under the impact of external environment (customer, enterprise, institute and government), connections among entities, between entity and sub-system. Among entity, sub-system and environment, are formed for coordinated innovation. The majority of connections are non-linear. The innovation ability is the outcome of each connection under non-linear mutual interactions. Therefore, the cultivation system of innovation ability is a complex system. The structural model can be constructed as shown in Fig. 1:

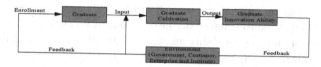

**Fig. 1.** Structural model of cultivation process of graduate innovation ability

It can be seen from this model that the graduate innovation ability is mainly decided by graduates, graduate cultivation and environment together; the cultivation of graduate innovation ability is a cyclic and dynamic process. The output of graduate innovation ability is fed to system environment. If it does not adapt to the environment and cannot satisfy the anticipated profit of customer demand and each environmental entity, the system input or structure is adjusted to accommodate the environmental change.

# Analysis of Influencing Factors on Cultivation of Graduate Innovation Ability

It can be known from above analysis that the cultivation system of graduate innovation ability is a complex system [10]. In this research, the influencing factors are analyzed from the perspectives of graduate enterprise, government-customer-industry-university-institute practical cultivation mode and innovation environment.

## .1    Influence of Enterprise on Cultivation of Graduate Innovation Ability

Graduate enterprise refers to a behavioral mode in which the graduates overcome various obstacles and pursue their own goals under self-initiation without the specific guidance from instructors [11]. In the cultivation process of graduate innovation ability, graduates are always at the core entity position. Students with strong enterprise have specific graduate study goals, complete knowledge system, excellent personality, clear behavioral features and active thinking features, which contribute to their innovation achievements. Students lacking in enterprise have become obstacles influencing graduate education effects [12]. They are characteristic of unspecific graduate study goal, messy knowledge system and bad team cooperation awareness. Besides, they easily give up in face of difficulties, cannot solve their problems independently, and lack innovation awareness and exploration spirit. The cultivation of graduate innovation ability cannot be separated from active participation and thinking of graduates. Although their enterprise might be somewhat inherited, it is more cultivated by later efforts. In later cultivation environment, the graduate enterprise is affected by knowledge system, incentive mechanism, common goal and teacher-student relation.

### 3.2    Influence of Government-Customer-Industry-University-Institute Practical Cultivation Mode on Cultivation of Graduate Innovation Ability

Government-customer-industry-university-institute practical cultivation mode is an innovative and practical cultivation mode. It can be known from above analysis that government-customer-industry-university-institute practical cultivation mode is aimed to improve graduate innovation ability and centered on universities. It actively guides customer, government, enterprise and institute to participate in graduate enrollment and cultivation as well as feedback, forming a coordinated cultivation innovation mode. In this innovation cultivation mode, according to customer demand, government plays the enabler role in practical cultivation mode of graduate innovation ability and should provide macroscopic service for graduate innovation ability, including funds and policy support. Enterprise and institute should together participate in formulation of graduate cultivation plan, curriculum setting, construction of professional instruction team and joint cultivation base, in order to carry out systematic research training on graduates. Government-customer-industry-university-institute practical cultivation mode mainly involves: enrollment plan, cultivation plan, curriculum learning, industry-university-institute cooperation (academic exchange, project research, practical base, instructor guidance and funds support). The customer mainly offers demand, the government mainly offers policy and funds assurance, the enterprise mainly offers funds, practical base and instructor cooperation, and the institute mainly offers technology, equipment, research platforms and instructor cooperation. Therefore, government-customer-industry-university-institute practical cultivation mode has provided excellent social networks and relations for graduate innovation, which contributes to graduates' finding the joint goal in innovation process.

### 3.3    Influence of Environment on Graduate Innovation Ability

As the cultivation of graduate innovation ability is a complex system, its environment refers to the total of things existing outside the system and specifically refers to the total of various conditions under which the graduates carry out innovative activities. The cultivation and environment of graduate innovation ability are co-existing. There is indispensable relation among the structure, function, entity and environment of cultivation system of innovation ability. The entity must accommodate to the change in environment. Here, the environment mainly refers to the customer demand environment, the governmental policy support environment, funds guarantee environment industry-university-institute cooperation and practical environment among enterprise research institute and university sharing close ties with graduate innovation.

# 4  Dynamics Model for Influencing Factors on Cultivation of Graduate Innovation Ability

## 4.1  Determination of System Boundary and Basic Hypothesis

**Determination of System Boundary**

The purpose of model construction in this research is mainly to analyze the influencing factors on graduate innovation ability, investigate the graduate enterprise, and the relation of structure and environment of "government-customer-industry-university-institute" cultivation mode with graduate innovation ability. According to the analysis on the influencing factors on graduate innovation ability, the graduate enrollment, cultivation mode and cultivation quality can be adjusted. It can be known from the modeling purpose that the system mainly involves 3 sub-systems, namely enrollment sub-system, cultivation sub-system and feedback sub-system. The boundary investigated only involves related factors.

**Basic Model Hypothesis**

H1: The coordinated cultivation of enterprise, university, research institute, government and customer is a continuous and gradual behavioral process. H2: The system breakdown caused by major governmental policy reform and abnormal circumstances is not considered. H3: Input of coordinated cultivation mainly includes input of coordinated government-customer-industry-university-institute cultivation among university, enterprise and research institute (including labor force, funds and materials, etc.), governmental funds and policy input of coordinated government-customer-industry-university-institute cultivation and input of the customer demand for coordinated government-customer-industry-university-institute cultivation. The output of coordinated cultivation includes the graduate innovation ability (mainly embodied in paper, patent and project) and the coordinated government-customer-industry-university-institute innovation ability. H4: University itself possesses certain research & development ability. Due to the cultivation demand for graduate innovation and its own resource limits, to accommodate to external environmental change, coordinated government-customer-industry-university-institute cultivation must be implemented.

## 4.2  Determination of System Boundary and Basic Hypothesis

It can be known from above analysis that the cultivation of graduate innovation ability is a dynamic and complex behavior under the joint impacts of multiple factors as well as a complex system problem involving dynamics, complexity, feedback and non-linearity. Such complex problem must be studied by combination of qualitative and quantitative methods. In this research, the System Dynamics modeling method is applied to fully reflect the non-linear and dynamic relation among different entities, between entity and sub-system, among different sub-systems and between system and environment. In addition, the cultivation of graduated innovation ability in university is regarded as one system, which is further divided into 3 sub-systems (enrollment sub-system, cultivation sub-system and feedback sub-system) according to the cultivation process. Graduates are the most core innovation entities in the cultivation system of graduate innovation ability and their innovation ability is mainly reflected by innovation achievements (considering paper, patent and monograph). In this paper, the graduate innovation ability is adopted as the first-level system variable, and the improvement and reduction of graduate innovation ability are adopted as the velocity variables. The improvement of graduate innovation ability cannot be separated from the cultivation structure and environmental support. Here, the government-customer-industry-institute environment is mainly considered. Therefore, the coordinated innovation structure between university and government-customer-industry-institute environment plays an important role in graduate cultivation. Besides, the coordinated innovation ability (mainly considering research, patent and monograph from coordinated innovation) serves as the second-level system variable, and the improvement and reduction of coordinated innovation ability serve as the velocity variables. The main variables in each sub-system serve as the parameters influencing the cultivation of graduate innovation ability, including enrollment examination involved in enrollment sub-system, knowledge system involved in teaching sub-system, technological exchange and cooperation, instruction relation, research project and fund, industry-university-institute cooperation and construction of research platform involved in innovation ability improvement sub-system as well as knowledge protection and incentive mechanism involved in management sub-system. In the meanwhile, it is assumed that the profit distribution mechanism and trust mechanism directly influence the government-customer-industry-university-institute cooperation intensity and indirectly influence the cultivation of graduate innovation ability. Therefore, these 10 variables serve as the parameters influencing the graduate innovation ability and the system flow chart can be constructed as shown in Fig. 2.

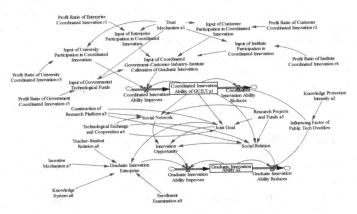

**Fig. 2.** Dynamics flow chart for influencing factors on cultivation of graduate innovation ability

## 4.3 Dynamics Equations for Cultivation System of Graduate Innovation Ability

To clearly describe the mutual relation among different variables involved in the model by reference to the research achievements of Chinese and foreign scholars, the major dynamics equations for influencing factors on cultivation of graduate innovation ability are established by adopting System Dynamics theory as follows:

System Equations:

$$\frac{dx_1}{dt} = (0.328a_1r_3 + 0.672(0.27a_1r_1 + 0.03a_1r_2 + 0.05a_1r_4 + 0.65a_1r_5))x_1(t)$$
$$- (1 - a_2)x_1(t) \tag{1}$$

$$\frac{dx_2}{dt} = ((0.147(0.15a_3 + 0.4a_4 + 0.2a_5 + 0.25x_1(t)) + 0.114(0.2a_3 + 0.15a_4 + 0.4a_5$$
$$+ 0.25x_1(t)) + 0.065(0.1a_3 + 0.4a_4 + 0.2a_5 + 0.3x_1(t)) + 0.674(0.25a_6 + 0.1a_7$$
$$+ 0.15a_8 + 0.4a_9 + 0.1(0.2a_3 + 0.15a_4 + 0.4a_5 + 0.25x_2(t))))x_2(t)$$
$$- (1 - a_2)(0.1a_3 + 0.4a_4 + 0.2a_5 + 0.3x_1(t))x_2(t) \tag{2}$$

where, $x_1(t)$ is the coordinated innovation ability of government-customer-industry-university-institute (GCIUI), and $x_2(t)$ is the graduate innovation ability. $r_1$, $r_2$, $r_3$, $r_4$ and $r_5$ are the coordinated innovation profit ratios of enterprise, customer, university, institute and government. $a_1$, $a_2$, $a_3$, $a_4$, $a_5$, $a_6$, $a_7$, $a_8$ and $a_9$ are values of trust mechanism, knowledge protection, construction of research platform, technological exchange and cooperation, research projects and funds, teacher-student relation, incentive mechanism, knowledge system and enrollment examination.

## 4.4   Model Output

This model is used to study the influencing factors on cultivation of graduate innovation ability. Therefore, the graduate innovation ability serves as the main output and reflects the direct achievement output of graduate innovation ability, and the coordinated government-customer-industry-university-institute innovation ability serves as the second output and indirectly reflects the graduate innovation ability.

## 5   Analysis of Simulation Result

### Setting of Model Parameters

In this study, the determination of initial model data and the mutual variable relations is based on the questionnaire survey in the Guangdong graduated education innovation plan, research on government-customer-industry-university-institute practical cultivation mode for graduate innovation ability based on system theory (2015JGXM-MS40), the Guangdong Tech Yearbook and the research achievements of other scholars. Specifically, 500 questionnaires were distributed and 485 effective questionnaires were recovered. According to the result and experience in questionnaires, the horizontal variables and parameters are graded on five levels, respectively very satisfied, satisfied, generally satisfied, unsatisfied and very unsatisfied. In addition, the parameter weights are graded on five levels, respectively very important, important, generally important, unimportant and very unimportant, corresponding to 9, 7, 5, 3 and 1. The weights of university participation input in coordinated innovation and the government-customer-industry-institute participation input in coordinated cultivation of graduate innovation ability are calculated according to the funds input; the weights of graduate innovation participation enterprise and innovation opportunity are obtained from reference, and the weights of other parameters are obtained by survey questionnaire. Considering the inconsistency in dimensions makes the calculation more difficult, the final output result of horizontal variables is expressed by relative value, and the range of parameter and weight is generally set as [0, 1].

## Model Effectiveness Examination

The model construction is based on massive actual related reference and practical investigations. In the model, the cultivation of graduate innovation ability in university is adopted as one system, and the government-customer-university-institute is regarded as the environment in which the system realizes its objective. It is assumed in System Dynamics that during model examination, the correctness of model structure is far more important than selection of parameters. Therefore, the theoretical model examination is dominant [13]. The major concepts and variables describing the problem are adopted as the endogenous variables. It is shown by repeated computer simulation that the design of model boundary is rational, each model equation is meaningful, the dimension of each equation is consistent and the relation among model variables is true. Therefore, the model is suitable.

Then, the historical effectiveness of the dynamics model for influencing factors on cultivation of graduate innovation ability is examined by mainly seeing if the simulation result can effectively reflect the objective system. The initial values of graduate innovation ability and coordinated government-customer-industry-university-institute innovation ability are inputted as shown in Table 1 to obtain the changes in horizontal variables at different stages as shown in Fig. 3.

It can be obtained from the 10-year simulation on graduate innovation ability and coordinated government-customer-industry-university-institute innovation ability that under excellent trust mechanism (0.9) and good knowledge protection (0.85), the coordinated government-customer-industry-university-institute innovation ability keeps improving, and the graduate innovation ability starts from improving at slower pace and then accelerates. It is indicated that the graduate innovation ability needs certain accumulation, including knowledge and experience accumulation, which takes a certain period. It can be seen from Figs. 3 and 4 that within 10 years, the graduate innovation ability is influenced by external environment and their innovation demand.

**Table 1.** Data comparison of horizontal variables at different stages.

| Horizontal variables | 1-year | 2-year | 3-year | 4-year | 5-year | 6-year | 7-year | 8-year | 9-year | 10-year |
|---|---|---|---|---|---|---|---|---|---|---|
| Coordinated innovation ability of GCIUI $x_1$ | 0.5 | 0.515 | 0.530 | 0.546 | 0.563 | 0.579 | 0.597 | 0.615 | 0.633 | 0.652 |
| Graduate innovation ability $x_2$ | 0.5 | 0.713 | 1.016 | 1.449 | 2.069 | 2.956 | 4.226 | 6.045 | 8.653 | 12.396 |

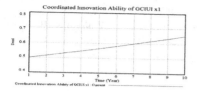
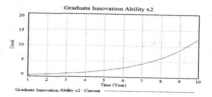

**Fig. 3.** Change trend of coordinated innovation ability of GCIUI and graduate innovation ability at different stages.

and their innovation ability keeps improving. In addition, the coordinated government-customer-industry-university-institute innovation ability improves steadily, and the inputs of graduate innovation ability and coordinated government-customer-industry-university-institute innovation ability mutually affect and promotes, thus improving in this good cycle.

It can be known from the theoretical and historical model examination that the characteristics of both the dynamics model for influencing factors on graduate innovation ability established in this research and the actual cultivation of graduate innovation ability in each stage are basically consistent. Therefore, this model is effective.

## Policy Regulation in Dynamics Model for Cultivation System of Graduate Innovation Ability

The policy regulation in System Dynamics model is to study the influences of some parameters on policies by model and seek the executable policy scheme which can solve the actual system problem by changing parameter values or initial values. In this research, Vensim version 5.8 is applied to adjust the profit distribution ratio or increase or reduce each parameter. Then, simulation is carried out in combination of these initial values, and the sensitivity of each parameter to graduate innovation ability is judged by the results. Through the analysis and simulation results, university can cooperate with government, institute, enterprise and customer to formulate countermeasures beneficial for cultivation of graduate innovation ability.

(1) Adjust the profit distribution ratio. The data in Table 1 serve as the initial simulation value. When other variables remain unchanged, the enterprise profit distribution ratio is increased to 80% and the profit distribution ratio of other entities is reduced to 5% in Scheme 1; the government profit distribution ratio is increased to 80% and the profit distribution ratio of other entities is reduced to 5% in Scheme 2; the university profit distribution ratio is increased to 80% and the profit distribution ratio of other entities is reduced to 5% in Scheme 3; the institute profit distribution ratio is increased to 80% and the profit distribution ratio of other entities is reduced to 5% in

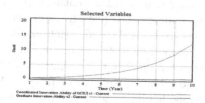

**Fig. 4.** Simulation figure under original data

Scheme 4; the customer profit distribution ratio is increased to 80% and the profit distribution ratio of other entities is reduced to 5% in Scheme 5. The change trends of graduate innovation ability and coordinated government-customer-industry-university-institute innovation ability are shown in Fig. 5a–e.

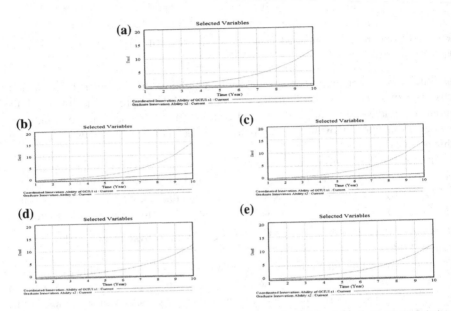

**Fig. 5.** (a) Simulation figure in Scheme 1. (b) Simulation figure in Scheme 2. (c) Simulation figure in Scheme 3. (d) Simulation figure in Scheme 4. (e) Simulation figure in Scheme 5

It can be found by comparing Fig. 4 with Fig. 5a–e that the graduate innovation ability and coordinated government-customer-industry-university-institute innovation ability improve as the government and university profit distribution ratios increase, but reduce as the institute and customer profit distribution ratios increase, and are slightly influenced by enterprise. Considering that the government assures the funds and policy formulation in actual coordinated innovation process, the government profit ratio can be properly increased in actual operation to make the government incline on policies and funds. In addition, university plays a core entity role in cultivation of graduate innovation ability. If the university profit ratio can be increased, both the graduate innovation ability and coordinated government-customer-industry-university-institute innovation ability can be improved. Besides, institute provides platform for cultivation of graduate innovation ability. However, due to its fund limitation, compared to university and enterprise, its innovation achievements (patent, paper and monograph) are not remarkable. Then, customer serves as demand orientation in cultivation of graduate innovation ability and focuses more on profit than quantity of innovation achievements. Although enterprise accounts for a large ratio in authorized domestic patents, the cultivation of graduate innovation ability is not its major function. Therefore, the influence which increasing its profit ratio has on the graduate innovation ability and coordinated government-customer-industry-university-institute innovation ability is not obvious.

(2) Adjust trust mechanism. The data in Table 1 serve as the initial simulation values. When other variables remain unchanged, if the trust intensity is increased or decreased by 20%, the simulation figures are shown in Fig. 6.

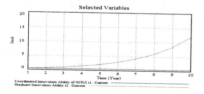 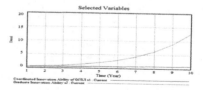

**Fig. 6.** Horizontal variable simulation figure under different trust intensity

It can be found by comparing Fig. 4 with Fig. 6 that the graduate innovation ability and coordinated government-customer-industry-university-institute innovation ability improve as the trust intensity increases, and reduce as the trust intensity decreases. The influence of trust intensity on coordinated government-customer-industry-university-institute innovation ability is more obvious. When the trust intensity is decreased by 20%, the coordinated government-customer-industry-university-institute innovation ability reduces, which indicates that the trust mechanism plays a very important role in coordinated government-customer-industry-university-institute innovation process. The trust of mutual trust among different entities lays the coordination foundation.

(3) Adjust the knowledge protection intensity. The data in Table 1 serve as the initial simulation values. When other variables remain unchanged, if the trust intensity is increased or decreased by 20%, the simulation figures are shown in Fig. 7.

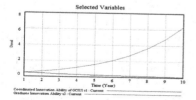 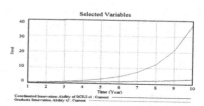

**Fig. 7.** Horizontal variable simulation figure under different knowledge system

It can be found by comparing Fig. 4 with Fig. 7 that the graduate innovation ability and coordinated government-customer-industry-university-institute innovation ability improve obviously as the knowledge protection intensity increases, but reduce obviously as the knowledge protection intensity decreases. When the knowledge protection intensity is decreased by 20%, the coordinated government-customer-industry-university-institute innovation ability reduces, which indicates that the knowledge protection intensity plays a very important role in coordinated government-customer-industry-university-institute innovation process. Improving knowledge protection can minimize the reliance and inertia among innovation entities.

(4) Adjust other parameters. The values of construction of research platform, technological exchange and cooperation, research projects and funds, teacher-student relation, incentive mechanism, knowledge system and enrollment examination are increased or decreased by 20%. The change trends of the graduate innovation ability and coordinated government-customer-industry-university-institute innovation ability are shown from Figs. 8, 9, 10, 11, 12, 13 and 14.

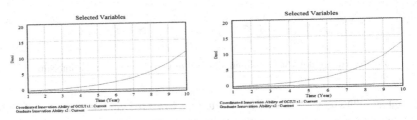

**Fig. 8.** Horizontal variable simulation figure under different construction of research platform

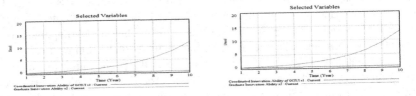

**Fig. 9.** Horizontal variable simulation figure under different technological exchange and cooperation

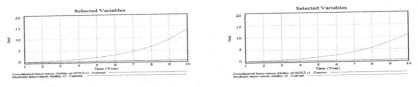

**Fig. 10.** Horizontal variable simulation figure under different research projects and funds

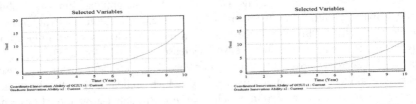

**Fig. 11.** Horizontal variable simulation figure under different teacher-student relation

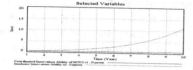
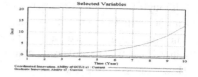

**Fig. 12.** Horizontal variable simulation figure under different incentive mechanism

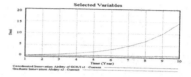

**Fig. 13.** Horizontal variable simulation figure under different knowledge system

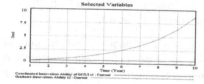
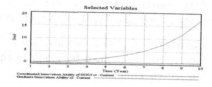

**Fig. 14.** Horizontal variable simulation figure under different enrollment examination

It can be known from comparison among Fig. 4, Figs. 8, 9, 10, 11, 12, 13 and 14 that the graduate innovation ability improves as the construction of research platform, technological exchange and cooperation, research projects and funds, teacher-student relation, incentive mechanism, knowledge system and enrollment examination get improved. The influential intensity of above parameters on graduate innovation ability ranks from enrollment examination, teacher-student relation, knowledge system, research projects and funds, incentive mechanism, construction of research platform to technological exchange and cooperation. The coordinated government-customer-industry-university-institute innovation ability is basically not affected by above parameters.

# Discussion

1) Considering that the influencing factors on graduate innovation ability are mainly considered in this paper and their change trend instead of value accuracy is focused on, for computational convenience, the model parameter estimation is simplified and those parameters which do not change obviously with time are calculated as constant values.

(2) Although the sample recovery rate of questionnaire survey in this paper is 97%, only a limited amount of 500 samples were distributed due to time limitation.

(3) The model established in this paper does not consider the system breakdown caused by major governmental policy reform and abnormal circumstances. However, governmental policies will change with external environment, and sometime the policy reform might be even revolutionary.

(4) The coordinated cultivation of enterprise, university, institute, government and customer is regarded as one continuous and gradual behavioral process. However, under many actual circumstances, the coordinated cultivation might be intermittent or one-time.

# 6  Conclusions

In this paper, systematic thinking, systematic dynamic theory and empirical study are combined to establish the System Dynamics model for influencing factors on graduate innovation ability. The influences of model parameters are studied by changing the model parameters, and a simulation analysis is carried out on the influencing factors under current policies. Finally, following conclusions are drawn:

(1) Cultivation of graduate innovation ability is a continuous, dynamic and cyclic process and is decided by graduate entity, graduate cultivation structure (coordinated government-customer-industry-university-institute cultivation) and environment (government-customer-university-institute) jointly. Therefore, during cultivation of graduate innovation ability, the formulation of various policies should consider the graduate entity, cultivation system structure of graduate innovation ability and cultivation environment of graduate innovation ability systematically.

(2) During cultivation of graduate innovation ability, the graduate enterprise, innovation opportunity and knowledge protection directly influence the graduate innovation ability. In the multiple influencing factors, the significances rank from knowledge protection, enrollment examination, teacher-student relation, knowledge system, research projects and funds, incentive mechanism, construction of research platform, technological exchange and cooperation. Therefore, government should firstly formulate auxiliary knowledge protection policies, prevent the reliance and inertia of graduates during innovation process and promote original innovation; university should pay high attention to graduate enterprise during graduate enrollment and cultivation process, and design related examination which can investigate the graduate enterprise during enrollment examination and interview process; after graduates are enrolled, instructor should firstly establish good teacher-student relation, and then carry out knowledge system integration and learning based on student personality, foundation, hobby and advantage; university should issue specific and feasible incentive mechanism combining penalty and award, and cooperate with government, enterprise, institute and instructor to try to provide various innovation opportunities for graduates under the orientation of customer demand.

(3) The structure of the cultivation system of graduate innovation ability indirectly influences the graduate innovation ability and is influenced by profit distribution mechanism, trust mechanism and knowledge protection. The coordinated government-customer-industry-university-institute innovation ability improves as the government and university profit distribution ratio increases, and reduces as the research institute and customer profit distribution ratio increases. However, the influence from enterprises is lower compared to other entities. Therefore, when university coordinates with government, enterprise, institute and customer in cultivation of graduate innovation ability, the establishment and perfection of profit distribution mechanism should be considered together and supplied according to different entity demands. Government should issue corresponding policies to reinforce knowledge protection and minimize the reliance and inertia among different innovation entities. The entities participating in coordinated innovation should construct complete trust mechanism and improve management on trust relation, thus maximizing the coordinated innovation ability of different entities.

(4) Innovation environment guarantees the cultivation of graduate innovation ability. The graduate innovation cannot be separated from customer demand and must be related to practices. On this basis, university should create a good innovation and research atmosphere, instructor should establish harmonious teacher-student relation, government should provide support on various policies and funds, enterprise and institute should provide various platforms and practical sites, and the university management on graduates should be flexible. These environments highly-related to government-customer-university-institute contribute to improving graduate innovation ability.

**Acknowledgments.** This research was financially supported by the Teaching Reform Project of Guangdong Province Department of Education (number: 2015JGXM-MS40) and the innovative project of Guangdong Province (number: 2016GXJK091) and the innovative project of Guangdong Polytechnic Normal University (number: 991460319).

# References

1. http://www.moe.edu.cn/publicfiles/business/htmlfiles/moe/A22_zcwj/201307/154118.html. Accessed 25 Feb 2017. (in Chinese)
2. Ma, Q., Cai, M.H.: Investigation and analysis of current graduates' learning. High. Educ. Explor. **2**, 95–98 (2013). (in Chinese)
3. Zheng, L.H., Chen, Ch.W.: Study on the influence of research opportunity on the cultivation of graduates' innovative ability. Degree Grad. Educ. **2**(7), 20–27 (2008). (in Chinese)
4. Zhang, Y.B., Liu, H.F., Gu, J.B.: Study on the influence of graduate enthusiasm and social capital on the cultivation of innovation ability. Degree Grad. Educ. **5**(14), 47–52 (2014). (in Chinese)
5. Wang, D.Zh., Hu, R.: Analysis of the environmental factors of the cultivation of graduate innovation ability. Degree Grad. Educ. **6**(6), 22–26 (2007). (in Chinese)

6. Chen, Ch., Cheng, T., Tian, Y., Tang, R., Xie, W.F.: Analysis of the influence of instructors on the cultivation of graduate innovative ability. China Mod. Educ. Equip. (243), 103–106 (2016). (in Chinese)
7. Zhao, W.F., Deng, R., Guo, X.M.: Research on the restricted factors and solutions for the cultivation of graduate innovative ability in local colleges. Sci. Manag. Res. 33(2), 89–92 (2015). (in Chinese)
8. Xiao, B., Chen, C.T.: Study on dynamic relationship of structure-function based on CAS theory. J. Syst. Sci. 23(2), 41–44 (2015). (in Chinese)
9. Yu, J.Y., Liu, Y.: Complexity research and system science. Stud. Sci. Sci. 20(5), 449–453 (2002). (in Chinese)
10. Dai, R.W.: Research on system science and system complexity. J. Syst. Simul. 14(11), 1411–1466 (2002). (in Chinese)
11. Fay, D., Frese, M.: The concept of personal initiative: an overview of validity studies. Hum. Perform. 14(1), 97–124 (2001)
12. Zhang, L.Y., Liang, Ch.J.: From the factor-driven to innovation-driven: the key to the reform of doctoral education. China High. Educ. Res. 1, 45–49 (2011). (in Chinese)
13. Sun, X.H.: Modeling and simulation of industrial dynamics model of industrial agglomeration effect. Sci. Technol. Manag., 71–76 (2008). (in Chinese)

# Author Index

Chankong, Vira  250

Dang, Tran-Thai  53
Dang, Yanzhong  227
Dong, Xuefan  101
Du, Zhonglian  150

Gu, Hualu  12
Guo, Chonghui  150

Ho, Tu-Bao  53
Htun, Htet Htet  76
Hung, Nguyen Duy  197
Huynh, Van-Nam  23, 186

Jin, ChengHao  117

Kaffashi, Farhad  66
Kou, Xinyue  150

Lhatoo, Samden  66
Li, Ding  101
Li, Huaiming  129
Li, Qianqian  136
Li, Ying  136
Lian, Ying  101
Liu, Haiyuan  129
Liu, Jiamou  38
Liu, Niu  38
Liu, Yijun  101
Loparo, Kenneth  66
Lu, Chenhao  38
Luo, Shuangling  1

McDonald, James  66

Mi, Lin  38

Pan, Donghua  162
Pan, Qian  1

Pannakkong, Warut  186
Pechsiri, Chaveevan  91

Qin, Xuezhi  242
Qu, Ling  162

Rong, LiLi  117

Sornlertlamvanich, Virach  76
Sukharomana, Renu  91
Sun, Kang  117
Suprasongsin, Sirin  23

Tang, Xijin  175
Theeranaew, Wanchat  66
Tian, Jing  212

Vasanthapriyan, Shanmuganathan  212

Wang, Ning  129
Wang, Shaoni  227
Wang, Shouyang  12
Wang, Wenhua  242
Wang, Xuehua  129
Wang, Yanzhang  129
Wu, Jiangning  227

Xia, Haoxiang  1
Xiang, Jianwen  212
Xiao, Bing  250
Xu, Nuo  175

Yang, Guangfei  162
Yang, Xianduan  12
Yenradee, Pisal  23

Zhong, Qiuyan  129
Zonjy, Bilal  66

Printed in the United States
By Bookmasters

Printed in the United States
By Bookmasters